NOMENCLATURE OF INORGANIC CHEMISTRY
IUPAC Recommendations 2005

IUPAC Periodic Table of the Elements

1	2	3	4	5	6	7	8	9	10	11	12	13	14	15	16	17	18
1 **H**																	2 **He**
3 **Li**	4 **Be**											5 **B**	6 **C**	7 **N**	8 **O**	9 **F**	10 **Ne**
11 **Na**	12 **Mg**											13 **Al**	14 **Si**	15 **P**	16 **S**	17 **Cl**	18 **Ar**
19 **K**	20 **Ca**	21 **Sc**	22 **Ti**	23 **V**	24 **Cr**	25 **Mn**	26 **Fe**	27 **Co**	28 **Ni**	29 **Cu**	30 **Zn**	31 **Ga**	32 **Ge**	33 **As**	34 **Se**	35 **Br**	36 **Kr**
37 **Rb**	38 **Sr**	39 **Y**	40 **Zr**	41 **Nb**	42 **Mo**	43 **Tc**	44 **Ru**	45 **Rh**	46 **Pd**	47 **Ag**	48 **Cd**	49 **In**	50 **Sn**	51 **Sb**	52 **Te**	53 **I**	54 **Xe**
55 **Cs**	56 **Ba**	*57–71 lanthanoids	72 **Hf**	73 **Ta**	74 **W**	75 **Re**	76 **Os**	77 **Ir**	78 **Pt**	79 **Au**	80 **Hg**	81 **Tl**	82 **Pb**	83 **Bi**	84 **Po**	85 **At**	86 **Rn**
87 **Fr**	88 **Ra**	‡89–103 actinoids	104 **Rf**	105 **Db**	106 **Sg**	107 **Bh**	108 **Hs**	109 **Mt**	110 **Ds**	111 **Rg**	112 Uub	113 Uut	114 Uuq	115 Uup	116 Uuh	117 Uus	118 Uuo

* 57 **La**	58 **Ce**	59 **Pr**	60 **Nd**	61 **Pm**	62 **Sm**	63 **Eu**	64 **Gd**	65 **Tb**	66 **Dy**	67 **Ho**	68 **Er**	69 **Tm**	70 **Yb**	71 **Lu**
‡ 89 **Ac**	90 **Th**	91 **Pa**	92 **U**	93 **Np**	94 **Pu**	95 **Am**	96 **Cm**	97 **Bk**	98 **Cf**	99 **Es**	100 **Fm**	101 **Md**	102 **No**	103 **Lr**

International Union of Pure and Applied Chemistry

Nomenclature of Inorganic Chemistry

IUPAC RECOMMENDATIONS 2005

Issued by the Division of Chemical Nomenclature and Structure Representation in collaboration with the Division of Inorganic Chemistry

Prepared for publication by

Neil G. Connelly
University of Bristol, UK

Ture Damhus
Novozymes A/S, Denmark

Richard M. Hartshorn
University of Canterbury, New Zealand

Alan T. Hutton
University of Cape Town, South Africa

RSCPublishing

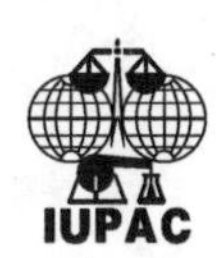

Cover images ©Murray Robertson/visual elements 1998–99, taken from the 109 Visual Elements Periodic Table, available at www.chemsoc.org/viselements

ISBN 0-85404-438-8

A catalogue record for this book is available from the British Library

Published for the International Union of Pure and Applied Chemistry by The Royal Society of Chemistry, Thomas Graham House, Science Park, Milton Road, Cambridge CB4 0WF, UK

Registered Charity Number 207890

For further information see our web site at www.rsc.org and the IUPAC site at www.iupac.org

Typeset by Alden Bookset, Northampton, UK
Printed by Biddles Ltd, King's Lynn, Norfolk, UK

Preface

Chemical nomenclature must evolve to reflect the needs of the community that makes use of it. In particular, nomenclature must be created to describe new compounds or classes of compounds; modified to resolve ambiguities which might arise; or clarified where there is confusion over the way in which nomenclature should be used. There is also a need to make nomenclature as systematic and uncomplicated as possible in order to assist less familiar users (for example, because they are only in the process of studying chemistry or are non-chemists who need to deal with chemicals at work or at home). A revision of *Nomenclature of Inorganic Chemistry, IUPAC Recommendations 1990* (Red Book I) was therefore initiated in 1998, under the guidance of the IUPAC Commission on Nomenclature of Inorganic Chemistry (CNIC) and then, on the abolition of CNIC in 2001 as part of the general restructuring of IUPAC, by a project group working under the auspices of the Division of Chemical Nomenclature and Structure Representation (Division VIII).

The need to ensure that inorganic and organic nomenclature systems are, as far as possible, consistent has resulted in extensive cooperation between the editors of the revised Red Book and the editors of *Nomenclature of Organic Chemistry, IUPAC Recommendations* (the revised 'Blue Book', in preparation). At present, the concept of preferred IUPAC names (PINs), an important element in the revision of the Blue Book, has not been extended to inorganic nomenclature (though preferred names are used herein for organic, *i.e.* carbon-containing, compounds when appropriate). A planned future project on inorganic PINs will need to face the problem of choice between the equally valid nomenclature systems currently in use.

The present book supersedes not only Red Book I but also, where appropriate, *Nomenclature of Inorganic Chemistry II, IUPAC Recommendations 2000* (Red Book II). One of the main changes from Red Book I is the different organization of material, adopted to improve clarity. Thus, Chapters IR-5 (Compositional Nomenclature, and Overview of Names of Ions and Radicals), IR-6 (Parent Hydride Names and Substitutive Nomenclature), and IR-7 (Additive Nomenclature) deal with the general characteristics of the three main nomenclature systems applied to inorganic compounds. (Note that the notation 'IR-' is used to distinguish chapters and sections in the current book from those in Red Book I, prefixed 'I-'). The next three chapters deal with their application, particularly that of additive nomenclature, to three large classes of compounds: inorganic acids and derivatives (Chapter IR-8), coordination compounds (Chapter IR-9) and organometallic compounds (Chapter IR-10). Overall, the emphasis on additive nomenclature (generalized from the classical nomenclature of coordination compounds) which was already apparent in Red Book I is reinforced here. Examples are even included of organic compounds, from the borderline between inorganic and organic chemistry, which may be conveniently named using additive nomenclature (although their PINs will be different).

One important addition in this book is Chapter IR-10 on Organometallic Compounds. The separation of this material from that on Coordination Compounds (Chapter IR-9) reflects the huge growth in importance of organometallic chemistry and the very different problems associated with the presence of π-bonded ligands. Chapter IR-9 is also considerably changed (*cf.* Red Book I, Chapter I-10). This revised chapter includes a clarification of the use of the η and κ conventions in coordination and organometallic compounds (Section IR-9.2.4.3); new rules for the ordering of central atoms in names of polynuclear compounds (Section IR-9.2.5.6); the bringing together of sections on configuration (Section IR-9.3) and their separation from those on constitution (Section IR-9.2); and the addition of polyhedral symbols for T-shaped (Section IR-9.3.3.7) and see-saw (Section IR-9.3.3.8) molecules, along with guidance on how to choose between these shapes and those of closely related structures (Section IR-9.3.2.2).

The chapter on Oxoacids and Derived Anions (Red Book I, Chapter I-9) has also been extensively modified. Now called Inorganic Acids and Derivatives (Chapter IR-8), it includes the slightly revised concept of 'hydrogen names' in Section IR-8.4 (and some traditional 'ous' and 'ic' names have been reinstated for consistency and because they are required for organic nomenclature purposes, *i.e.* in the new Blue Book).

The reader facing the problem of how to name a given compound or species may find help in several ways. A flowchart is provided in Section IR-1.5.3.5 which will in most cases guide the user to a Section or Chapter where rules can be found for generating at least one possible name; a second flowchart is given in Section IR-9.2.1 to assist in the application of additive nomenclature specifically to coordination and organometallic compounds. A more detailed subject index is also provided, as is an extended guide to possible alternative names of a wide range of simple inorganic compounds, ions and radicals (in Table IX).

For most compounds, formulae are another important type of compositional or structural representation and for some compounds a formula is perhaps easier to construct. In Chapter IR-4 (Formulae) several changes are made in order to make the presentation of a formula and its corresponding name more consistent, *e.g.* the order of ligand citation (which does not now depend on the charge on the ligand) (Section IR-4.4.3.2) and the order and use of enclosing marks (simplified and more consistent with the usage proposed for the nomenclature of organic compounds) (Section IR-4.2.3). In addition, the use of ligand abbreviations can make formulae less cumbersome. Thus, recommendations for the construction and use of abbreviations are provided in Section IR-4.4.4, with an extensive list of established abbreviations given in Table VII (and with structural formulae for the ligands given in Table VIII).

Two chapters of Red Book I have been shortened or subsumed since in both areas extensive revision is still necessary. First, the chapter on Solids (IR-11) now describes only basic topics, more recent developments in this area tending to be covered by publications from the International Union of Crystallography (IUCr). It is to be hoped that future cooperation between IUPAC and IUCr will lead to the additional nomenclature required for the rapidly expanding field of solid-state chemistry.

Second, boron chemistry, particularly that of polynuclear compounds, has also seen extensive development. Again, therefore, only the basics of the nomenclature of boron-containing compounds are covered here (*cf.* the separate, more comprehensive but dated, chapter on boron nomenclature, I-11, in Red Book I), within Chapter IR-6 (Parent Hydride Names and Substitutive Nomenclature), while more advanced aspects are left for elaboration in a future project.

Other changes include a section on new elements and the procedure by which they are now named (Section IR-3.1) and a simplified coverage of the systematic method for naming chains and rings (adapted from Chapter II-5 of Red Book II). Lesser omissions include the section on single strand polymers (now updated as Chapter II-7 in Red Book II) and the several different outdated versions of the periodic table. (That on the inside front cover is the current IUPAC-agreed version.)

Some new recommendations represent breaks with tradition, in the interest of increased clarity and consistency. For example, the application of the ending 'ido' to all anionic ligands with 'ide' names in additive nomenclature (*e.g.* chlorido and cyanido instead of chloro and cyano, and hydrido throughout, *i.e.* no exception in boron nomenclature) is part of a general move to a more systematic approach.

Acknowledgements

It is important to remember that the current volume has evolved from past versions of the Red Book and it is therefore appropriate first to acknowledge the efforts of previous editors and contributors. However, we would also like to thank the many people without whose help this revision would not have come to fruition. Members of CNIC were involved in the early stages of the revision (including Stanley Kirschner who began the task of compiling ligand abbreviations and what has become Tables VII and VIII), and members of the IUPAC Division VIII Advisory Subcommittee (particularly Jonathan Brecher, Piroska Fodor-Csányi, Risto Laitinen, Jeff Leigh and Alan McNaught) and the editors of the revised Blue Book (Warren Powell and Henri Favre) have made extremely valuable comments. However, the bulk of the work has been carried out by a project group comprising the two Senior Editors, Richard Hartshorn and Alan Hutton.

NEIL G. CONNELLY and TURE DAMHUS
(Senior Editors)

Contents

TABLES

IR-1 General Aims, Functions and Methods of Chemical Nomenclature

CONTENTS

IR-1.1 INTRODUCTION

This Chapter provides a brief historical overview of chemical nomenclature (Section IR-1.2) followed by summaries of its aims, functions and methods (Sections IR-1.3 to IR-1.5). There are several systems of nomenclature that can be applied to inorganic compounds, briefly described in Section IR-1.5.3.5 as an introduction to the later, more detailed, chapters. Because each system can provide a valid name for a compound, a flowchart is presented in Section IR-1.5.3 which should help identify which is the most appropriate for the type of compound of interest. Section IR-1.6 summarises the major changes from previous recommendations and, finally, reference is made in Section IR-1.7 to nomenclature in other areas of chemistry, underlining that inorganic chemistry is part of an integrated whole.

IR-1.2 HISTORY OF CHEMICAL NOMENCLATURE

The activities of alchemy and of the technical arts practised prior to the founding of what we now know as the science of chemistry produced a rich vocabulary for describing chemical substances although the names for individual species gave little indication of composition. However, almost as soon as the true science of chemistry was established a 'system' of chemical nomenclature was developed by Guyton de Morveau in 1782.[1] Guyton's statement of the need for a 'constant method of denomination, which helps the intelligence and relieves the memory' clearly defines the basic aims of chemical nomenclature. His system was extended by a joint contribution[2] with Lavoisier, Berthollet, and de Fourcroy and was popularized by Lavoisier.[3] Later, Berzelius championed Lavoisier's ideas, adapting the nomenclature to the Germanic languages,[4] expanding the system and adding many new terms. This system, formulated before the enunciation of the atomic theory by Dalton, was based upon the concept of elements forming compounds with oxygen, the oxides in turn reacting with each other to form salts; the two-word names in some ways resembled the binary system introduced by Linnaeus (Carl von Linné) for plant and animal species.

When atomic theory developed to the point where it was possible to write specific formulae for the various oxides and other binary compounds, names reflecting composition more or less accurately then became common; no names reflecting the composition of the oxosalts were ever adopted, however. As the number of inorganic compounds rapidly grew, the essential pattern of nomenclature was little altered until near the end of the 19th century. As a need arose, a name was proposed and nomenclature grew by accretion rather than by systematization.

When Arrhenius focused attention on ions as well as molecules, it became necessary to name charged particles in addition to neutral species. It was not deemed necessary to develop a new nomenclature for salts; cations were designated by the names of the appropriate metal and anions by a modified name of the non-metal portion.

Along with the theory of coordination, Werner proposed[5] a system of nomenclature for coordination entities which not only reproduced their compositions but also indicated many of their structures. Werner's system was completely additive in that the names of the ligands were cited, followed by the name of the central atom (modified by the ending 'ate' if the complex was an anion). Werner also used structural descriptors and locants. The additive nomenclature system was capable of expansion and adaptation to new compounds and even to other fields of chemistry.

IR-1.2.1 International cooperation on inorganic nomenclature

In 1892 a conference in Geneva[6] laid the basis for an internationally accepted system of organic nomenclature but at that time there was nothing comparable for inorganic nomenclature. Thus, many *ad hoc* systems had developed for particular rather than general purposes, and two or more methods often evolved for naming a given compound belonging to a given class. Each name might have value in a specific situation, or be preferred by some users, but there was then the possibility of confusion.

The need for uniform practice among English-speaking chemists was recognized as early as 1886 and resulted in agreements on usage by the British and American Chemical Societies. In 1913, the Council of the International Association of Chemical Societies appointed a commission of inorganic and organic nomenclature, but World War I abruptly ended its activities. Work was resumed in 1921 when IUPAC, at its second conference, appointed commissions on the nomenclature of inorganic, organic, and biological chemistry.

The first comprehensive report of the inorganic commission, in 1940,[7] had a major effect on the systematization of inorganic nomenclature and made many chemists aware of the necessity for developing a more fully systematic nomenclature. Among the significant features of this initial report were the adoption of the Stock system for indicating oxidation states, the establishment of orders for citing constituents of binary compounds in formulae and in names, the discouragement of the use of bicarbonate, *etc*. in the names of acid salts, and the development of uniform practices for naming addition compounds.

These IUPAC recommendations were then revised and issued as a small book in 1959[8] followed by a second revision in 1971[9] and a supplement, entitled *How to Name an Inorganic Substance*, in 1977.[10] In 1990 the IUPAC recommendations were again fully revised[11] in order to bring together the many and varied changes which had occurred in the previous 20 years.

More specialized areas have also been considered, concerning polyanions,[12] metal complexes of tetrapyrroles (based on Ref. 13), inorganic chain and ring compounds,[14] and graphite intercalation compounds.[15] These topics, together with revised versions of papers on isotopically modified inorganic compounds,[16] hydrides of nitrogen and derived cations, anions and ligands,[17] and regular single-strand and quasi single-strand inorganic and coordination polymers,[18] comprise the seven chapters of *Nomenclature of Inorganic Chemistry II, IUPAC Recommendations 2000*.[19] A paper entitled *Nomenclature of Organometallic Compounds of the Transition Elements*[20] forms the basis for Chapter IR-10 of this book.

IR-1.3 AIMS OF CHEMICAL NOMENCLATURE

The primary aim of chemical nomenclature is to provide methodology for assigning descriptors (names and formulae) to chemical species so that they can be identified without ambiguity, thereby facilitating communication. A subsidiary aim is to achieve standardization. Although this need not be so absolute as to require only one name for a substance, the number of 'acceptable' names needs to be minimized.

When developing a system of nomenclature, public needs and common usage must also be borne in mind. In some cases, the only requirement may be to identify a substance, essentially the requirement prior to the late 18th century. Thus, local names and abbreviations are still used by small groups of specialists. Such local names suffice as long as the specialists understand the devices used for identification. However, this is not nomenclature as defined above since local names do not necessarily convey structural and compositional information to a wider audience. To be widely useful, a nomenclature system must be recognisable, unambiguous, and general; the unnecessary use of local names and abbreviations in formal scientific language should therefore be discouraged.

IR-1.4 FUNCTIONS OF CHEMICAL NOMENCLATURE

The first level of nomenclature, beyond the assignment of totally trivial names, gives some systematic information about a substance but does not allow the inference of composition. Most of the common names of the oxoacids (*e.g.* sulfuric acid, perchloric acid) and of their salts are of this type. Such names may be termed semi-systematic and as long as they are used for common materials and understood by chemists, they are acceptable. However, it should be recognized that they may hinder compositional understanding by those with limited chemical training.

When a name itself allows the inference of the stoichiometric formula of a compound according to general rules, it becomes truly systematic. Only a name at this second level of nomenclature becomes suitable for retrieval purposes.

The desire to incorporate information concerning the three-dimensional structures of substances has grown rapidly and the systematization of nomenclature has therefore had to expand to a third level of sophistication. Few chemists want to use such a degree of sophistication every time they refer to a compound, but they may wish to do so when appropriate.

A fourth level of nomenclature may be required for the compilation and use of extensive indexes. Because the cost to both compiler and searcher of multiple entries for a given substance may be prohibitive, it becomes necessary to develop systematic hierarchical rules that yield a unique name for a given substance.

IR-1.5 METHODS OF INORGANIC NOMENCLATURE

IR-1.5.1 Formulation of rules

The revision of nomenclature is a continuous process as new discoveries make fresh demands on nomenclature systems. IUPAC, through the Division of Chemical Nomenclature and Structure Representation (formed in 2001), studies all aspects of the nomenclature of inorganic and other substances, recommending the most desirable practices to meet specific problems, for example for writing formulae and generating names. New nomenclature rules need to be formulated precisely, to provide a systematic basis for assigning names and formulae within the defined areas of application. As far as possible, such rules should be consistent with existing recommended nomenclature, in both inorganic and other areas of chemistry, and take into account emerging chemistry.

IR-1.5.2 Name construction

The systematic naming of an inorganic substance involves the construction of a name from entities which are manipulated in accordance with defined procedures to provide compositional and structural information. The element names (or roots derived from them or from their Latin equivalents) (Tables I and II*, see also Chapter IR-3) are combined with affixes in order to construct systematic names by procedures which are called systems of nomenclature.

* Tables numbered with a Roman numeral are collected together at the end of this book.

There are several accepted systems for the construction of names, as discussed in Section IR-1.5.3. Perhaps the simplest is that used for naming binary substances. This set of rules leads to a name such as iron dichloride for the substance $FeCl_2$; this name involves the juxtaposition of element names (iron, chlorine), their ordering in a specific way (electropositive before electronegative), the modification of an element name to indicate charge (the 'ide' ending designates an elementary anion and, more generally, an element being treated formally as an anion), and the use of the multiplicative prefix 'di' to indicate composition.

Whatever the pattern of nomenclature, names are constructed from entities such as:

element name roots,
multiplicative prefixes,
prefixes indicating atoms or groups – either substituents or ligands,
suffixes indicating charge,
names and endings denoting parent compounds,
suffixes indicating characteristic substituent groups,
infixes,
locants,
descriptors (structural, geometric, spatial, *etc.*),
punctuation.

IR-1.5.3 Systems of nomenclature

IR-1.5.3.1 *General*

In the development of nomenclature, several systems have emerged for the construction of chemical names; each system has its own inherent logic and set of rules (grammar). Some systems are broadly applicable whereas practice has led to the use of specialized systems in particular areas of chemistry. The existence of several distinct nomenclature systems leads to logically consistent alternative names for a given substance. Although this flexibility is useful in some contexts, the excessive proliferation of alternatives can hamper communication and even impede trade and legislation procedures. Confusion can also occur when the grammar of one nomenclature system is mistakenly used in another, leading to names that do not represent any given system.

Three systems are of primary importance in inorganic chemistry, namely compositional, substitutive and additive nomenclature; they are described in more detail in Chapters IR-5, IR-6 and IR-7, respectively. Additive nomenclature is perhaps the most generally applicable in inorganic chemistry, but substitutive nomenclature may be applied in appropriate areas. These two systems require knowledge of the constitution (connectivity) of the compound or species being named. If only the stoichiometry or composition of a compound is known or to be communicated, compositional nomenclature is used.

IR-1.5.3.2 *Compositional nomenclature*

This term is used in the present recommendations to denote name constructions which are based solely on the composition of the substances or species being named, as opposed to systems involving structural information. One such construction is that of a generalized *stoichiometric name*. The names of components which may themselves be elements or composite entities (such as polyatomic ions) are listed with multiplicative prefixes giving the overall stoichiometry of the compound. If there are two or more components, they are formally divided into two classes, the electropositive and the electronegative components. In this respect, the names are like traditional salt names although there is no implication about the chemical nature of the species being named.

Grammatical rules are then required to specify the ordering of components, the use of multiplicative prefixes, and the proper endings for the names of the electronegative components.

Examples:

1. trioxygen, O_3
2. sodium chloride, NaCl
3. phosphorus trichloride, PCl_3
4. trisodium pentabismuthide, Na_3Bi_5
5. magnesium chloride hydroxide, MgCl(OH)
6. sodium cyanide, NaCN
7. ammonium chloride, NH_4Cl
8. sodium acetate, NaO_2CMe

IR-1.5.3.3 *Substitutive nomenclature*

Substitutive nomenclature is used extensively for organic compounds and is based on the concept of a parent hydride modified by substitution of hydrogen atoms by atoms and/or groups.[21] (In particular it is used for naming organic ligands in the nomenclature of coordination and organometallic compounds, even though this is an overall additive system.)

It is also used for naming compounds formally derived from the hydrides of certain elements in groups 13–17 of the periodic table. Like carbon, these elements form chains and rings which can have many derivatives, and the system avoids the necessity for specifying the location of the hydrogen atoms of the parent hydride.

Rules are required to name parent compounds and substituents, to provide an order of citation of substituent names, and to specify the positions of attachment of substituents.

Examples:

1. 1,1-difluorotrisilane, $SiH_3SiH_2SiHF_2$
2. trichlorophosphane, PCl_3

Operations in which certain non-hydrogen atoms of parents are replaced by different atoms or groups, *e.g.* the skeletal replacements leading to ‘a’ names in organic chemistry (see Sections P-13.2 and P-51.3 of Ref. 21), are usually considered as part of substitutive nomenclature and are also used in certain parts of inorganic chemistry.

Examples:

3. 1,5-dicarba-*closo*-pentaborane(5), $B_3C_2H_5$ (CH replacing BH)
4. stiborodithioic acid, $H_3SbO_2S_2$

Subtractive operations are also regarded as part of the machinery of substitutive nomenclature.

Example:

5. 4,5-dicarba-9-debor-*closo*-nonaborate(2−), $[B_6C_2H_8]^{2-}$ (loss of BH)

IR-1.5.3.4 *Additive nomenclature*

Additive nomenclature treats a compound or species as a combination of a central atom or central atoms with associated ligands. The particular additive system used for coordination compounds (see Chapter IR-9) is sometimes known as coordination nomenclature although it may be used for much wider classes of compounds, as demonstrated for inorganic acids (Chapter IR-8) and organometallic compounds (Chapter IR-10) and for a large number of simple molecules and ions named in Table IX. Another additive system is well suited for naming chains and rings (Section IR-7.4; see Example 6 below).

Rules within these systems provide ligand names and guidelines for the order of citation of ligand names and central atom names, designation of charge or unpaired electrons on species, designation of point(s) of ligation in complicated ligands, designation of spatial relationships, *etc.*

Examples:

1. PCl_3, trichloridophosphorus
2. $[CoCl_3(NH_3)_3]$, triamminetrichloridocobalt
3. $H_3SO_4^+$ ($= [SO(OH)_3]^+$), trihydroxidooxidosulfur(1+)
4. $[Pt(\eta^2\text{-}C_2H_4)Cl_3]^-$, trichlorido($\eta^2$-ethene)platinate(1−)
5. $HONH^{\bullet}$, hydridohydroxidonitrogen(•)
6.

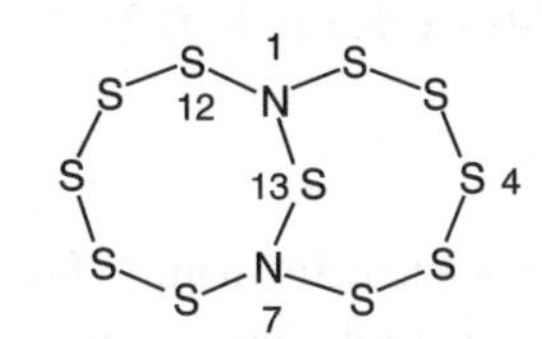

1,7-diazyundecasulfy-[012.1^{1,7}]dicycle

IR-1.5.3.5 *General naming procedures*

The three basic nomenclature systems may provide different but unambiguous names for a given compound, as demonstrated for PCl_3 above.

The choice between the three systems depends on the class of inorganic compound under consideration and the degree of detail one wishes to communicate. The following examples further illustrate typical aspects that need to be considered before deciding on a name.

Examples:

1. NO_2
   Would you like simply to specify a compound with this *empirical* formula, or a compound with this *molecular* formula? Would you like to stress that it is a radical? Would you like to specify the connectivity ONO?
2. $Al_2(SO_4)_3 \cdot 12H_2O$
   Would you like simply to indicate that this is a compound composed of dialuminium trisulfate and water in the proportion 1:12, or would you like to specify explicitly that it contains hexaaquaaluminium(3+) ions?
3. $H_2P_3O_{10}{}^{3-}$
   Would you like to specify that this is triphosphoric acid (as defined in Table IR-8.1) from which three hydrogen(1+) ions have been removed? Would you like to specify from where they have been removed?

The flowchart shown in Figure IR-1.1 (see page 9) proposes general guidelines for naming compounds and other species.

IR-1.6 CHANGES TO PREVIOUS IUPAC RECOMMENDATIONS

This section highlights significant changes made in the present recommendations relative to earlier IUPAC nomenclature publications. In general, these changes have been introduced to provide a more logical and consistent nomenclature, aligned with that of *Nomenclature of Organic Chemistry, IUPAC Recommendations*, Royal Society of Chemistry, in preparation (Ref. 21), as far as possible.

IR-1.6.1 **Names of cations**

Certain cations derived from parent hydrides were given names in Refs. 11 and 19 which appear to be substitutive but which do not follow the rules of substitutive nomenclature. For example, according to Refs. 11 and 19, $N_2H_6{}^{2+}$ may be named hydrazinium(2+). However, the ending 'ium' in itself denotes addition of hydrogen(1+) and thus implies the charge. Consequently this cation is named hydrazinediium or diazanediium, with no charge number, both in Section IR-6.4.1 and in Ref. 21.

IR-1.6.2 **Names of anions**

When constructing systematic names for anions, consistency is achieved by adhering without exception to the following rules:

(i) Compositional names of homopolyatomic anions end in 'ide'.

Examples:

1. $I_3{}^-$, triiodide(1−)
2. $O_2{}^{2-}$, dioxide(2−)

(ii) Parent hydride-based names of anions based on the formal removal of hydrogen(1+) end in 'ide'.

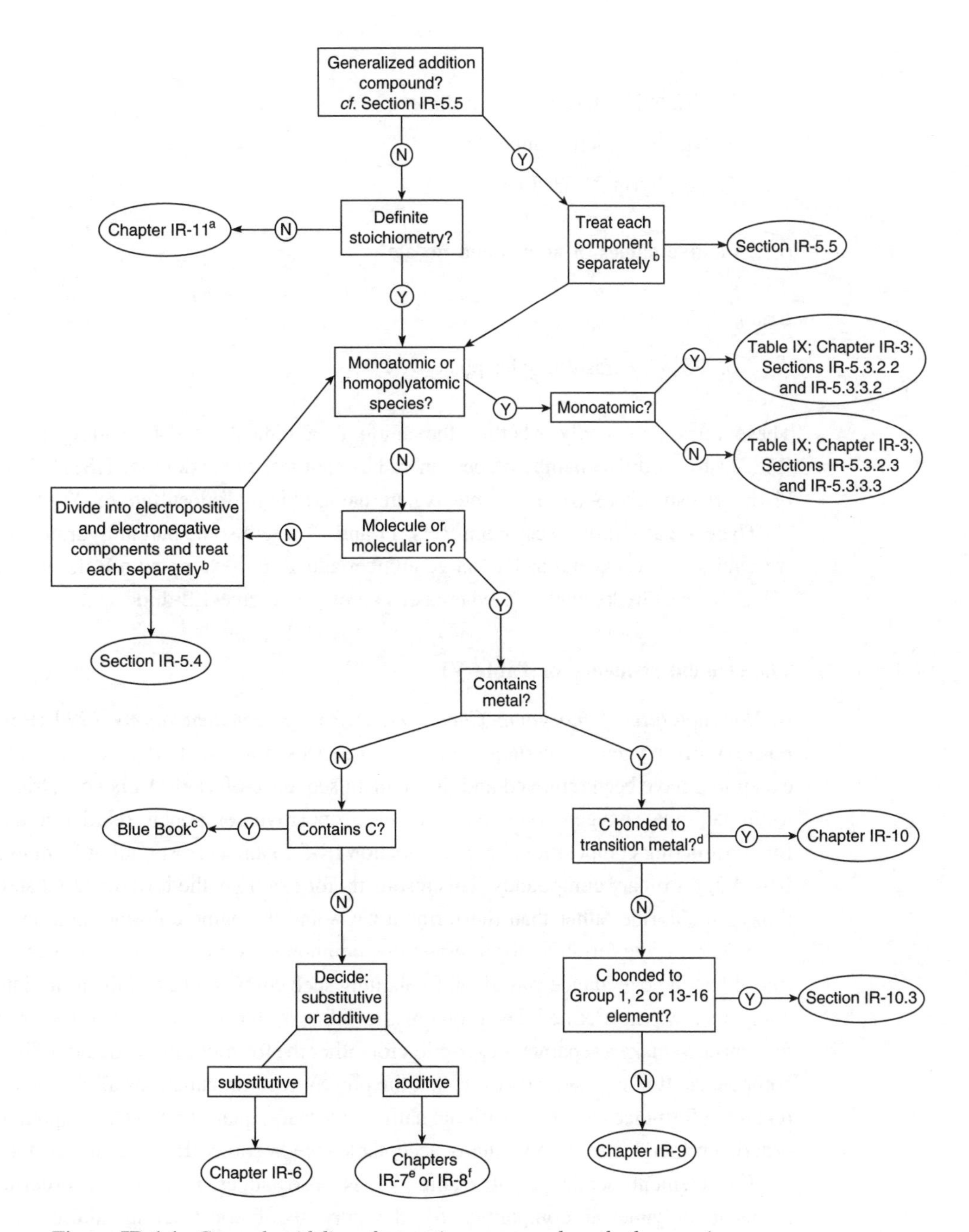

Figure IR-1.1. *General guidelines for naming compounds and other species.*

[a] Chapter IR-11 deals with nomenclature of the solid state.
[b] Each individual component is named by following the pathway indicated. The complete name is then assembled according to the recommendations in the Section of Chapter IR-5 indicated.
[c] In principle, the compound is outside the scope of this book. A few carbon compounds are named in Tables IR-8.1, IR-8.2 and IX, but otherwise the reader is referred to the Blue Book.[21]
[d] *C*-bonded cyanides are treated as coordination compounds, see Chapter IR-9.
[e] The species may be named as a coordination-type compound (Sections IR-7.1 to IR-7.3) or as a chain or ring (Section IR-7.4).
[f] For inorganic acids.

Examples:

3. $^{-}HNNH^{-}$, hydrazine-1,2-diide
4. $MeNH^{-}$, methanaminide
5. porphyrin-21,23-diide

(iii) Additive names of anions end in 'ate'.

Example:

6. PS_4^{3-}, tetrasulfidophosphate(3−)

These rules now apply whether the anion is a radical or not, leading to changes to Ref. 22 for additive names of certain radical anions. For example, $HSSH^{\bullet-}$ was named bis(hydridosulfide)(*S–S*)(•1−)[22] but is here named bis(hydridosulfate)(*S–S*)(•1−).

There are also differences from Refs. 11 and 19 where some parent hydride-based anions were missing locants and had a charge number added. For example, in Ref. 19 one name for $^{-}HNNH^{-}$ was hydrazide(2−), whereas it is now hydrazine-1,2-diide.

IR-1.6.3 The element sequence of Table VI

In *Nomenclature of Inorganic Chemistry, IUPAC Recommendations 1990* (Ref. 11), the position of oxygen in certain element sequences was treated as an exception. Such exceptions have been removed and the element sequence of Table VI is now strictly adhered to. In particular, oxygen is treated as the electropositive component relative to any halogen for constructing compositional names (Section IR-5.2) and corresponding formulae (Section IR-4.4.3) for binary compounds. This results in, for example, the formula O_2Cl and the name dioxygen chloride rather than the formula ClO_2 and the name chlorine dioxide.

In Ref. 11, the formulae for intermetallic compounds were also subject to an exceptional rule although no guidance was given for naming such compounds, and the term 'intermetallic compound' was not defined. The problem is to define the term 'metal'. Therefore, no attempt is now made to make a separate prescription for either the formulae or the names of intermetallic compounds. It is stressed, however, that the present recommendations allow some flexibility regarding formulae and compositional names of ternary, quaternary, *etc.* compounds. Several ordering principles are often equally acceptable (see Sections IR-4.4.2 and IR-4.4.3).

The element sequence of Table VI is also adhered to when ordering central atoms in polynuclear compounds for the purpose of constructing additive names (see Section IR-1.6.6).

IR-1.6.4 Names of anionic ligands in (formal) coordination entities

The rule now used, without exception, is that anion names ending in 'ide', 'ite' and 'ate', respectively, are changed to end in 'ido', 'ito' and 'ato', respectively, when modifying the ligand name for use in additive nomenclature (Sections IR-7.1.3, and IR-9.2.2.3). This entails several changes from Refs. 11 and 22.

Certain simple ligands have historically (and in Ref. 11) been represented in names by abbreviated forms: fluoro, chloro, bromo, iodo, hydroxo, hydro, cyano, oxo, *etc.* Following the rule stated above, these are now fluorido, chlorido, bromido, iodido, hydroxido, hydrido, cyanido, oxido, *etc*. In particular, thio is now reserved for functional replacement nomenclature (see Section IR-8.6), and the ligand S^{2-} is named sulfido.

In a number of cases the names of (formally) anionic ligands have changed as a result of modifications to the nomenclature of the anions themselves (see Section IR-1.6.2). For example, the ligand $^{-}HNNH^{-}$ is now named hydrazine-1,2-diido (Example 3 in Section IR-1.6.2), and $HNCO^{\bullet-}$ was (hydridonitrido)oxidocarbonate(•1−) in Ref. 22 but is now named (hydridonitrato)oxidocarbonate(•1−).

Particular attention has been given to providing the correct names and endings for organic ligands. Thus, with reference to Examples 4 and 5 in Section IR-1.6.2, methanaminido is now used rather than methaminato, and a porphyrin ligand is named porphyrin-21,23-diido rather than the name porphyrinato(2−) (which is used in Ref. 11).

The systematic organic ligand names given in Table VII are now in accord with anion names derived by the rules of Ref. 21. In a number of cases they differ from the names given as systematic in Ref. 11.

IR-1.6.5 Formulae for (formal) coordination entities

In the formulae for coordination entities, ligands are now ordered alphabetically according to the abbreviation or formula used for the ligand, irrespective of charge (Sections IR-4.4.3.2 and IR-9.2.3.1).

In Ref. 11, charged ligands were cited before neutral ligands. Thus, two ordering principles were in use for no obvious reason other than tradition, and the person devising the formula needed to decide whether a particular ligand was charged. Such a decision is not always straightforward.

Thus, for example, the recommended formula for the anion of Zeise's salt is now $[Pt(\eta^2\text{-}C_2H_4)Cl_3]^-$ whereas in Ref. 11 it was $[PtCl_3(\eta^2\text{-}C_2H_4)]^-$ because chloride is anionic.

IR-1.6.6 Additive names of polynuclear entities

The system developed in Ref. 11 for additive names of dinuclear and polynuclear entities has been clarified and to some extent changed for reasons of consistency: the order of citation of central atoms in names is now always the order in which they appear in Table VI, the element occurring later being cited first (see Sections IR-7.3.2 and IR-9.2.5.6).

The system can be used for polynuclear entities with any central atoms. In this system, the order of the central atoms in the name reflects the order in which they are assigned locants to be used in the kappa convention (Section IR-9.2.4.2) for specifying which ligator atoms coordinate to which central atoms. The atom symbols used at the end of the name to indicate metal-metal bonding are similarly ordered. Thus, for example, $[(CO)_5ReCo(CO)_4]$ is now named nonacarbonyl-1κ^5*C*,2κ^4*C*-rheniumcobalt(*Re*—*Co*) rather than nonacarbonyl-1κ^5*C*,2κ^4*C*-cobaltrhenium(*Co*—*Re*) (as in Ref. 11).

IR-1.6.7 Names of inorganic acids

The names of inorganic acids are dealt with separately in Chapter IR-8.

Names described in Ref. 11 under the heading 'acid nomenclature', *e.g.* tetraoxosulfuric acid, trioxochloric(V) acid, have been abandoned. In addition, the format of the names described in Ref. 11 under the heading 'hydrogen nomenclature' has been changed so that 'hydrogen' is always attached directly to the second part of the name, and this part is always in enclosing marks. The charge number at the end of the name is the total charge.

Examples:

1. $HCrO_4^-$, hydrogen(tetraoxidochromate)(1−)
2. $H_2NO_3^+$, dihydrogen(trioxidonitrate)(1+)

A restricted list of names of this type where the enclosing marks and charge number may be omitted is given in Section IR-8.5 (hydrogencarbonate, dihydrogenphosphate and a few others). (These names do not differ from those in Ref. 11.)

The main principle, however, is to use additive nomenclature for deriving systematic names for inorganic acids. For example, the systematic name for dihydrogenphosphate, $H_2PO_4^-$, is dihydroxidodioxidophosphate(1−).

For a number of inorganic acids, used as functional parents in organic nomenclature, the parent names used are now consistently allowed in the present recommendations, although fully systematic additive names are also given in all cases in Chapter IR-8. Examples are phosphinous acid, bromic acid and peroxydisulfuric acid. (Some of these names were absent from Ref. 11.)

IR-1.6.8 Addition compounds

The formalism for addition compounds, and other compounds treated as such, has been rationalized (see Sections IR-4.4.3.5 and IR-5.5) so as to remove the exceptional treatment of component boron compounds and to make the construction of the name self-contained rather than dependent on the formula. Thus, the double salt carnallite, when considered formally as an addition compound, is given the formula:

$KCl{\cdot}MgCl_2{\cdot}6H_2O$
(*formulae* of compounds ordered alphabetically, water still placed last),

and the name:

magnesium chloride—potassium chloride—water (1/1/6)
(*names* of components ordered alphabetically).

IR-1.6.9 Miscellaneous

(i) In the present recommendations the radical dot is regarded as optional in formulae and names whereas in Ref. 22 the dot is not omitted in any systematic names. [For example, in Ref. 22, NO is shown as $NO^{\bullet}$ with the name oxidonitrogen(•).]

(ii) The order of enclosing marks (Section IR-2.2.1) has been changed from that in Ref. 11 in order to ensure consistency with Ref. 21.

(iii) Certain names were announced as 'preferred' in Refs. 20 and 22. This announcement was premature and, as explained in the preface, no preferred names are selected in the present recommendations.

IR-1.7 NOMENCLATURE RECOMMENDATIONS IN OTHER AREAS OF CHEMISTRY

Inorganic chemical nomenclature, as inorganic chemistry itself, does not develop in isolation from other fields, and those working in interdisciplinary areas will find useful IUPAC texts on the general principles of chemical nomenclature[23] as well as the specific topics of organic,[21] biochemical,[24] analytical[25] and macromolecular chemistry.[26] Other IUPAC publications include a glossary of terms in bioinorganic chemistry,[27] a compendium of chemical terminology[28] and quantities, units and symbols in physical chemistry.[29] Other texts concerning chemical nomenclature are given in Ref. 30.

IR-1.8 REFERENCES

1. L.B. Guyton de Morveau, *J. Phys.*, **19**, 310 (1782); *Ann. Chim. Phys.*, **1**, 24 (1798).
2. L.B. Guyton de Morveau, A.L. Lavoisier, C.L. Berthollet and A.F. de Fourcroy, *Méthode de Nomenclature Chimique*, Paris, 1787.
3. A.L. Lavoisier, *Traité Elémentaire de Chimie*, Third Edn., Deterville, Paris, 1801, Vol. I, pp. 70–81, and Vol. II.
4. J.J. Berzelius, *Journal de Physique, de Chimie, et d'Histoire Naturelle*, **73**, 253 (1811).
5. A. Werner, *Neuere Anschauungen auf den Gebieten der Anorganischen Chemie*, Third Edn., Vieweg, Braunschweig, 1913, pp. 92–95.
6. *Bull. Soc. Chem. (Paris)*, **3**(7), XIII (1892).
7. W.P. Jorissen, H. Bassett, A. Damiens, F. Fichter and H. Remy, *Ber. Dtsch. Chem. Ges. A*, **73**, 53–70 (1940); *J. Chem. Soc.*, 1404–1415 (1940); *J. Am. Chem. Soc.*, **63**, 889–897 (1941).
8. *Nomenclature of Inorganic Chemistry*, 1957 Report of CNIC, IUPAC, Butterworths Scientific Publications, London, 1959; *J. Am. Chem. Soc.*, **82**, 5523–5544 (1960).
9. *Nomenclature of Inorganic Chemistry. Definitive Rules 1970*, Second Edn., Butterworths, London, 1971.
10. *How to Name an Inorganic Substance, 1977. A Guide to the Use of Nomenclature of Inorganic Chemistry: Definitive Rules 1970*, Pergamon Press, Oxford, 1977.
11. *Nomenclature of Inorganic Chemistry, IUPAC Recommendations 1990*, ed. G.J. Leigh, Blackwell Scientific Publications, Oxford, 1990.
12. Nomenclature of Polyanions, Y. Jeannin and M. Fournier, *Pure Appl. Chem.*, **59**, 1529–1548 (1987).
13. Nomenclature of Tetrapyrroles, Recommendations 1986, G.P. Moss, *Pure Appl. Chem.*, **59**, 779–832 (1987); Nomenclature of Tetrapyrroles, Recommendations 1978, J.E. Merritt and K.L. Loening, *Pure Appl. Chem.*, **51**, 2251–2304 (1979).
14. Nomenclature of Inorganic Chains and Ring Compounds, E.O. Fluck and R.S. Laitinen, *Pure Appl. Chem.*, **69**, 1659–1692 (1997).

15. Nomenclature and Terminology of Graphite Intercalation Compounds, H.-P. Boehm, R. Setton and E. Stumpp, *Pure Appl. Chem.*, **66**, 1893–1901 (1994).
16. Isotopically Modified Compounds, W.C. Fernelius, T.D. Coyle and W.H. Powell, *Pure Appl. Chem.*, **53**, 1887–1900 (1981).
17. The Nomenclature of Hydrides of Nitrogen and Derived Cations, Anions, and Ligands, J. Chatt, *Pure Appl. Chem.*, **54**, 2545–2552 (1982).
18. Nomenclature for Regular Single-strand and Quasi Single-strand Inorganic and Coordination Polymers, L.G. Donaruma, B.P. Block, K.L. Loening, N. Platé, T. Tsuruta, K.Ch. Buschbeck, W.H. Powell and J. Reedijk, *Pure Appl. Chem.* **57**, 149–168 (1985).
19. *Nomenclature of Inorganic Chemistry II, IUPAC Recommendations 2000*, eds. J.A. McCleverty and N.G. Connelly, Royal Society of Chemistry, 2001. (Red Book II.)
20. Nomenclature of Organometallic Compounds of the Transition Elements, A. Salzer, *Pure Appl. Chem.*, **71**, 1557–1585 (1999).
21. *Nomenclature of Organic Chemistry, IUPAC Recommendations*, eds. W.H. Powell and H. Favre, Royal Society of Chemistry, in preparation. [See also, *Nomenclature of Organic Chemistry*, Pergamon Press, Oxford, 1979; *A Guide to IUPAC Nomenclature of Organic Compounds, Recommendations 1993*, eds. R. Panico, W.H. Powell and J.-C. Richer, Blackwell Scientific Publications, Oxford, 1993; and corrections in *Pure Appl. Chem.*, **71**, 1327–1330 (1999)].
22. Names for Inorganic Radicals, W.H. Koppenol, *Pure Appl. Chem.*, **72**, 437–446 (2000).
23. *Principles of Chemical Nomenclature, A Guide to IUPAC Recommendations*, G.J. Leigh, H.A. Favre and W.V. Metanomski, Blackwell Scientific Publications, Oxford, 1998.
24. *Biochemical Nomenclature and Related Documents*, for IUBMB, C. Liébecq, Portland Press Ltd., London, 1992. (The White Book.)
25. *Compendium of Analytical Nomenclature, IUPAC Definitive Rules*, 1997, Third Edn., J. Inczedy, T. Lengyel and A.M. Ure, Blackwell Scientific Publications, Oxford, 1998. (The Orange Book.)
26. *Compendium of Macromolecular Nomenclature*, ed. W.V. Metanomski, Blackwell Scientific Publications, Oxford, 1991. (The Purple Book. The second edition is planned for publication in 2005). See also Glossary of Basic Terms in Polymer Science, A.D. Jenkins, P. Kratochvíl, R.F.T. Stepto and U.W. Suter, *Pure Appl. Chem.*, **68**, 2287–2311 (1996); Nomenclature of Regular Single-strand Organic Polymers, J. Kahovec, R.B. Fox and K. Hatada, *Pure Appl. Chem.*, **74**, 1921–1956 (2002).
27. Glossary of Terms used in Bioinorganic Chemistry, M.W.G. de Bolster, *Pure Appl. Chem.*, **69**, 1251–1303 (1997).
28. *Compendium of Chemical Terminology, IUPAC Recommendations*, Second Edn., eds. A.D. McNaught and A. Wilkinson, Blackwell Scientific Publications, Oxford, 1997. (The Gold Book.)
29. *Quantities, Units and Symbols in Physical Chemistry*, Second Edn., eds. I. Mills, T. Cvitas, K. Homann, N. Kallay and K. Kuchitsu, Blackwell Scientific Publications, Oxford, 1993. (The Green Book. The third edition is planned for publication in 2005).
30. *Nomenclature of Coordination Compounds*, T.E. Sloan, Vol. 1, Chapter 3, *Comprehensive Coordination Chemistry*, Pergamon Press, 1987; *Inorganic Chemical Nomenclature, Principles and Practice*, B.P. Block, W.H. Powell and W.C. Fernelius, American Chemical Society, Washington, DC, 1990; *Chemical Nomenclature*, K.J. Thurlow, Kluwer Academic Pub., 1998.

IR-2 Grammar

CONTENTS

IR-2.1 INTRODUCTION

Chemical nomenclature may be considered to be a language. As such, it consists of words and it should obey the rules of syntax.

In the language of chemical nomenclature, the simple names of atoms are the words. As words are assembled to form a sentence, so names of atoms are assembled to form names of chemical compounds. Syntax is the set of grammatical rules for building sentences out of words. In nomenclature, syntax includes the use of symbols, such as dots, commas and hyphens, the use of numbers for appropriate reasons in given places, and the order of citation of various words, syllables and symbols.

Generally, nomenclature systems require a root on which to construct the name. This root can be an element name (*e.g.* 'cobalt' or 'silicon') for use in additive nomenclature, or can be derived from an element name (*e.g.* 'sil' from 'silicon', 'plumb' from 'plumbum' for lead) and elaborated to yield a parent hydride name (*e.g.* 'silane' or 'plumbane') for use in substitutive nomenclature.

Names are constructed by joining other units to these roots. Among the most important units are affixes. These are syllables added to words or roots and can be suffixes, prefixes or infixes according to whether they are placed after, before or within a word or root.

Suffixes and endings are of many different kinds (Table III)*, each of which conveys specific information. The following examples illustrate particular uses. They may specify the degree of unsaturation of a parent compound in substitutive nomenclature: hex*ane*, hex*ene*; and phosph*ane*, diphosph*ene*, diphosph*yne*. Other endings indicate the nature of the charge carried by the whole compound; cobalt*ate* refers to an anion. Further suffixes can indicate that a name refers to a group, as in hex*yl*.

Prefixes indicate, for example, substituents in substitutive nomenclature, as in the name *chloro*trisilane, and ligands in additive nomenclature, as in the name *aqua*cobalt. Multiplicative prefixes (Table IV) can be used to indicate the number of constituents or ligands, *e.g.* *hexa*aquacobalt. Prefixes may also be used to describe the structural types or

* Tables numbered with a Roman numeral are collected together at the end of this book.

other structural features of species; geometrical and structural prefixes are listed in Table V. The ordering of prefixes in substitutive nomenclature is dealt with in Chapter IR-6, and in additive nomenclature in Chapters IR-7, IR-9 and IR-10.

Other devices may be used to complete the description of the compound. These include the charge number to indicate the ionic charge, *e.g.* hexaaquacobalt(2+), and, alternatively, the oxidation number to indicate the oxidation state of the central atom, *e.g.* hexaaquacobalt(II).

The designation of central atom and ligands, generally straightforward in mononuclear complexes, is more difficult in polynuclear compounds where there are several central atoms in the compound to be named, *e.g.* in polynuclear coordination compounds, and chain and ring compounds. In each case, a priority order or hierarchy has to be established. A hierarchy of functional groups is an established feature of substitutive nomenclature; Table VI shows an element sequence used in compositional and additive nomenclature.

The purpose of this Chapter is to guide the users of nomenclature in building the name or formula of an inorganic compound and to help them verify that the derived name or formula fully obeys the accepted principles. The various devices used in names (or formulae) are described successively below, together with their meanings and fields of application.

IR-2.2 ENCLOSING MARKS

IR-2.2.1 General

Chemical nomenclature employs three types of enclosing mark, namely: braces { }, square brackets [ ], and parentheses ( ).

In *formulae*, these enclosing marks are used in the following nesting order: [ ], [( )], [{( )}], [({( )})], [{({( )})}], *etc.* Square brackets are normally used only to enclose entire formulae; parentheses and braces are then used alternately (see also Sections IR-4.2.3 and IR-9.2.3.2). There are, however, some specific uses of square brackets in formulae, *cf.* Section IR-2.2.2.1.

In *names*, the nesting order is: ( ), [( )], {[( )]}, ({[( )]}), *etc.* This ordering is that used in substitutive nomenclature, see Section P-16.4 of Ref. 1. (See also Section IR-9.2.2.3 for the use of enclosing marks with ligand names.)

Example:

1. $[Rh_3Cl(\mu\text{-}Cl)(CO)_3\{\mu_3\text{-}Ph_2PCH_2P(Ph)CH_2PPh_2\}_2]^+$

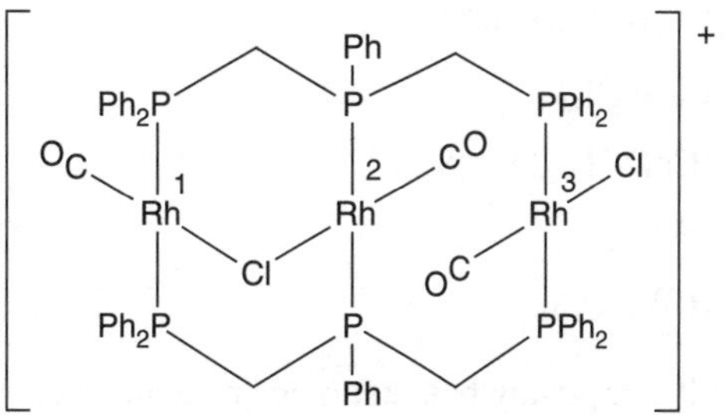

tricarbonyl-1κ*C*,2κ*C*,3κ*C*-μ-chlorido-1:2κ^2*Cl*-chlorido-3κ*Cl*-bis{μ$_3$-bis[(diphenylphosphanyl)methyl]-1κ*P*:3κ*P*′-phenylphosphane-2κ*P*}trirhodium(1+)

IR-2.2.2 Square brackets

IR-2.2.2.1 *Use in formulae*

Square brackets are used in *formulae* in the following ways.

(a) To enclose the whole coordination entity of a neutral (formal) coordination compound.

Examples:

1. $[Fe(\eta^5\text{-}C_5H_5)_2]$ (for use of the symbol η see Sections IR-9.2.4.3 and IR-10.2.5.1)
2. $[Pt(\eta^2\text{-}C_2H_4)Cl_2(NH_3)]$
3. $[PH(O)(OH)_2]$

No numerical subscript should follow the square bracket used in this context. For example, where the molecular formula is double the empirical formula, this should be indicated inside the square bracket.

Example:

4\.

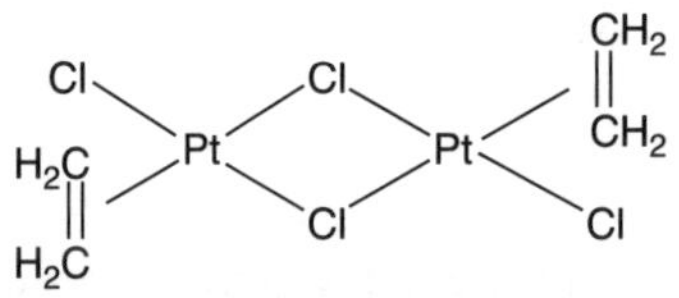

$[\{Pt(\eta^2\text{-}C_2H_4)Cl(\mu\text{-}Cl)\}_2]$ is more informative than $[Pt_2(\eta^2\text{-}C_2H_4)_2Cl_4]$; the representation $[Pt(\eta^2\text{-}C_2H_4)Cl_2]_2$ is incorrect.

(b) To enclose a charged (formal) coordination entity. In this case, the superscript showing the charge appears outside the square bracket as do any subscripts indicating the number of ions in the salt.

Examples:

5. $[BH_4]^-$
6. $[Al(OH)(OH_2)_5]^{2+}$
7. $[Pt(\eta^2\text{-}C_2H_4)Cl_3]^-$
8. $Ca[AgF_4]_2$
9. $[Co(NH_3)_5(N_3)]SO_4$
10. $[S_2O_5]^{2-}$
11. $[PW_{12}O_{40}]^{3-}$

(c) In a salt comprising both cationic and anionic coordination entities, each ion is separately enclosed in square brackets. (The cation is placed before the anion and no individual charges are shown.) Any subscripts indicating the number of complex ions in the salt are shown outside the square brackets.

Examples:

12. $[Co(NH_3)_6][Cr(CN)_6]$ (comprising the ions $[Co(NH_3)_6]^{3+}$ and $[Cr(CN)_6]^{3-}$)
13. $[Co(NH_3)_6]_2[Pt(CN)_4]_3$ (comprising the ions $[Co(NH_3)_6]^{3+}$ and $[Pt(CN)_4]^{2-}$)

(d) To enclose structural formulae.

Example:

14.

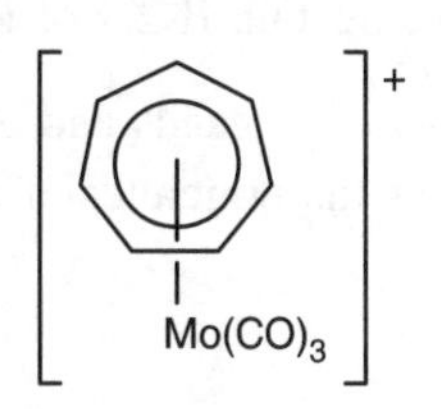

$[Mo(\eta^7\text{-}C_7H_7)(CO)_3]^+$

(e) In solid-state chemistry, to indicate an atom or a group of atoms in an octahedral site. (See Section IR-11.4.3.)

Example:

15. $(Mg)[Cr_2]O_4$

(f) In specifically labelled compounds (see also Section II-2.4.2.2 of Ref. 2).

Example:

16. $H_2[^{15}N]NH_2$

Note that this distinguishes the specifically labelled compound from the isotopically substituted compound $H_2{}^{15}NNH_2$.

(g) In selectively labelled compounds (see also Section II-2.4.3.2 of Ref. 2).

Example:

17. $[^{18}O,^{32}P]H_3PO_4$

(h) To indicate repeating units in chain compounds.

Example:

18. $SiH_3[SiH_2]_8SiH_3$

IR-2.2.2.2 *Use in names*

Square brackets are used in *names* in the following ways.

(a) In specifically and selectively labelled compounds the nuclide symbol is placed in square brackets before the name of the part of the compound that is isotopically modified.

(Compare with the use of parentheses for isotopically substituted compounds in Section IR-2.2.3.2, and also see Sections II-2.4.2.3, II-2.4.2.4 and II-2.4.3.3 of Ref. 2.)

Examples:

1. $[^{15}N]H_2[^2H]$ $[^2H_1,^{15}N]$ammonia
2. $HO[^{18}O]H$ dihydrogen $[^{18}O_1]$peroxide

For more details, see Section II-2.4 of Ref. 2.

(b) When naming organic ligands and organic parts of coordination compounds the use of square brackets obeys the principles of organic nomenclature.[1]

Example:

3.

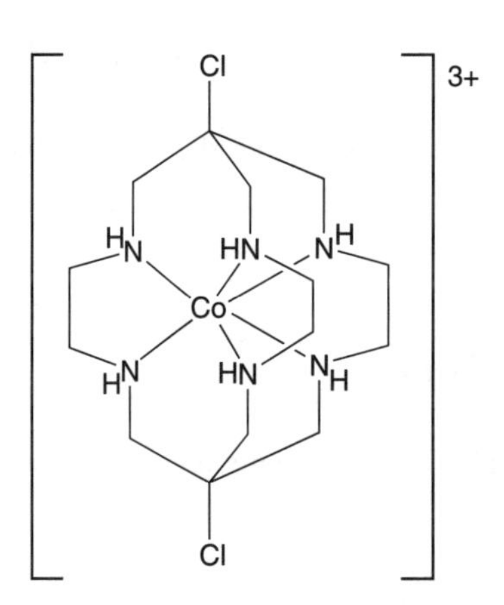

1,8-dichloro-3,6,10,13,16,19-hexaazabicyclo[6.6.6]icosanecobalt(3+)

(c) In chain and ring nomenclature, square brackets are used to enclose the nodal descriptor (Section IR-7.4.2 and Chapter II-5 of Ref. 2).

Examples:

4. $HSSH^{\bullet-}$ 1,4-dihydrony-2,3-disulfy-[4]catenate(•1−)
5.

1,7-diazyundecasulfy-[012.1^{1,7}]dicycle

IR-2.2.3 Parentheses

IR-2.2.3.1 *Use in formulae*

Parentheses are used in *formulae* in the following ways.

(a) To enclose formulae for groups of atoms (the groups may be ions, substituent groups, ligands or molecules), to avoid ambiguity or when the group is being multiplied. In the latter case, a multiplicative subscript numeral follows the closing parenthesis. In the case of common ions such as nitrate and sulfate, parentheses are recommended but not mandatory.

Examples:

1. $Ca_3(PO_4)_2$
2. $[Te(N_3)_6]$
3. $(NO_3)^-$ or NO_3^-
4. $[FeH(H_2)(Ph_2PCH_2CH_2PPh_2)_2]^+$
5. $PH(O)(OH)_2$
6. $[Co(NH_3)_5(ONO)][PF_6]_2$

(b) To enclose the abbreviation of a ligand name in formulae. (Recommended ligand abbreviations are given in Tables VII and VIII. See also Sections IR-4.4.4 and IR-9.2.3.4.)

Example:

7. $[Co(en)_3]^{3+}$

(c) To enclose the superscripted radical dot and its multiplier for polyradicals, in order to avoid ambiguity in relation to multiplying the charge symbol.

Example:

8. $NO^{(2\bullet)-}$

(d) In solid-state chemistry, to enclose symbols of atoms occupying the same type of site in a random fashion. The symbols themselves are separated by a comma, with no space.

Example:

9. K(Br,Cl)

(e) In solid-state chemistry, to indicate an atom or a group of atoms in a tetrahedral site.

Example:

10. $(Mg)[Cr_2]O_4$

(f) To indicate the composition of a non-stoichiometric compound.

Examples:

11. $Fe_{3x}Li_{4-x}Ti_{2(1-x)}O_6$ $(x = 0.35)$
12. $LaNi_5H_x$ $(0 < x < 66.7)$

(g) In the Kröger–Vink notation (see Section IR-11.4), to indicate a complex defect.

Example:

13. $(Cr_{Mg}V_{Mg}Cr_{Mg})^{x}$

(h) For crystalline substances, to indicate the type of crystal formed (see Chapter IR-11).

Examples:

14. ZnS(*c*)
15. AuCd (*CsCl* type)

(i) To enclose a symbol representing the state of aggregation of a chemical species.

Example:

16. HCl(g) hydrogen chloride in the gaseous state

(j) In optically active compounds, to enclose the signs of rotation.

Example:

17. $(+)_{589}$-$[Co(en)_3]Cl_3$

(k) To enclose stereodescriptors, such as chirality descriptors and configuration indexes (see Section IR-9.3.3.2).

Examples:

18. (2*R*,3*S*)-$SiH_2ClSiHClSiHClSiH_2SiH_3$
19. (*OC*-6-22)-$[Co(NH_3)_3(NO_2)_3]$

(l) In polymers, the repeating unit is enclosed in strike-through parentheses, with the dash superimposed on the parentheses representing the bond.[3]

Example:

20. $\text{-(}S\text{)-}_n$

IR-2.2.3.2 *Use in names*

Parentheses are used in *names* in the following ways.

(a) To enclose substituent group or ligand names in order to avoid ambiguity, for example if the substituent group or ligand names contain multiplicative prefixes, such as (dioxido) or (triphosphato), or if substitution patterns would otherwise not be unambiguously specified, or if the substituent group or ligand name contains numerical or letter descriptors. It may be necessary to use different enclosing marks if the ligand names or substituent groups themselves include parentheses, *cf.* the nesting rule of Section IR-2.2.1.

Examples:

1. $[Pt(\eta^2\text{-}C_2H_4)Cl_3]^-$ trichlorido(η^2-ethene)platinate(II)
2. $[Hg(CHCl_2)Ph]$ (dichloromethyl)(phenyl)mercury

(b) Following multiplicative prefixes of the series bis, tris, *etc.*, unless other enclosing marks are to be used because of the nesting order (see Section IR-2.2.1).

Examples:

3. $[CuCl_2(NH_2Me)_2]$ dichloridobis(methylamine)copper(II)
4. Fe_2S_3 diiron tris(sulfide)

(c) To enclose oxidation and charge numbers.

Example:

5. $Na[B(NO_3)_4]$ sodium tetranitratoborate(III), or sodium tetranitratoborate(1−)

(d) For radicals, to enclose the radical dot, and the charge number if appropriate.

Examples:

6. $ClOO^{\bullet}$ chloridodioxygen(•)
7. $Cl_2^{\bullet-}$ dichloride(•1−)

(e) To enclose stoichiometric ratios for formal addition compounds.

Example:

8. $8H_2S{\cdot}46H_2O$ hydrogen sulfide—water (8/46)

(f) To enclose italic letters representing bonds between two (or more) metal atoms in coordination compounds.

Example:

9. $[Mn_2(CO)_{10}]$ bis(pentacarbonylmanganese)(*Mn*—*Mn*)

(g) To enclose stereochemical descriptors (see Section IR-9.3)

Examples:

10.

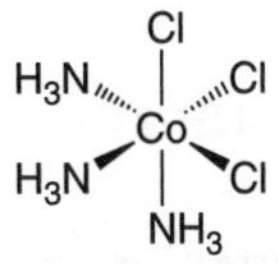

Cl
H3N Cl
Co
H3N Cl
NH3

$[CoCl_3(NH_3)_3]$ (*OC*-6-22)-triamminetrichloridocobalt(III)

11. $(+)_{589}$-$[Co(en)_3]Cl_3$ $(+)_{589}$-tris(ethane-1,2-diamine)cobalt(III) trichloride

12. $(2R,3S)$-$\overset{1}{Cl}\overset{2}{SiH_2}\overset{3}{SiHCl}\overset{4}{SiHCl}\overset{5}{SiH_2}SiH_3$
(2*R*,3*S*)-1,2,3-trichloropentasilane

(h) In isotopically substituted compounds, the appropriate nuclide symbol(s) is placed in parentheses before the name of the part of the compound that is isotopically substituted (see Section II-2.3.3 of Ref. 2). Compare with the use of square brackets for specifically and selectively labelled compounds in Section IR-2.2.2.2(a).

Example:

13. H^3HO $(^3H_1)$water

(i) To enclose the number of hydrogen atoms in boron compounds.

Example:

14. B_6H_{10} hexaborane(10)

(j) In hydrogen names (Section IR-8.4), to enclose the part of the name following the word hydrogen.

Example:

15. $[HMo_6O_{19}]^-$ hydrogen(nonadecaoxidohexamolybdate)(1−)

IR-2.2.4 Braces

Braces are used in *names* and *formulae* within the hierarchical sequence outlined and exemplified in Section IR-2.2.1.

IR-2.3 HYPHENS, PLUS AND MINUS SIGNS, 'EM' DASHES AND BOND INDICATORS

IR-2.3.1 Hyphens

Hyphens are used in *formulae* and in *names*. Note that there is no space on either side of a hyphen.

(a) To separate symbols such as μ (mu), η (eta) and κ (kappa) from the rest of the formula or name.

Example:

1. $[\{Cr(NH_3)_5\}_2(\mu\text{-}OH)]^{5+}$ μ-hydroxido-bis(pentaamminechromium)(5+)

(b) To separate geometrical or structural and stereochemical designators such as *cyclo, catena, triangulo, quadro, tetrahedro, octahedro, closo, nido, arachno, cis* and *trans* from the rest of the formula or name. In dealing with aggregates or clusters, locant designators are similarly separated.

Example:

2.

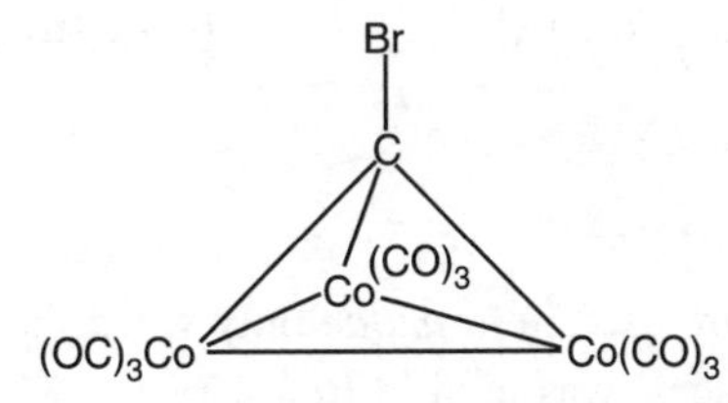

μ_3-(bromomethanetriyl)-*cyclo*-tris(tricarbonylcobalt)(3 *Co—Co*)

(c) To separate locant designators from the rest of the name.

Example:

3. $SiH_2ClSiHClSiH_2Cl$ 1,2,3-trichlorotrisilane

(d) To separate the labelling nuclide symbol from its locant in the formula of a selectively labelled compound.

Example:

4. $[1\text{-}^{2}H_{1;2}]SiH_3OSiH_2OSiH_3$

(e) To separate the name of a bridging ligand from the rest of the name.

Example:

5.

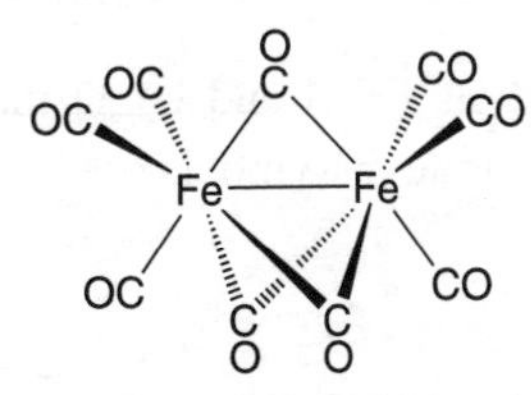

$[Fe_2(\mu\text{-}CO)_3(CO)_6]$ tri-μ-carbonyl-bis(tricarbonyliron)(*Fe—Fe*)

IR-2.3.2 Plus and minus signs

The signs + and − are used to indicate the charge on an ion in a formula or name.

Examples:

1. Cl^-
2. Fe^{3+}
3. $[SO_4]^{2-}$
4. $[Co(CO)_4]^-$ tetracarbonylcobaltate(1−)

They can also indicate the sign of optical rotation in the formula or name of an optically active compound.

Example:

5. $(+)_{589}$-$[Co(en)_3]^{3+}$ $(+)_{589}$-tris(ethane-1,2-diamine)cobalt(3+)

IR-2.3.3 'Em' dashes

'Em' dashes are used in *formulae* only when the formulae are structural. (The less precise term 'long dashes' was used in Ref. 4.)

In *names*, 'em' dashes are used in two ways.

(a) To indicate metal–metal bonds in polynuclear compounds. They separate the italicized symbols of the bond partners which are contained in parentheses at the end of the name.

Example:

1. $[Mn_2(CO)_{10}]$ bis(pentacarbonylmanganese)(*Mn*—*Mn*)

(b) To separate the individual constituents in names of (formal) addition compounds.

Examples:

2. $3CdSO_4{\cdot}8H_2O$ cadmium sulfate—water (3/8)
3. $2CHCl_3{\cdot}4H_2S{\cdot}9H_2O$ chloroform—hydrogen sulfide—water (2/4/9)

IR-2.3.4 Special bond indicators for line formulae

The structural symbols ┌──┐ and └──┘ may be used in line formulae to indicate bonds between non-adjacent atom symbols.

Examples:

1.

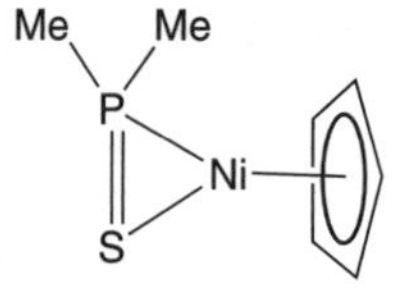

$[Ni(S{=}PMe_2)(\eta^5\text{-}C_5H_5)]$

2.

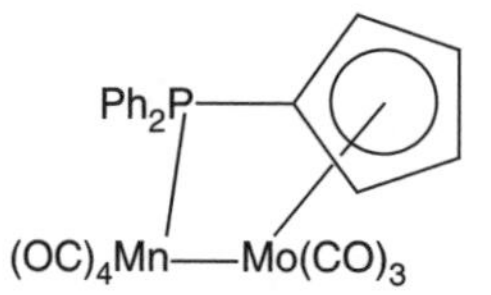

$[(CO)_4MnMo(CO)_3(\eta^5\text{-}C_5H_4PPh_2)]$

3.

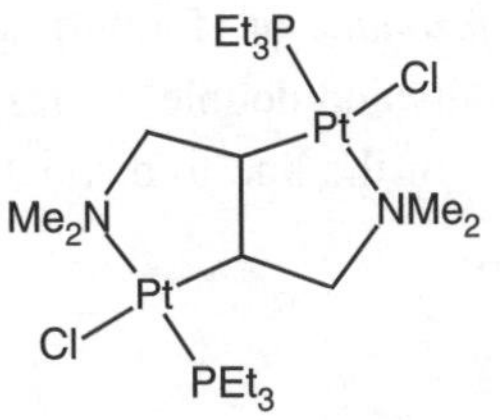

$[(Et_3P)ClPt(Me_2NCH_2CHCHCH_2NMe_2)PtCl(PEt_3)]$

4.

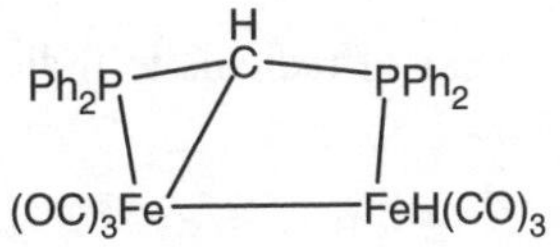

$[(OC)_3Fe(\mu\text{-}Ph_2PCHPPh_2)FeH(CO)_3]$

IR-2.4 SOLIDUS

The solidus (/) is used in names of formal addition compounds to separate the arabic numerals which indicate the proportions of individual constituents in the compound.

Examples:

1. $BF_3{\cdot}2H_2O$ boron trifluoride—water (1/2)
2. $BiCl_3{\cdot}3PCl_5$ bismuth trichloride—phosphorus pentachloride (1/3)

IR-2.5 DOTS, COLONS, COMMAS AND SEMICOLONS

IR-2.5.1 Dots

Dots are used in *formulae* in various positions.

(a) As right superscripts they indicate unpaired electrons in radicals (see Section IR-4.6.2).

Examples:

1. $HO^{\bullet}$
2. $O_2^{2\bullet}$

(b) As right superscripts in the Kröger–Vink notation of solid-state chemistry, they indicate the unit of positive effective charge (see Section IR-11.4.4).

Example:

3. $Li^{x}_{Li,1-2x}Mg^{\bullet}_{Li,x}V'_{Li,x}Cl^{x}_{Cl}$

(c) Centre dots in *formulae* of (formal) addition compounds, including hydrates, adducts, clathrates, double salts and double oxides, separate the individual constituents. The dot is written in the centre of the line to distinguish it from a full stop (period).

Examples:

4. $BF_3{\cdot}NH_3$
5. $ZrCl_2O{\cdot}8H_2O$
6. $CuCl_2{\cdot}3Cu(OH)_2$
7. $Ta_2O_5{\cdot}4WO_3$

Dots are used in *names* of radicals to indicate the presence of unpaired electrons.

Examples:

8. $ClO^{\bullet}$ oxidochlorine(•)
9. $Cl_2^{\bullet-}$ dichloride(•1−)

IR-2.5.2 Colons

Colons are used in *names* in the following ways.

(a) In coordination and organometallic compounds, to separate the ligating atoms of a ligand which bridges central atoms.

Example:

1. $[\{Co(NH_3)_3\}_2(\mu\text{-}NO_2)(\mu\text{-}OH)_2]^{3+}$
   di-μ-hydroxido-μ-nitrito-κ*N*:κ*O*-bis(triamminecobalt)(3+)

(See Sections IR-9.2.4.2 and IR-10.2.3.3 for the use of κ, and Sections IR-9.2.5.2 and IR-10.2.3.1 for the use of μ.)

(b) In polynuclear coordination and organometallic compounds, to separate the central atom locants when single ligating atoms or unsaturated groups bind to two or more central atoms. Thus, a chloride ligand bridging between central atoms 1 and 2 would be indicated by μ-chlorido-1:2κ^2*Cl*, and a carbonyl group terminally bonded to atom 1 and bridging atoms 2 and 3 *via* its π electrons would be indicated by μ_3-2η^2:3η^2-carbonyl-1κ*C*.

(c) In boron compounds, to separate the sets of locants of boron atoms which are connected by bridging hydrogen atoms.

Example:

2.

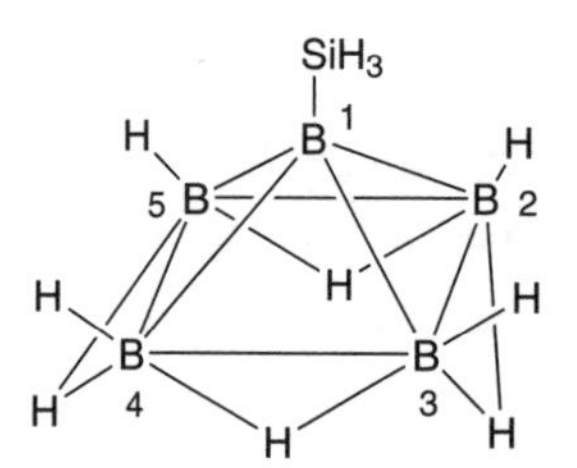

1-silyl-2,3:2,5:3,4:4,5-tetra-μ*H*-pentaborane(9)

(d) In chains and rings nomenclature, to separate nodal descriptors of individual modules of an assembly (see Section IR-7.4.2).

IR-2.5.3 **Commas**

Commas are used in the following ways.

(a) To separate locants.

Example:

1. $SiH_2ClSiHClSiH_2Cl$ 1,2,3-trichlorotrisilane

(b) To separate the symbols of the ligating atoms of a polydentate ligand.

Example:

2. *cis*-bis(glycinato-κ*N*,κ*O*)platinum

(c) In solid-state chemistry, to separate symbols of atoms occupying the same type of site in a random fashion.

Example:

3. $(Mo,W)_nO_{3n-1}$

(d) To separate oxidation numbers in a mixed valence compound.

Example:

4. $[(H_3N)_5Ru{-}N\langle C_4H_4\rangle N{-}Ru(NH_3)_5]^{5+}$

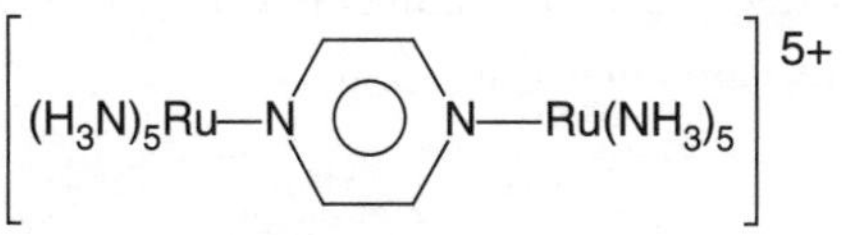

$[(H_3N)_5Ru(\mu\text{-pyz})Ru(NH_3)_5]^{5+}$

μ-pyrazine-bis(pentaammineruthenium)(II,III)

(e) To separate symbols of labelled atoms in selectively labelled compounds. (See Section II-2.4.3.3 of Ref. 2.)

Example:

5. $[^{18}O,^{32}P]H_3PO_4$ $[^{18}O,^{32}P]$phosphoric acid

IR-2.5.4 **Semicolons**

Semicolons are used in the following ways.

(a) In the names of coordination compounds, to order locants already separated by commas, as in the kappa convention. (See examples in Section IR-9.2.5.6.)

(b) To separate the subscripts that indicate the possible numbers of labelling nuclides in selectively labelled compounds.

Example:

1. $[1\text{-}^{2}H_{1;2}]SiH_3OSiH_2OSiH_3$

IR-2.6 SPACES

In inorganic nomenclature, spaces are used in *names* in the following ways in English; the rules may differ in other languages. Spaces are never used within formulae.

(a) To separate the names of ions in salts.

Examples:

1. NaCl	sodium chloride
2. $NaTl(NO_3)_2$	sodium thallium(I) dinitrate

(b) In names of binary compounds, to separate the electropositive part from the electronegative part.

Example:

3. P_4O_{10}	tetraphosphorus decaoxide

(c) To separate the arabic numeral from the symbols of central atoms in the bonding descriptor in the name of a polynuclear entity with several direct bonds between central atoms.

Example:

4. $[Os_3(CO)_{12}]$	*cyclo*-tris(tetracarbonylosmium)(3 *Os*—*Os*)

(d) In names of (formal) addition compounds, to separate the stoichiometric descriptor from the remainder of the name.

Example:

5. $3CdSO_4{\cdot}8H_2O$	cadmium sulfate—water (3/8)

(e) In solid-state nomenclature, to separate formula and structural type.

Example:

6. $TiO_2(o)$ (*brookite* type)

IR-2.7 ELISIONS

In general, in compositional and additive nomenclature no elisions are made when using multiplicative prefixes.

Example:

1. tetraaqua (*not* tetraqua)
2. monoooxygen (*not* monoxygen)
3. tetraarsenic hexaoxide

However, monoxide, rather than monooxide, is an allowed exception through general use.

IR-2.8 NUMERALS

IR-2.8.1 Arabic numerals

Arabic numerals are crucially important in nomenclature; their placement in a formula or name is especially significant.

They are used in *formulae* in many ways.

(a) As right subscripts, to indicate the number of individual constituents (atoms or groups of atoms). Unity is not indicated.

Examples:

1. $CaCl_2$
2. $[Co(NH_3)_6]Cl_3$

(b) As a right superscript, to indicate the charge. Unity is not indicated.

Examples:

3. Cl^-
4. NO^+
5. Cu^{2+}
6. $[Al(H_2O)_6]^{3+}$

(c) To indicate the composition of (formal) addition compounds or non-stoichiometric compounds. The numeral is written on the line before the formula of each constituent except that unity is omitted.

Examples:

7. $Na_2CO_3{\cdot}10H_2O$
8. $8WO_3{\cdot}9Nb_2O_5$

(d) To designate the mass number and/or the atomic number of nuclides represented by their symbols. The mass number is written as a left superscript, and the atomic number as a left subscript.

Examples:

9. ${}^{18}_{8}O$
10. ${}^{3}_{1}H$

(e) As a right superscript to the symbol η, to indicate the hapticity of a ligand (see Sections IR-9.2.4.3 and IR-10.2.5.1). As a right subscript to the symbol μ, to indicate the bridging multiplicity of a ligand (see Section IR-9.2.5.2).

Example:

11. $[\{Ni(\eta^5\text{-}C_5H_5)\}_3(\mu_3\text{-}CO)_2]$

Arabic numerals are also used as locants in *names* (see Section IR-2.14.2), and in the following ways.

(a) To indicate the number of metal–metal bonds in polynuclear compounds.

Example:

12.

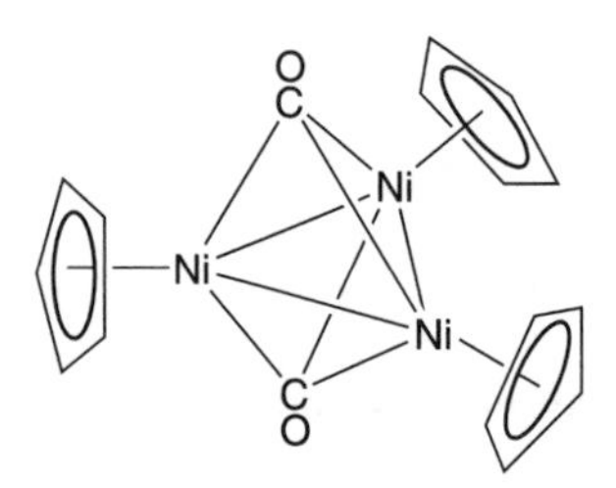

di-μ_3-carbonyl-*cyclo*-tris(cyclopentadienylnickel)(3 *Ni*—*Ni*)

(b) To indicate charge.

Examples:

13. $[CoCl(NH_3)_5]^{2+}$ pentaamminechloridocobalt(2+)
14. $[AlCl_4]^-$ tetrachloridoaluminate(1−)

Note that the number '1' must be included in order to avoid ambiguity in relation to symbols for optical rotation [see Section IR-2.2.3.1(j)].

(c) As a right subscript to the symbol μ, to indicate bridging multiplicity of a ligand (see Section IR-9.2.5.2).

Example:

15.

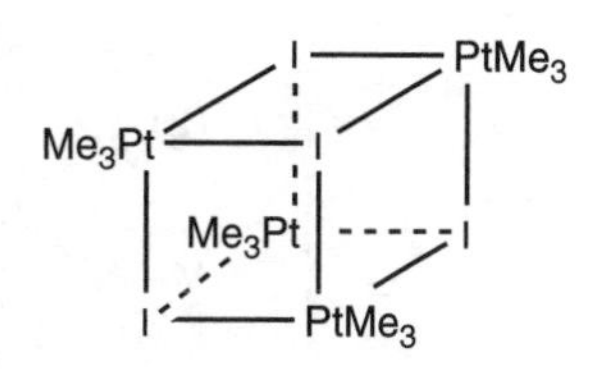

$[\{Pt(\mu_3\text{-}I)Me_3\}_4]$ tetra-μ_3-iodido-tetrakis[trimethylplatinum(IV)]

(d) In the nomenclature of boron compounds (see Chapter IR-6.2.3), to indicate the number of hydrogen atoms in the parent borane molecule. The arabic numeral is enclosed in parentheses immediately following the name.

Examples:

16. B_2H_6 diborane(6)
17. $B_{10}H_{14}$ decaborane(14)

(e) As a right superscript to the symbol κ, to indicate the number of donor atoms of a particular type bound to a central atom or central atoms (see Sections IR-9.2.4.2 and IR-10.2.3.3).

(f) As a right superscript to the symbol η, to indicate the hapticity of a ligand. (See Sections IR-9.2.4.3 and IR-10.2.5.1.)

(g) In polynuclear structures, arabic numerals are part of the CEP descriptor[5] used to identify polyhedral shapes. (See also Section IR-9.2.5.6.)

(h) In the stoichiometric descriptor terminating the name of a (formal) addition compound (see Section IR-5.5).

Example:

18. $8H_2S{\cdot}46H_2O$ hydrogen sulfide—water (8/46)

(i) As a right superscript, to indicate the non-standard bonding number in the λ convention. (See Section IR-6.2.1.)

Example:

19. IH_5 λ^5-iodane

(j) To describe the coordination geometry and configuration of ligands around a central atom using polyhedral symbols and configuration indexes (see Sections IR-9.3.2 and IR-9.3.3).

Example:

20.

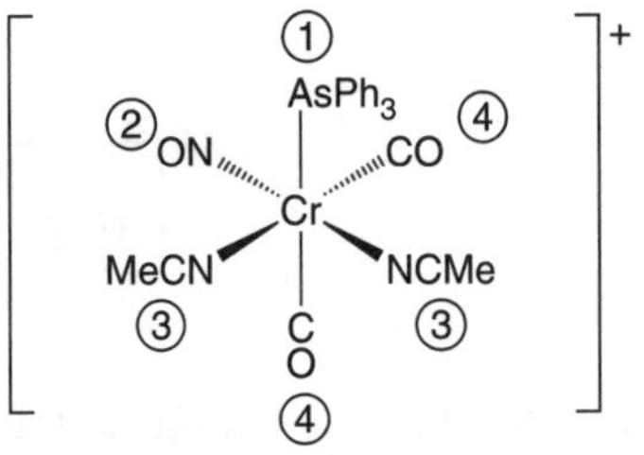

$$\left[\begin{array}{c} \text{①} \\ \text{AsPh}_3 \\ \text{②ON} \cdots \text{Cr} \cdots \text{CO ④} \\ \text{MeCN ③} \quad \text{NCMe ③} \\ \text{CO ④} \end{array} \right]^+$$

(*OC*-6-43)-bis(acetonitrile)dicarbonylnitrosyl(triphenylarsane)chromium(1+)

IR-2.8.2 Roman numerals

Roman numerals are used in *formulae* as right superscripts to designate the formal oxidation state.

Examples:

1. $[Co^{II}Co^{III}W_{12}O_{42}]^{7-}$
2. $[Mn^{VII}O_4]^-$
3. $Fe^{II}Fe^{III}{}_2O_4$

In *names* they indicate the formal oxidation state of an atom, and are enclosed in parentheses immediately following the name of the atom being qualified.

Examples:

4. $[Fe(H_2O)_6]^{2+}$ hexaaquairon(II)
5. $[FeO_4]^{2-}$ tetraoxidoferrate(VI)

IR-2.9 ITALIC LETTERS

Italic letters are used in *names* as follows.

(a) For geometrical and structural prefixes such as *cis, cyclo, catena, triangulo, nido, etc.* (see Table V).

(b) To designate symbols of central atoms in the bonding descriptor in polynuclear compounds.

Example:

1. $[Mn_2(CO)_{10}]$ bis(pentacarbonylmanganese)(*Mn*—*Mn*)

(c) In double oxides and hydroxides when the structural type is to be indicated.

Example:

2. $MgTiO_3$ (*ilmenite* type)

(d) In coordination compounds, to designate the symbols of the atom or atoms of a ligand (usually polydentate) to which the central atom is bound, whether the kappa convention is used or not. (See Section IR-9.2.4.4.)

Example:

3.

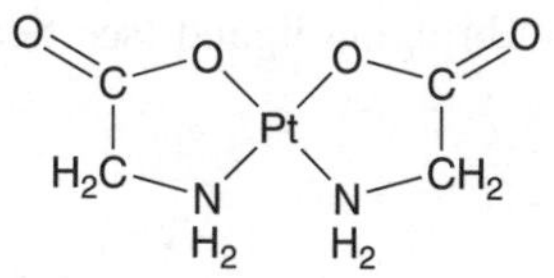

cis-bis(glycinato-κ*N*,κ*O*)platinum

(e) In solid-state chemistry, in Pearson and crystal system symbols. (See Sections IR-3.4.4 and IR-11.5.)

(f) Italicized capital letters are used in polyhedral symbols. (See Section IR-9.3.2.1.)

Example:

4.

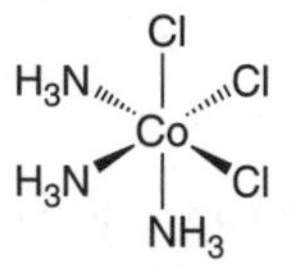

$[CoCl_3(NH_3)_3]$ (*OC*-6-22)-triamminetrichloridocobalt(III)

(g) Other uses of italicized capital letters are as locants in substitutive nomenclature (see, for example, Section IR-6.2.4.1), and the letter *H* for indicated hydrogen (see, for example, Section IR-6.2.3.4). Italic lower case letters are used to represent numbers, especially in formulae where the numbers are undefined.

Examples:

5. $(HBO_2)_n$
6. Fe^{n+}

IR-2.10 GREEK ALPHABET

Greek letters (in Roman type) are used in systematic inorganic nomenclature as follows:

Δ to show absolute configuration, or as a structural descriptor to designate deltahedra (see Section IR-9.3.4);

δ to denote the absolute configuration of chelate ring conformations (see Section IR-9.3.4); in solid-state chemistry to indicate small variations of composition (see Section IR-11.3.2); to designate cumulative double bonds in rings or ring systems (see Section P-25.7 of Ref. 1);

η to designate the hapticity of a ligand (see Sections IR-9.2.4.3 and IR-10.2.5.1);

κ as a ligating atom designator in the kappa convention (see Sections IR-9.2.4.2 and IR-10.2.3.3);

Λ to show absolute configuration (see Section IR-9.3.4);

λ to indicate non-standard bonding number in the lambda convention (see Section IR-6.2.1 and Section P-14.1 of Ref.1); to denote the absolute configuration of chelate ring conformations (see Section IR-9.3.4);

μ to designate a bridging ligand (see Sections IR-9.2.5.2 and IR-10.2.3.1).

IR-2.11 ASTERISKS

The asterisk (*) is used in *formulae* as a right superscript to the symbol of an element, in the following ways:

(a) To highlight a chiral centre.

Example:

1.

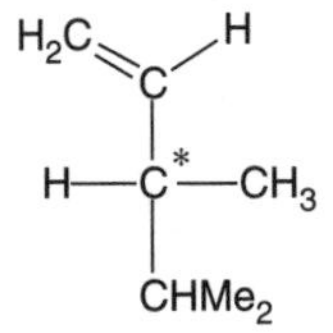

This usage has been extended to label a chiral ligand or a chiral centre in coordination chemistry.

Example:

2.

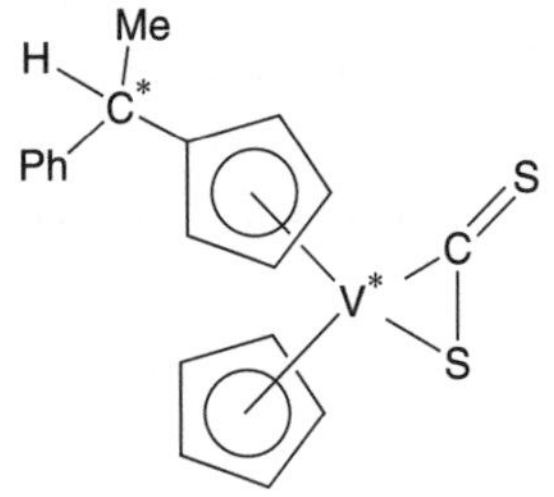

(b) To designate excited molecular or nuclear states.

Example:

3. NO*

IR-2.12 PRIMES

(a) Primes (′), double primes (″), triple primes (‴), *etc.* may be used in the names and formulae of coordination compounds in the following ways:

(i) within ligand names, in order to differentiate between sites of substitution;
(ii) when specifying donor atoms (IR-9.2.4.2), in order to differentiate between donor atoms;

(iii) when specifying configuration using configuration indexes (IR-9.3.5.3), in order to differentiate between donor atoms of the same priority, depending on whether they are located within the same ligand or portion of the ligand.

Example:

1. $[Rh_3Cl(\mu\text{-}Cl)(CO)_3\{\mu_3\text{-}Ph_2PCH_2P(Ph)CH_2PPh_2\}_2]^+$

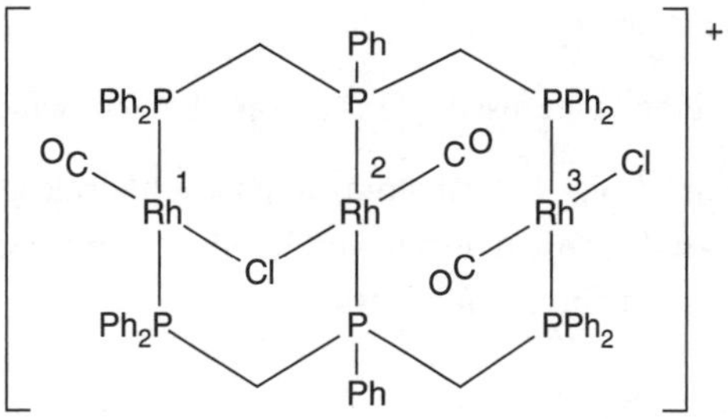

tricarbonyl-1κ*C*,2κ*C*,3κ*C*-μ-chlorido-1:2κ²*Cl*-chlorido-3κ*Cl*-bis{μ₃-bis[(diphenylphosphanyl)methyl]-1κ*P*:3κ*P′*-phenylphosphane-2κ*P*}trirhodium(1+)

(b) Primes, double primes, triple primes, *etc.* are also used as right superscripts in the Kröger–Vink notation (see Section IR-11.4) where they indicate a site which has one, two, three, *etc.* units of negative effective charge.

Example:

2. $Li^{x}_{Li,1-2x}Mg^{\bullet}_{Li,x}V'_{Li,x}Cl^{x}_{Cl}$

IR-2.13 MULTIPLICATIVE PREFIXES

The number of identical chemical entities in a name is expressed by a multiplicative prefix (see Table IV).

In the case of simple entities such as monoatomic ligands the multiplicative prefixes di, tri, tetra, penta, *etc.*, are used.

The multiplicative prefixes bis, tris, tetrakis, pentakis, *etc.* are used with composite ligand names or in order to avoid ambiguity. The modified entity is placed within parentheses.

Examples:

1. Fe_2O_3	diiron trioxide
2. $[PtCl_4]^{2-}$	tetrachloridoplatinate(2−)
3. $[Fe(CCPh)_2(CO)_4]$	tetracarbonylbis(phenylethynyl)iron
4. TlI_3	thallium tris(iodide) (*cf.* Section IR-5.4.2.3)
5. $Ca_3(PO_4)_2$	tricalcium bis(phosphate)
6. $[Pt(PPh_3)_4]$	tetrakis(triphenylphosphane)platinum(0)

Composite multiplicative prefixes are built up by citing units first, then tens, hundreds and so on, *e.g.* 35 is written pentatriaconta (or pentatriacontakis).

IR-2.14 LOCANTS

IR-2.14.1 Introduction

Locants are used to indicate the position of a substituent on, or a structural feature within, a parent molecule. The locants can be arabic numerals or letters.

IR-2.14.2 Arabic numerals

Arabic numerals are used as locants in the following ways.

(a) For numbering skeletal atoms in parent hydrides, to indicate: the placement of hydrogen atoms when there are non-standard bonding numbers; unsaturation; the positions of bridging hydrogen atoms in a borane structure.

Examples:

1. $H_5\overset{1}{S}\overset{2}{S}\overset{3}{S}H_4\overset{4}{S}H$ $1\lambda^6,3\lambda^6$-tetrasulfane (*not* $2\lambda^6,4\lambda^6$)

2. $H_2\overset{1}{N}\overset{2}{N}{=}\overset{3}{N}H\overset{4}{N}\overset{5}{N}H_2$ pentaaz-2-ene

3.

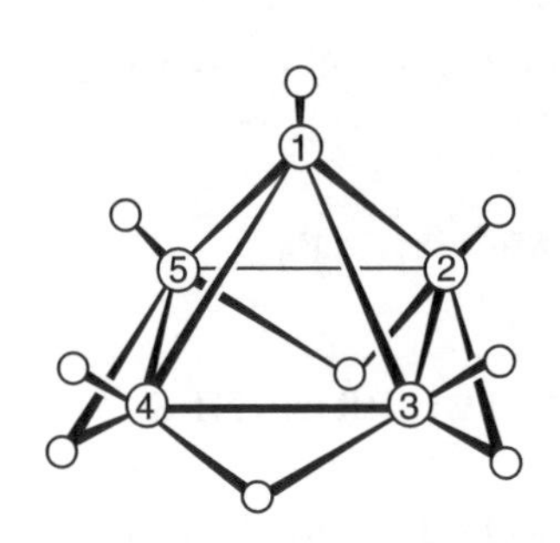

2,3:2,5:3,4:4,5-tetra-μ*H*-*nido*-pentaborane(9)

(b) In replacement nomenclature.

Example:

4.
$\overset{1}{C}H_3\overset{2}{S}\overset{3}{C}H_2\overset{4}{Si}H_2\overset{5}{C}H_2\overset{6}{C}H_2\overset{7}{O}\overset{8}{C}H_2\overset{9}{C}H_2\overset{10}{O}\overset{11}{C}H_3$
7,10-dioxa-2-thia-4-silaundecane

(c) In additive nomenclature.

Example:

5.
$\overset{1}{Si}H_3\overset{2}{Ge}H_2\overset{3}{Si}H_2\overset{4}{Si}H_2\overset{5}{Si}H_3$
1,1,1,2,2,3,3,4,4,5,5,5-dodecahydrido-2-germy-1,3,4,5-tetrasily-[5]catena

(d) In the Hantzsch–Widman nomenclature (Section IR-6.2.4.3), to indicate positions of skeletal atoms.

Example:

6.

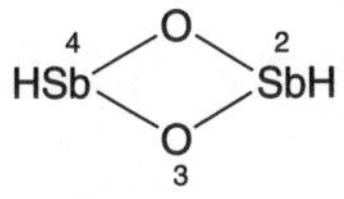

1,3,2,4-dioxadistibetane

(e) In the Hantzsch–Widman nomenclature (Section IR-6.2.4.3), to denote indicated hydrogen.

Example:

7.

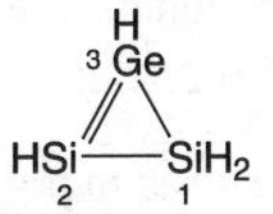

1*H*-1,2,3-disilagermirene

(f) In substitutive nomenclature, to specify the positions of substituent groups.

Example:

8.

$\overset{1}{HOSi}H_2\overset{2}{Si}H_2\overset{3}{Si}H_2\overset{4}{Si}HCl\overset{5}{Si}H_2Cl$

4,5-dichloropentasilan-1-ol

(g) In substitutive nomenclature, to specify the skeletal atom at which an additive or substractive operation is performed.

Example:

9. •HNNH• and –HNNH– hydrazine-1,2-diyl

(h) In von Baeyer names, to indicate the topology of a polycyclic ring system.

Example:

10.

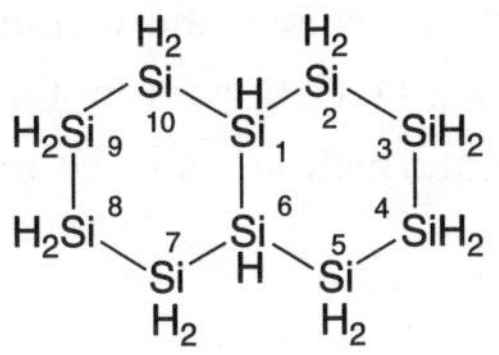

bicyclo[4.4.0]decasilane

(i) In polynuclear coordination compounds, for numbering the central atoms (see Section IR-9.2.5).

Example:

11.

$[(OC)_5\overset{1}{Re}\overset{2}{Co}(CO)_4]$ nonacarbonyl-1κ^5*C*,2κ^4*C*-rheniumcobalt(*Re*—*Co*)

(j) To indicate stereochemistry at particular atoms in structures where arabic numerals have been used for numbering those atoms.

Example:

12.

$\overset{1}{Cl}SiH_2\overset{2}{Si}HCl\overset{3}{Si}HCl\overset{4}{Si}H_2\overset{5}{Si}H_3$

(2*R*,3*S*)-1,2,3-trichloropentasilane

IR-2.14.3 Letter locants

Italicized upper case letters are used as locants in certain substitutive names. (See, for example, Section IR-6.2.4.1.)

Lower case letters are used in polyoxometallate nomenclature to designate the vertices of the coordination polyhedra around the central atoms. They are attached to the number of the central atom to which a particular vertex refers. A detailed treatment is given in Chapter II-1 of Ref. 2.

IR-2.15 ORDERING PRINCIPLES

IR-2.15.1 Introduction

Chemical nomenclature deals with names of elements and their combinations. Whereas writing the symbol or the name of an element is straightforward, a choice of which element to write first in the formula and in the name has to be made as soon as an element is associated with one or more other elements to form, for example, a binary compound. The order of citation of elements in formulae and names is based upon the methods outlined below. Furthermore, groups of atoms, such as ions, ligands in coordination compounds and substituent groups in derivatives of parent hydrides, are ordered according to specified rules.

IR-2.15.2 Alphabetical order

Alphabetical order is used in *formulae* as follows.

(a) Within the group of cations and within the group of anions, respectively, in formulae of salts and double salts. Deviations from this rule may be acceptable if it is desired to convey specific structural information, as in Example 5 below.

Examples:

1. BiClO (anions Cl^- and O^{2-})
2. NaOCl (the anion is OCl^-, *cf.* Section IR-4.4.3.1)
3. $KNa_4Cl(SO_4)_2$
4. $CaTiO_3$ (*perovskite* type)
5. $SrFeO_3$ (*perovskite* type)

(b) In formulae of coordination compounds and species formally treated as coordination compounds, ligands are cited in alphabetical order of their formulae or abbreviations (*cf.* Section IR-2.15.3.4). Where possible, the donor atom symbol in ligand formulae should be placed nearest the symbol of the central atom to which it is coordinated. (See Section IR-9.2.3.1.)

Example:

6. $[CrCl_2(NH_3)_2(OH_2)_2]$

(c) In the construction of the formula for a (formal) addition compound, the formulae of the individual components are ordered first by number of each component, then alphabetically. (See Section IR-4.4.3.5.)

Alphabetical order is used in *names* as follows.

(d) In compositional names, the names of the formally electropositive components and the names of the formally electronegative components are each arranged alphabetically with the former group of components preceding the latter. Note that this order of components may therefore deviate from the order of the corresponding components in the formula, as in Examples 7, 9 and 10 below.

Examples:

7. $KMgF_3$ — **m**agnesium **p**otassium fluoride
8. BiClO — bismuth **c**hloride **o**xide
9. ZnI(OH) — zinc **h**ydroxide **i**odide
10. $SrFeO_3$ — **ir**on **s**trontium oxide (*cf.* Example 5 above)

(e) In the citation of ligands in additive names. The alphabetical citation of ligand names is maintained regardless of the number of each ligand, or whether the compounds are mononuclear or polynuclear (*cf.* Section IR-2.15.3.4).

Examples:

11. $K[AuS(S_2)]$ — potassium (**d**isulfido)**s**ulfidoaurate(1−)
12. $[CrCl_2(NH_3)_4]^+$ — tetra**a**mminedi**c**hloridochromium(1+)

A similar rule applies when citing names of substituent groups in substitutive nomenclature (see Section IR-6.3.1).

(f) For citation of the names of the skeletal atoms in the chains and rings additive nomenclature (*cf.* Section IR-7.4.3).

Example:

13. $HOS(O)_2SeSH$
1,4-dihydrido-2,2-dioxido-1-**o**xy-3-**s**eleny-2,4-di**s**ulfy-[4]catena

(g) In the construction of the name for a (formal) addition compound, the names of the individual components are ordered first by number of each component, then alphabetically. (See Section IR-5.5.)

IR-2.15.3 Other ordering rules

IR-2.15.3.1 *Element ordering on the basis of the periodic table*

One important element sequence based on the periodic table is shown in Table VI. The element columns (1 to 18) are connected by arrows leading in a direction starting from the less metallic elements and moving towards the more metallic elements. Only H has a unique position. This order has its origin in electronegativity considerations even though O is now placed at its usual position in group 16. It is used for ordering element symbols and element names in the following cases.

(a) In compositional names of binary compounds and corresponding formulae, the element encountered *last* when following the arrow in Table VI is represented *first* in the formula as well as the name.

Examples:

1. S_2Cl_2 disulfur dichloride
2. O_2Cl dioxygen chloride
3. H_2Te dihydrogen telluride
4. AlH_3 aluminium trihydride

(b) In additive names of polynuclear compounds, the central atom encountered *last* when following the arrow is listed first, *cf.* Sections IR-7.3.2 and IR-9.2.5.1.

(c) In additive names for chains and rings, to determine the numbering of the skeletal atoms if this is not defined fully by the structure of the skeleton. The element encountered *first* when following the arrows in Table VI is given the *lowest* number. Note that the element 'y' terms (Table X) are cited alphabetically.

Example:

5. $HOS(O)_2SeSH$
   1,4-dihydrido-2,2-dioxido-1-oxy-3-seleny-2,4-disulfy-[4]catena

(d) In Hantzsch–Widman names, the element encountered *first* when following the arrows in Table VI is given the *lowest* number. The element 'a' terms (Table X) are cited in the same order.

Examples:

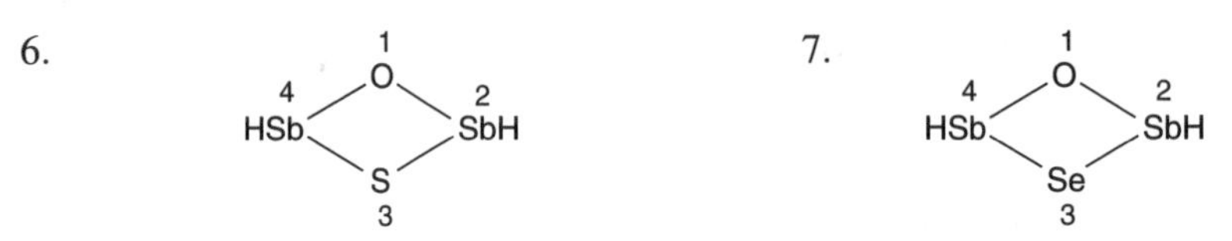

1,3,2,4-oxathiadistibetane 1,3,2,4-oxaselenadistibetane

(e) In names where heteroatoms replacing skeleton atoms are indicated by 'a' prefixes, the element encountered *first* when following the arrows in Table VI is given the *lowest* number. The element 'a' terms (Table X) are cited in the same order.

Examples:

8. 1-oxa-3-thia-2,4-distibacyclobutane

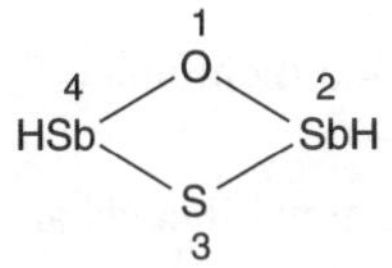

9. 1-oxa-3-selena-2,4-distibacyclobutane

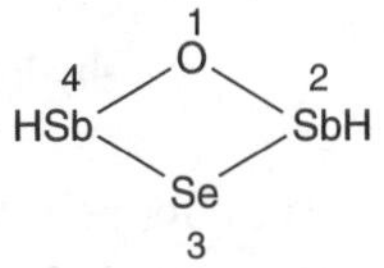

IR-2.15.3.2 *Ordering of parent hydrides*

Where there is a choice of parent hydrides among those listed in Table IR-6.1 (or corresponding hydrides with non-standard bonding numbers, *cf.* Section IR-6.2.2.2), the name is based on the parent hydride of the element occurring first in the sequence:

N > P > As > Sb > Bi > Si > Ge > Sn > Pb > B > Al > Ga > In > Tl > O >

S > Se > Te > C > F > Cl > Br > I.

This applies in particular to the naming of organometallic compounds of elements of groups 13–16 when a choice has to be made between several parent hydrides (Section IR-10.3.4).

Example:

1. $AsCl_2GeH_3$ dichloro(germyl)arsane

Note that due to the rules of substitutive nomenclature[1] the above does not necessarily come into play even if two or more elements appearing in the sequence are present in the compound. For example, the substitutive name for HTeOH is tellanol, *i.e.* based on tellane, not oxidane, because the characteristic group OH must be cited as a suffix.

IR-2.15.3.3 *Ordering characteristic groups for substitutive nomenclature*

In substitutive nomenclature, an order for the choice of principal functional group is defined (see Section P-41 of Ref.1).

IR-2.15.3.4 *Ordering ligands in formulae and names*

In *formulae* of coordination compounds, the formulae or abbreviations representing the ligands are cited in alphabetical order as the general rule. Bridging ligands are cited immediately after terminal ligands of the same kind, if any, and in increasing order of bridging multiplicity. (See also Sections IR-9.2.3 and IR-9.2.5.)

In *names* of coordination compounds, the ligand names precede the name of the central atom and are cited in alphabetical order. For each ligand type, bridging ligands are cited immediately before terminal ligands of the same kind, if any, *e.g.* di-μ-chlorido-tetrachlorido, and in decreasing order of bridging multiplicity, *e.g.* μ_3-oxido-di-μ-oxido.... (See also Sections IR-9.2.2 and IR-9.2.5.1.)

Example:

1. $[Cr_2(\mu\text{-}O)(OH)_8(\mu\text{-}OH)]^{5-}$
   μ-hydroxido-octahydroxido-μ-oxido-dichromate(5−)

Thus, for both *formulae* and *names* the terminal ligands are closer to the central atom, with the multiplicity of the bridging ligands increasing further away from the metal.

IR-2.15.3.5 *Ordering components in salt formulae and names*

In formulae and names of salts, double salts and coordination compounds, cations precede anions. Ordering within each of these groups is alphabetical, *cf.* Section IR-2.15.2.

IR-2.15.3.6 *Isotopic modification*

In isotopically modified compounds, a principle governs the order of citation of nuclide symbols. (See Section II-2.2.5 of Ref. 2.)

IR-2.15.3.7 *Stereochemical priorities*

In the stereochemical nomenclature of coordination compounds, the procedure for assigning priority numbers to the ligating atoms of a mononuclear coordination system is based upon the standard sequence rules developed for chiral carbon compounds (the Cahn, Ingold, Prelog or CIP rules[6], see Section IR-9.3.3.2).

IR-2.15.3.8 *Hierarchial ordering of punctuation marks*

In the names of coordination compounds and boron compounds, the punctuation marks used to separate the symbols of atoms from the numerical locants, the locants indicating bridging atoms, and the various other sets of locants which may be present, are arranged in the following hierarchy:

semicolon > colon > comma.

The colon is only used for bridging ligands, so that the more restricted general hierarchy is simply comma < semicolon. The sequence when bridging ligands are being specified is comma < colon. (See Example 2 in Section IR-2.5.2, and Section IR-9.2.5.5.)

IR-2.16 FINAL REMARKS

In this chapter, the various uses of letters, numerals and symbols in names and formulae have been gathered under common headings to provide an easy check to ensure that the constructed name or formula is in accord with agreed practice. However, this chapter is not sufficient to make clear all the rules needed to build a name or a formula, and the reader is therefore advised to consult other appropriate chapters for the more detailed treatment.

IR-2.17 REFERENCES

1. *Nomenclature of Organic Chemistry, IUPAC Recommendations,*, eds. W.H. Powell and H. Favre, Royal Society of Chemistry, in preparation: [See also, *Nomenclature of Organic Chemistry*, Pergamon Press, Oxford, 1979; *A Guide to IUPAC Nomenclature of Organic Compounds, Recommendations 1993*, eds. R. Panico, W.H. Powell and J.-C. Richer, Blackwell Scientific Publications, Oxford, 1993; and corrections in *Pure Appl. Chem.*, **71**, 1327–1330 (1999)].
2. *Nomenclature of Inorganic Chemistry II, IUPAC Recommendations 2000*, eds. J.A. McCleverty and N.G. Connelly, Royal Society of Chemistry, 2001. (Red Book II.)
3. *Compendium of Macromolecular Nomenclature*, ed. W.V. Metanomski, Blackwell Scientific Publications, Oxford, 1991. (The Purple Book. The second edition is planned for publication in 2005).
4. *Nomenclature of Inorganic Chemistry, IUPAC Recommendations 1990*, ed. G.J. Leigh, Blackwell Scientific Publications, Oxford, 1990.
5. J.B. Casey, W.J. Evans and W.H. Powell, *Inorg. Chem.*, **20**, 1333–1341 (1981).
6. R.S. Cahn, C. Ingold and V. Prelog, *Angew. Chem., Int. Ed. Engl.*, **5**, 385–415 (1966); V. Prelog and G. Helmchen, *Angew. Chem., Int. Ed. Engl.*, **21**, 567–583 (1982).

IR-3 Elements

CONTENTS

IR-3.1 NAMES AND SYMBOLS OF ATOMS

The origins of the names of some chemical elements, for example antimony, are lost in antiquity. Other elements recognized (or discovered) during the past three centuries were named according to various associations of origin, physical or chemical properties, *etc*., and more recently to commemorate the names of eminent scientists.

In the past, some elements were given two names because two groups claimed to have discovered them. To avoid such confusion it was decided in 1947 that after the existence of a new element had been proved beyond reasonable doubt, discoverers had the right to *suggest* a name to IUPAC, but that only the Commission on Nomenclature of Inorganic Chemistry (CNIC) could make a recommendation to the IUPAC Council to make the final decision.

Under the present procedure,[1] claims of the discovery of a new element are first investigated by a joint IUPAC-IUPAP (International Union of Pure and Applied Physics) committee which then assigns priority. The acknowledged discoverers are then invited to *suggest* a name to the Division of Inorganic Chemistry, which then makes a formal recommendation to the IUPAC Council. It is emphasized that the name for any new element resulting from the above process, once approved by Council, is not intended to carry any implication regarding priority of discovery. The same is understood to apply to the names approved by IUPAC in the past, whatever the prehistory.

The IUPAC-approved names of the atoms of atomic numbers 1-111 for use in the English language are listed in alphabetical order in Table I*. It is obviously desirable that the names used in any language resemble these names as closely as possible, but it is recognized that for elements named in the past there are often well-established and very different names in other languages. In the footnotes of Table I, certain names are cited which are not used now in English, but which either provide the basis of the atomic symbol, or the basis of certain affixes used in nomenclature.

For use in chemical formulae, each atom is represented by a unique symbol in upright type as shown in Table I. In addition, the symbols D and T may be used for the hydrogen isotopes of mass numbers two and three, respectively (see Section IR-3.3.2).

IR-3.1.1 Systematic nomenclature and symbols for new elements

Newly discovered elements may be referred to in the scientific literature but until they have received permanent names and symbols from IUPAC, temporary designators are required. Such elements may be referred to by their atomic numbers, as in 'element 120' for example, but IUPAC has approved a systematic nomenclature and series of three-letter symbols (see Table II).[2]

The name is derived directly from the atomic number of the element using the following numerical roots:

0 = nil	3 = tri	6 = hex	9 = enn
1 = un	4 = quad	7 = sept	
2 = bi	5 = pent	8 = oct	

The roots are put together in the order of the digits which make up the atomic number and terminated by 'ium' to spell out the name. The final 'n' of 'enn' is elided when it occurs before 'nil', and the final 'i' of 'bi' and of 'tri' when it occurs before 'ium'.

The symbol for the element is composed of the initial letters of the numerical roots which make up the name.

Example:

1. element 113 = ununtrium, symbol Uut

IR-3.2 INDICATION OF MASS, CHARGE AND ATOMIC NUMBER USING INDEXES (SUBSCRIPTS AND SUPERSCRIPTS)

The mass, charge and atomic number of a nuclide are indicated by means of three indexes (subscripts and superscripts) placed around the symbol. The positions are occupied as follows:

left upper index	mass number
left lower index	atomic number
right upper index	charge

A charge placed on an atom of symbol A is indicated as A^{n+} or A^{n-}, not as A^{+n} or A^{-n}.

* Tables numbered with a Roman numeral are collected together at the end of this book.

The right lower position of an atomic symbol is reserved for an index (subscript) indicating the number of such atoms in a formula. For example, S_8 is the formula of a molecule containing eight sulfur atoms (see Section IR-3.4). For formalisms when oxidation states or charges are also shown, see Section IR-4.6.1.

Example:

1. $^{32}_{16}S^{2+}$ represents a doubly ionized sulfur atom of atomic number 16 and mass number 32.

The nuclear reaction between $^{26}_{12}Mg$ and $^{4}_{2}He$ nuclei to yield $^{29}_{13}Al$ and $^{1}_{1}H$ nuclei is written as follows[3]:

$$^{26}Mg(\alpha, p)^{29}Al$$

For the use of atomic symbols to indicate isotopic modification in chemical formulae and the nomenclature of isotopically modified compounds see Section IR-4.5 and Chapter II-2 of Ref. 4 respectively.

IR-3.3 ISOTOPES

IR-3.3.1 Isotopes of an element

The isotopes of an element all bear the same name (but see Section IR-3.3.2) and are designated by mass numbers (see Section IR-3.2). For example, the atom of atomic number 8 and mass number 18 is named oxygen-18 and has the symbol ^{18}O.

IR-3.3.2 Isotopes of hydrogen

Hydrogen is an exception to the rule in Section IR-3.3.1 in that the three isotopes ^{1}H, ^{2}H and ^{3}H can have the alternative names protium, deuterium and tritium, respectively. The symbols D and T may be used for deuterium and tritium but ^{2}H and ^{3}H are preferred because D and T can disturb the alphabetical ordering in formulae (see Section IR-4.5). The combination of a muon and an electron behaves like a light isotope of hydrogen and is named muonium, symbol Mu.[5]

These names give rise to the names proton, deuteron, triton and muon for the cations $^{1}H^{+}$, $^{2}H^{+}$, $^{3}H^{+}$ and Mu^{+}, respectively. Because the name proton is often used in contradictory senses, *i.e.* for isotopically pure $^{1}H^{+}$ ions on the one hand, and for the naturally occurring undifferentiated isotope mixture on the other, it is recommended that the undifferentiated mixture be designated generally by the name hydron, derived from hydrogen.

IR-3.4 ELEMENTS (or elementary substances)

IR-3.4.1 Name of an element of indefinite molecular formula or structure

A sample of an element that has an undefined formula, or is a mixture of allotropes (see Sections IR-3.4.2 to IR-3.4.5), bears the same name as the atom.

IR-3.4.2 Allotropes (allotropic modifications) of elements

Allotropic modifications of an element bear the name of the atom from which they are derived, together with a descriptor to specify the modification. Common descriptors are Greek letters (α, β, γ, *etc.*), colours and, where appropriate, mineral names (*e.g.* graphite and diamond for the well known forms of carbon). Such names should be regarded as provisional, to be used only until structures have been established, after which a rational system based on molecular formula (see Section IR-3.4.3) or crystal structure (see Section IR-3.4.4) is recommended. Common names will continue to be used for amorphous modifications of an element and for those which are mixtures of closely related structures in their commonly occurring forms (such as graphite) or have an ill-defined disordered structure (such as red phosphorus) (see Section IR-3.4.5).

IR-3.4.3 Names of allotropes of definite molecular formula

Systematic names are based on the number of atoms in the molecule, indicated by a multiplicative prefix from Table IV. The prefix 'mono' is only used when the element does not normally occur in a monoatomic state. If the number is large and unknown, as in long chains or large rings, the prefix 'poly' may be used. Where necessary, appropriate prefixes (Table V) may be used to indicate structure. When it is desired to specify a particular polymorph of an element with a defined structure (such as the α-, β- or γ-forms of S_8) the method of Section IR-3.4.4 should be used (see Examples 13–15 in Section IR-3.4.4).

Examples:

	Formula	*Systematic name*	*Acceptable alternative name*
1.	Ar	argon	
2.	H	monohydrogen	
3.	N	mononitrogen	
4.	N_2	dinitrogen	
5.	$N_3^{\bullet}$	trinitrogen(•)	
6.	O_2	dioxygen	oxygen
7.	O_3	trioxygen	ozone
8.	P_4	tetraphosphorus	white phosphorus
9.	S_6	hexasulfur	ε-sulfur
10.	S_8	*cyclo*-octasulfur	α-sulfur, β-sulfur, γ-sulfur
11.	S_n	polysulfur	μ-sulfur (or plastic sulfur)
12.	C_{60}	hexacontacarbon	[60]fullerene

In Example 12, the name [60]fullerene is to be regarded as an acceptable non-systematic name for a particular C_{60} structure. For more details see Section P-27 of Ref. 6.

IR-3.4.4 Crystalline allotropic modifications of an element

Crystalline allotropic modifications are polymorphs of the elements. Each can be named by adding the Pearson symbol (see Section IR-11.5.2)[7] in parentheses after the name of the atom. This symbol defines the structure of the allotrope in terms of its Bravais lattice (crystal class and type of unit cell, see Table IR-3.1) and the number of atoms in its unit cell. Thus, iron(*cF*4) is the allotropic modification of iron (γ-iron) with a cubic (*c*), all-face-centred (*F*) lattice containing four atoms of iron in the unit cell.

Table IR-3.1 *Pearson symbols used for the fourteen Bravais lattices*

System	*Lattice symbol*[a]	*Pearson symbol*
Triclinic	*P*	*aP*
Monoclinic	*P*	*mP*
	S[b]	*mS*
Orthorhombic	*P*	*oP*
	S	*oS*
	F	*oF*
	I	*oI*
Tetragonal	*P*	*tP*
	I	*tI*
Hexagonal (and trigonal *P*)	P	*hP*
Rhombohedral	*R*	*hR*
Cubic	*P*	*cP*
	F	*cF*
	I	*cI*

[a] *P*, *S*, *F*, *I* and *R* are primitive, side-face-centred, all-face-centred, body-centred and rhombohedral lattices, respectively. The letter *C* was formerly used in place of *S*.
[b] Second setting, *y*-axis unique.

Examples:

	Symbol	*Systematic name*	*Acceptable alternative name*
1.	P_n	phosphorus(*oS*8)	black phosphorus
2.	C_n	carbon(*cF*8)	diamond
3.	C_n	carbon(*hP*4)	graphite (common form)
4.	C_n	carbon(*hR*6)	graphite (less common form)
5.	Fe_n	iron(*cI*2)	α-iron
6.	Fe_n	iron(*cF*4)	γ-iron
7.	Sn_n	tin(*cF*8)	α- or grey tin
8.	Sn_n	tin(*tI*4)	β- or white tin
9.	Mn_n	manganese(*cI*58)	α-manganese
10.	Mn_n	manganese(*cP*20)	β-manganese
11.	Mn_n	manganese(*cF*4)	γ-manganese
12.	Mn_n	manganese(*cI*2)	δ-manganese

13.	S_8	sulfur(*oF*128)	α-sulfur
14.	S_8	sulfur(*mP*48)	β-sulfur
15.	S_8	sulfur(*mP*32)	γ-sulfur

In a few cases, the Pearson symbol fails to differentiate between two crystalline allotropes of the element. In such an event the space group is added to the parentheses. If this still fails to distinguish the allotropes, the characteristically different lattice parameters will have to be cited. An alternative notation involving compound type may also be useful (see Section IR-4.2.5 and Chapter IR-11).

IR-3.4.5 Solid amorphous modifications and commonly recognized allotropes of indefinite structure

Solid amorphous modifications and commonly recognized allotropes of indefinite structure are distinguished by customary descriptors such as a Greek letter, names based on physical properties, or mineral names.

Examples:

1. C_n vitreous carbon
2. C_n graphitic carbon (carbon in the form of graphite, irrespective of structural defects)
3. P_n red phosphorus [a disordered structure containing parts of phosphorus(*oS*8) and parts of tetraphosphorus]
4. As_n amorphous arsenic

IR-3.5 ELEMENTS IN THE PERIODIC TABLE

The groups of elements in the periodic table (see inside front cover) are numbered from 1 to 18. The elements (except hydrogen) of groups 1, 2 and 13–18 are designated as main group elements and, except in group 18, the first two elements of each main group are termed typical elements. Optionally, the letters s, p, d and f may be used to distinguish different blocks of elements. For example, the elements of groups 3–12 are the d-block elements. These elements are also commonly referred to as the transition elements, though the elements of group 12 are not always included; the f-block elements are sometimes referred to as the inner transition elements. If appropriate for a particular purpose, the various groups may be named from the first element in each, for example elements of the boron group (B, Al, Ga, In, Tl), elements of the titanium group (Ti, Zr, Hf, Rf), *etc.*

The following collective names for like elements are IUPAC-approved: alkali metals (Li, Na, K, Rb, Cs, Fr), alkaline earth metals (Be, Mg, Ca, Sr, Ba, Ra), pnictogens[8] (N, P, As, Sb, Bi), chalcogens (O, S, Se, Te, Po), halogens (F, Cl, Br, I, At), noble gases (He, Ne, Ar, Kr, Xe, Rn), lanthanoids (La, Ce, Pr, Nd, Pm, Sm, Eu, Gd, Tb, Dy, Ho, Er, Tm, Yb, Lu), rare earth metals (Sc, Y and the lanthanoids) and actinoids (Ac, Th, Pa, U, Np, Pu, Am, Cm, Bk, Cf, Es, Fm, Md, No, Lr).

The generic terms pnictide, chalcogenide and halogenide (or halide) are commonly used in naming compounds of the pnictogens, chalcogens and halogens.

Although lanthanoid means 'like lanthanum' and so should not include lanthanum, lanthanum has become included by common usage. Similarly, actinoid. The ending 'ide' normally indicates a negative ion, and therefore lanthanoid and actinoid are preferred to lanthanide and actinide.

IR-3.6 REFERENCES

1. Naming of New Elements, W.H. Koppenol, *Pure Appl. Chem.*, **74**, 787–791 (2002).
2. Recommendations for the Naming of Elements of Atomic Numbers Greater Than 100, J. Chatt, *Pure Appl. Chem.*, **51**, 381–384 (1979).
3. *Quantities, Units and Symbols in Physical Chemistry,* Second Edn., eds. I. Mills, T. Cvitas, K. Homann, N. Kallay and K. Kuchitsu, Blackwell Scientific Publications, Oxford, 1993. (The Green Book. The third edition is in preparation.)
4. *Nomenclature of Inorganic Chemistry II, IUPAC Recommendations 2000*, eds. J.A. McCleverty and N.G. Connelly, Royal Society of Chemistry, 2001. (Red Book II.)
5. Names for Muonium and Hydrogen Atoms and Their Ions, W.H. Koppenol, *Pure Appl. Chem.*, **73**, 377–379 (2001).
6. *Nomenclature of Organic Chemistry IUPAC Recommendations*, W.H. Powell and H. Favre, Royal Society of Chemistry, in preparation.
7. W.B. Pearson, *A Handbook of Lattice Spacings and Structures of Metals and Alloys*, Vol. 2, Pergamon Press, Oxford, 1967, pp. 1,2. For tabulated lattice parameters and data on elemental metals and semi-metals, see pp. 79–91. See also, P. Villars and L.D. Calvert, *Pearson's Handbook of Crystallographic Data for Intermetallic Phases*, Vols. 1–3, American Society for Metals, Metals Park, Ohio, USA, 1985.
8. The alternative spelling 'pnicogen' is also used.

IR-4 Formulae

CONTENTS

IR-4.1 INTRODUCTION

Formulae (empirical, molecular and structural formulae as described below) provide a simple and clear method of designating compounds. They are of particular importance in chemical equations and in descriptions of chemical procedures. In order to avoid ambiguity and for many other purposes, *e.g.* in databases, indexing, *etc.*, standardization is recommended.

IR-4.2 DEFINITIONS OF TYPES OF FORMULA

IR-4.2.1 Empirical formulae

The empirical formula of a compound is formed by juxtaposition of the atomic symbols with appropriate (integer) subscripts to give the simplest possible formula expressing the composition. For the order of citation of symbols in formulae, see Section IR-4.4, but, *in the absence of any other ordering criterion* (for example, if little structural information is available), the alphabetical order of atomic symbols should be used in an empirical formula, except that in carbon-containing compounds, C and H are usually cited first and second, respectively.[1]

Examples:

1. $BrClH_3N_2NaO_2Pt$
2. $C_{10}H_{10}ClFe$

IR-4.2.2 Molecular formulae

For compounds consisting of discrete molecules, the *molecular formula*, as opposed to the empirical formula, may be used to indicate the actual composition of the molecules. For the order of citation of symbols in molecular formulae, see Section IR-4.4.

The choice of formula depends on the chemical context. In some cases, the empirical formula may also correspond to a molecular composition, in which case the only possible difference between the two formulae is the ordering of the atomic symbols. If it is not desirable or possible to specify the composition, *e.g.* in the case of polymers, a letter subscript such as *n* may be used.

Examples:

	Molecular formula	*Empirical formula*
1.	S_8	S
2.	S_n	S
3.	SF_6	F_6S
4.	S_2Cl_2	ClS
5.	$H_4P_2O_6$	H_2O_3P
6.	Hg_2Cl_2	ClHg
7.	N_2O_4	NO_2

IR-4.2.3 Structural formulae and the use of enclosing marks in formulae

A structural formula gives partial or complete information about the way in which the atoms in a molecule are connected and arranged in space. In simple cases, a line formula that is just a sequence of atomic symbols gives structural information provided the reader knows that the formula represents the order of the atoms in the linear structure.

Examples:

1. HOCN (empirical formula CHNO)
2. HNCO (empirical formula also CHNO)
3. HOOH (empirical formula HO)

As soon as the compound has even a slightly more complex structure, it becomes necessary to use enclosing marks in line formulae to separate subgroups of atoms. Different enclosing marks must be used for repeating units and sidechains in order to avoid ambiguity.

The basic rules for applying enclosing marks in structural formulae are as follows:

(i) Repeating units in chain compounds are enclosed in square brackets.

(ii) Side groups to a main chain and groups (ligands) attached to a central atom are enclosed in parentheses (except single atoms when there is no ambiguity regarding their attachment in the structure, *e.g.* hydrogen in hydrides with a chain structure).

(iii) A formula or part of a formula which represents a molecular entity may be placed in enclosing marks. If an entire formula is enclosed, square brackets must be used, except if rule (v) applies.

(iv) A part of a formula which is to be multiplied by a subscript may also be enclosed in parentheses or braces, except in the case of repeating units in chain compounds, *cf.* rule (i).

(v) In the case of polymers, if the bonds between repeating units are to be shown, the repeating unit is enclosed in strike-through parentheses, with the dash superimposed on the parentheses representing the bond. (If this is typographically inconvenient, dashes can be placed before and after the parentheses.)

(vi) Inside square brackets, enclosing marks are nested as follows:
( ), {( )}, ({( )}), {({( )})}, *etc.*

(vii) Atoms or groups of atoms which are represented together with a prefixed symbol, *e.g.* a structural modifier such as 'μ', are placed within enclosing marks, using the nesting order of (vi).

The use of enclosing marks for the specification of isotopic modification is described in Section IR-4.5.

Compared to line formulae, displayed formulae (Examples 12 and 13 below) give more (or full) information about the structure.

(The rules needed for ordering the symbols in some of the example formulae below are given in Section IR-4.4.3.)

Examples:

4. $SiH_3[SiH_2]_8SiH_3$ [rule (i)]
5. $SiH_3[SiH_2]_5SiH(SiH_3)SiH_2SiH_3$ [rules (i) and (ii)]

6. $Ca_3(PO_4)_2$ [rule (iv)]
7. $[Co(NH_3)_6]_2(SO_4)_3$ [rules (iii), (iv), (vi)]
8. $[\{Rh(\mu\text{-}Cl)(CO)_2\}_2]$ [rules (iii), (vi), (vii)]
9. $K[Os(N)_3]$ [rules (ii), (iii)]
10. $-\!(\mathrm{S})\!-_n$ [rule (v)]
11. $(HBO_2)_n$, or $-\!(\mathrm{B(OH)O})\!-_n$ [rules (ii) and (v)]
12. 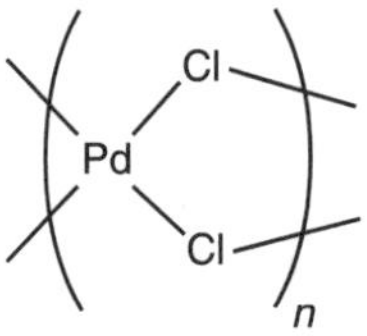
13. 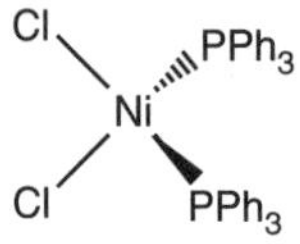
14. NaCl
15. [NaCl]

The first formula in Example 11 may be considered to be a molecular formula (Section IR-4.2.2) with no implications about the structure of the polymer in question.

In Examples 14 and 15, the formula [NaCl] may be used to distinguish the molecular compound consisting of one sodium atom and one chlorine atom from the solid with the composition NaCl.

IR-4.2.4 Formulae of (formal) addition compounds

In the formulae of addition compounds and compounds which can formally be regarded as such, including clathrates and multiple salts, a special format is used. The proportions of constituents are indicated by arabic numerals preceding the formulae of the constituents, and the formulae of the constituents are separated by a centre dot. The rules for ordering the constituent formulae are described in Section IR-4.4.3.5.

Examples:

1. $Na_2CO_3{\cdot}10H_2O$
2. $8H_2S{\cdot}46H_2O$
3. $BMe_3{\cdot}NH_3$

IR-4.2.5 Solid state structural information

Structural information can also be given by indicating structural type as a qualification of a formula. For example, polymorphs may be indicated by adding in parentheses an abbreviated expression for the crystal system (see Sections IR-11.5.2 and IR-11.7.2, and Table IR-3.1). Structures may also be designated by adding the name of a type-compound in italics in parentheses, but such usage may not be unambiguous. There are at least ten varieties of ZnS(*h*).

Where several polymorphs crystallise in the same crystal system they may be differentiated by the Pearson symbol (see Sections IR-3.4.4 and IR-11.5.2). Greek letters are frequently employed to designate polymorphs, but their use is often confused and contradictory and is not generally recommended.

Examples:

1. $TiO_2(t)$ (*anatase* type)
2. $TiO_2(t)$ (*rutile* type)
3. $AuCd(c)$, or AuCd (*CsCl* type)

For the formulae of solid solutions and non-stoichiometric phases, see Chapter IR-11.

IR-4.3 INDICATION OF IONIC CHARGE

Ionic charge is indicated by means of a right upper index, as in A^{n+} or A^{n-} (*not* A^{+n} or A^{-n}). If the formula is placed in enclosing marks, the right upper index is placed outside the enclosing marks. For polymeric ions, the charge of a single repeating unit should be placed inside the parentheses that comprise the polymeric structure or the total charge of the polymeric species should be placed outside the polymer parentheses. (The rules needed for ordering the symbols in some of the example formulae below are given in Section IR-4.4.3.)

Examples:

1. Cu^{+}
2. Cu^{2+}
3. NO^{+}
4. $[Al(OH_2)_6]^{3+}$
5. $H_2NO_3{}^{+}$
6. $[PCl_4]^{+}$
7. As^{3-}
8. $HF_2{}^{-}$
9. CN^{-}
10. $S_2O_7{}^{2-}$
11. $[Fe(CN)_6]^{4-}$
12. $[PW_{12}O_{40}]^{3-}$
13. $[P_3O_{10}]^{5-}$, or $[O_3POP(O)_2OPO_3]^{5-}$, or

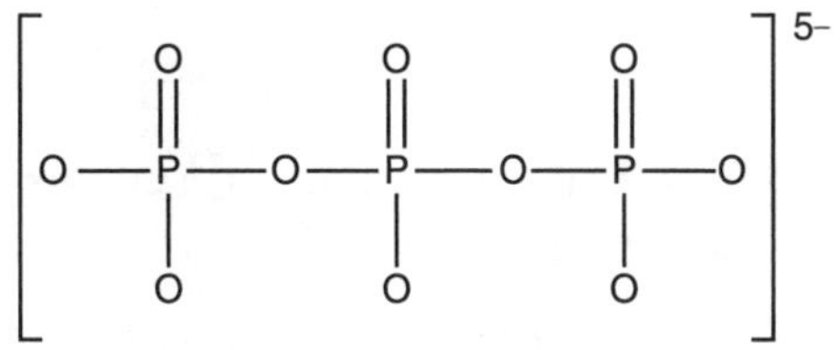

14. $([CuCl_3]^-)_n$, or $([CuCl_3])_n{}^{n-}$, or

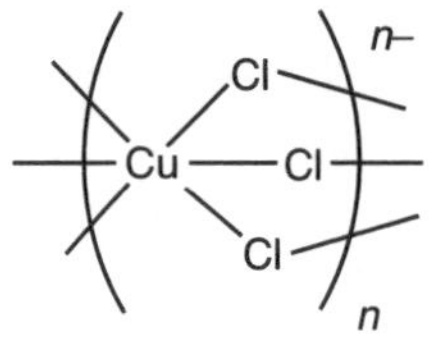

IR-4.4 SEQUENCE OF CITATION OF SYMBOLS IN FORMULAE

IR-4.4.1 **Introduction**

Atomic symbols in formulae may be ordered in various ways. Section IR-4.4.3 describes the conventions usually adopted for some important classes of compounds. As a prerequisite, Section IR-4.4.2 explains what is meant by the two ordering principles 'electronegativity' and 'alphabetical ordering'.

IR-4.4.2 **Ordering principles**

IR-4.4.2.1 *Electronegativity*

If electronegativity is taken as the ordering principle in a formula or a part of a formula, the atomic symbols are cited according to *relative* electronegativities, the least electronegative element being cited first. For this purpose, Table VI* is used as a guide. By convention, the later an element occurs when the table is traversed following the arrows, the more electropositive is the element.

IR-4.4.2.2 *Alphanumerical order*

Atomic symbols within line formulae are ordered alphabetically. A single-letter symbol always precedes a two-letter symbol with the same initial letter, *e.g.* B before Be, and two-letter symbols are themselves ordered alphabetically, *e.g.* Ba before Be.

Line formulae for different species can be ordered alphanumerically, *e.g.* in indexes and registries, according to the order of the atomic symbols and the right subscripts to these, *e.g.* $B < BH < BO < B_2O_3$. The group NH_4 is often treated as a single symbol and so listed after Na, for example.

To exemplify, the order of citation of some nitrogen- and sodium-containing entities is:

$$N^{3-}, NH_2^-, NH_3, NO_2^-, NO_2^{2-}, NO_3^-, N_2O_2^{2-}, N_3^-, Na, NaCl, NH_4Cl$$

Such ordering may be applied to entire formulae in indexes and registries *etc.*, but may also be used for ordering parts of a given formula, sometimes in connection with the ordering principle of Section IR-4.4.2.1, as decribed below for various specific classes of compounds and ions.

IR-4.4.3. **Formulae for specific classes of compounds**

IR-4.4.3.1 *Binary species*

In accordance with established practice, the electronegativity criterion (Section IR-4.4.2.1) is most often used in binary species.[2]

* Tables numbered with a Roman numeral are collected together at the end of this book.

Examples:

1. NH_3
2. H_2S
3. OF_2
4. O_2Cl
5. OCl^-
6. PH_4^+
7. $P_2O_7^{4-}$
8. $[SiAs_4]^{8-}$
9. RbBr
10. $[Re_2Cl_9]^-$
11. HO^- or OH^-
12. $Rb_{15}Hg_{16}$
13. Cu_5Zn_8 and Cu_5Cd_8

Note that the formula for the hydroxide ion should be HO^- to be consistent with the above convention.

Ordering by electronegativity could, in principle, be applied to ternary, quaternary, *etc.* species. For most species consisting of more than two elements, however, other criteria for ordering the element symbols in the formula are more often used (see Sections IR-4.4.3.2 to IR-4.4.3.4).

IR-4.4.3.2 *Formal treatment as coordination compounds*

The nomenclature of coordination compounds is described in detail in Chapter IR-9. A brief summary of the construction of *formulae* of coordination compounds is given here. Many polyatomic compounds may conveniently be treated as coordination compounds for the purpose of constructing a formula.

In the formula of a coordination entity, the symbol of the central atom(s) is/are placed first, followed by the symbols or formulae of the ligands, unless additional structural information can be presented by changing the order (see, for example, Section IR-4.4.3.3).

The order of citation of central atoms is based on electronegativity as described in Section IR-4.4.2.1. Ligands are cited alphabetically (Section IR-4.4.2.2) according to the first symbol of the ligand formula or ligand abbreviation (see Section IR-4.4.4) *as written.* Where possible, the ligand formula should be written in such a way that a/the donor atom symbol is closest to the symbol of the central atom to which it is attached.

Square brackets may be used to enclose the whole coordination entity whether charged or not. Established practice is always to use square brackets for coordination entities with a transition metal as the central atom (*cf.* Sections IR-2.2.2 and IR-9.2.3.2).

Examples:

1. $PBrCl_2$
2. $SbCl_2F$ or $[SbCl_2F]$

3. $[Mo_6O_{18}]^{2-}$
4. $[CuSb_2]^{5-}$
5. $[UO_2]^{2+}$
6. $[SiW_{12}O_{40}]^{4-}$
7. $[BH_4]^-$
8. $[ClO_4]^-$ or ClO_4^-
9. $[PtCl_2\{P(OEt)_3\}_2]$
10. $[Al(OH)(OH_2)_5]^{2+}$
11. $[PtBrCl(NH_3)(NO_2)]^-$
12. $[PtCl_2(NH_3)(py)]$
13. $[Co(en)F_2(NH_3)_2]^+$, but $[CoF_2(NH_2CH_2CH_2NH_2)(NH_3)_2]^+$
14. $[Co(NH_3)_5(N_3)]^{2-}$

In a few cases, a moiety which comprises different atoms and which occurs in a series of compounds is considered as an entity that acts as a central atom and is cited as such, even if this violates the alphabetical order of ligands. For example, PO and UO_2 are regarded as single entities in Examples 15 and 16.

Examples:

15. $POBr_3$ (alphabetically, PBr_3O)
16. $[UO_2Cl_2]$ (alphabetically, $[UCl_2O_2]$)

For derivatives of parent hydrides (see Chapter IR-6), the alphabetical order of ligands is traditionally disobeyed in that remaining hydrogen atoms are listed first among the ligands in the formula.

Examples:

17. GeH_2F_2
18. SiH_2BrCl
19. B_2H_5Cl

For carbaboranes, there has previously been some uncertainty over the order of B and C.[3] The order 'B before C' recommended here conforms to both electronegativity and alphabetical order (*i.e.* it is an exception to the Hill order[1] in Section IR-4.2.1). In addition, carbon atoms that replace skeletal boron atoms are cited immediately after boron, regardless of what other elements are present. (See also Section IR-6.2.4.4.)

Examples:

20. $B_3C_2H_5$ (recommended)
21. $B_3C_2H_4Br$ (recommended)

For inorganic oxoacids, there is a traditional ordering of formulae in which the 'acid' or 'replaceable' hydrogen atoms (hydrogen atoms bound to oxygen) are listed first, followed by the central atom, then 'non-replaceable' hydrogen atoms (hydrogen atoms bound directly to the central atom), and finally oxygen. This format is an alternative to writing the formulae as coordination compound formulae (see Section IR-8.3).

Examples:

22. HNO_3 (traditional) or $[NO_2(OH)]$ (coordination)
23. H_2PHO_3 (traditional) or $[PHO(OH)_2]$ (coordination)
24. $H_2PO_4^-$ (traditional) or $[PO_2(OH)_2]^-$ (coordination)
25. $H_5P_3O_{10}$ (traditional) or $[(HO)_2P(O)OP(O)(OH)OP(O)(OH)_2]$ (coordination)
26. $(HBO_2)_n$ (traditional) or $\{B(OH)O\}_n$ (coordination)

IR-4.4.3.3 *Chain compounds*

For chain compounds containing three or more different elements, the sequence of atomic symbols should generally be in accord with the order in which the atoms are bound in the molecule or ion, rather than using alphabetical order or order based on electronegativity. However, if one wishes to view a compound formally as a coordination compound, *e.g.* in connection with a discussion of additive naming of the compound, one may use a coordination-compound type of formula, as in Example 1 below.

Examples:

1. NCS^-or SCN^- (*not* CNS^-) = $[C(N)S]^-$, nitridosulfidocarbonate(1−)
2. BrSCN (*not* BrCNS)
3. HOCN (cyanic acid)
4. HNCO (isocyanic acid)

IR-4.4.3.4 *Generalized salt formulae*

If the formula of a compound containing three or more elements is not naturally assigned using the preceding two sections, the compound can be treated as a generalized salt. This term is taken to mean any compound in which it is possible to identify at least one constituent which is a positive ion or can be classified as electropositive or more electropositive than the other constituents, and at least one constituent which is a negative ion or can be classified as electronegative or more electronegative than the rest of the constitutents. The ordering principle is then:

(i) all electropositive constituents precede all electronegative constituents;
(ii) within each of the two groups of constituents, alphabetical order is used.

Examples:

1. $KMgF_3$
2. MgCl(OH)

3. FeO(OH)
4. $NaTl(NO_3)_2$
5. $Li[H_2PO_4]$
6. $NaNH_4[HPO_4]$
7. $Na[HPHO_3]$
8. CuK_5Sb_2 or K_5CuSb_2
9. $K_5[CuSb_2]$
10. $H[AuCl_4]$
11. $Na(UO_2)_3[Zn(H_2O)_6](O_2CMe)_9$

The first formula in Example 8 was arrived at by considering K and Cu to be electropositive constituents and Sb to be electronegative, the second by considering K to be electropositive and Cu and Sb to be electronegative. No structural information is conveyed by these formulae. The formula in Example 9, on the other hand, implies the presence of the coordination entity $[CuSb_2]^{5-}$.

Deviation from alphabetical order of constituents in the same class is allowed to emphasize similarities between compounds.

Example:

12. $CaTiO_3$ and $ZnTiO_3$ (rather than $TiZnO_3$)

Some generalized salts may also be treated as addition compounds, see Section IR-4.4.3.5.

IR-4.4.3.5 *(Formal) addition compounds*

In the formulae of addition compounds or compounds which can formally be regarded as such, including clathrates and multiple salts, the formulae of the component molecules or entities are cited in order of increasing number; if they occur in equal numbers, they are cited in alphabetical order in the sense of Section IR-4.4.2.2. In addition compounds containing water, the water remains conventionally cited last. However, component boron compounds are no longer treated as exceptions.

Examples:

1. $3CdSO_4{\cdot}8H_2O$
2. $Na_2CO_3{\cdot}10H_2O$
3. $Al_2(SO_4)_3{\cdot}K_2SO_4{\cdot}24H_2O$
4. $AlCl_3{\cdot}4EtOH$
5. $8H_2S{\cdot}46H_2O$
6. $C_6H_6{\cdot}NH_3{\cdot}Ni(CN)_2$
7. $BF_3{\cdot}2H_2O$
8. $BF_3{\cdot}2MeOH$

IR-4.4.4 *Ligand abbreviations*

Since abbreviations are widely used in the chemical literature, agreement on their use and meaning is desirable. This Section provides guidelines for the selection of ligand abbreviations for application in the formulae of coordination compounds (Section IR-9.2.3.4). Some commonly used ligand abbreviations are listed in Table VII with diagrams of most of the ligands shown in Table VIII.

An abbreviation for an organic ligand should be derived from a name consistent with the current rules for the systematic nomenclature of organic compounds.[4] (For some ligands a non-systematic name is included in Table VII if it was the source of the abbreviation and if that abbreviation is still commonly used.) New abbreviations should further be constructed according to the following recommendations:

(i) Ligand abbreviations should be constructed so as to avoid confusion and misunderstanding. Since a reader may not be familiar with an abbreviation, it should be explained when first used in a publication.

(ii) New meanings should not be suggested for abbreviations or acronyms that have generally accepted meanings, *e.g.* DNA, NMR, ESR, HPLC, Me (for methyl), Et (for ethyl), *etc.*

(iii) An abbreviation should readily suggest the ligand name, *e.g.* ida for iminodiacetato. (Ligand names may, however, eventually violate nomenclature rules as these are modified, for example iminodiacetate will be replaced by azanediyldiacetate in Ref. 4, but the ligand abbreviations need not be changed every time the naming rules change.)

(iv) Abbreviations should be as short as practicable, but should contain more than one letter or symbol.

(v) The use of non-systematic names for deriving new ligand abbreviations is discouraged.

(vi) Abbreviations should normally use only lower-case letters, with several well-established exceptions:

(a) abbreviations for alkyl, aryl and similar groups, which have the first letter capitalized with the remaining letters in lower case, *e.g.* Me (for methyl), Ac (for acetyl), Cp (for cyclopentadienyl), *etc.*;
(b) abbreviations containing atomic symbols, *e.g.* [12]aneS_4;
(c) abbreviations containing Roman numerals, *e.g.* H_2ppIX for protoporphyrin IX;
(d) abbreviations for ligands containing readily removable hydrons (see vii).

(N.B. Abbreviations for solvents that behave as ligands should also be in lower case letters [*e.g.* dmso for dimethyl sulfoxide {(methylsulfinyl)methane}, thf for tetrahydrofuran]; the practice of capitalizing the abbreviation of a solvent when it does not behave as a ligand is strongly discouraged as an unnecessary distinction.)

(vii) Hydronation of anionic ligands, *e.g.* ida, leads to acids which may be abbreviated by the addition of H, *e.g.* Hida, H_2ida.

(viii) Ligands which are normally neutral, but which continue to behave as ligands on losing one or more hydrons, are abbreviated by adding −1H, −2H, *etc.* (including the numeral 1) after the usual abbreviation of the ligand. For example, if $Ph_2PCH_2PPh_2$ (dppm) loses one hydron to give $[Ph_2PCHPPh_2]^-$ its abbreviation is dppm−1H; if it loses two hydrons, its abbreviation is dppm−2H, *etc.*

IR-4.5 ISOTOPICALLY MODIFIED COMPOUNDS

IR-4.5.1 General formalism

The mass number of any specific nuclide can be indicated in the usual way with a left superscript preceding the appropriate atomic symbol (see Section IR-3.2).

When it is necessary to cite different nuclides at the same position in a formula, the nuclide symbols are written in alphabetical order; when their atomic symbols are identical the order is that of increasing mass number. Isotopically modified compounds may be classified as *isotopically substituted* compounds and *isotopically labelled* compounds.

IR-4.5.2 Isotopically substituted compounds

An isotopically substituted compound has a composition such that all the molecules of the compound have only the indicated nuclide(s) at each designated position. The substituted nuclides are indicated by insertion of the mass numbers as left superscripts preceding the appropriate atom symbols in the normal formula.

Examples:

1. $H^{3}HO$
2. $H^{36}Cl$
3. $^{235}UF_6$
4. $^{42}KNa^{14}CO_3$
5. $^{32}PCl_3$
6. $K[^{32}PF_6]$
7. $K_3{}^{42}K[Fe(^{14}CN)_6]$

IR-4.5.3 Isotopically labelled compounds

IR-4.5.3.1 *Types of labelling*

An isotopically labelled compound may be considered formally as a mixture of an isotopically unmodified compound and one or more analogous isotopically substituted compounds. They may be divided into several different types. Specifically labelled compounds and selectively labelled compounds are treated briefly here and described in more detail in Ref. 5.

IR-4.5.3.2 *Specifically labelled compounds*

An isotopically labelled compound is called a specifically labelled compound when a unique isotopically substituted compound is added formally to the analogous isotopically unmodified compound. A specifically labelled compound is indicated by enclosing the appropriate nuclide symbol(s) and multiplying subscript (if any) in square brackets.

Examples:

1. $H[^{36}Cl]$
2. $[^{32}P]Cl_3$
3. $[^{15}N]H_2[^{2}H]$
4. $[^{13}C]O[^{17}O]$
5. $[^{32}P]O[^{18}F_3]$
6. $Ge[^{2}H_2]F_2$

IR-4.5.3.3 *Selectively labelled compounds*

A selectively labelled compound may be considered as a mixture of specifically labelled compounds. It is indicated by prefixing the formula by the nuclide symbol(s) preceded by any necessary locant(s) (but without multiplying subscripts) enclosed in square brackets.

Examples:

1. $[^{36}Cl]SOCl_2$
2. $[^{2}H]PH_3$
3. $[^{10}B]B_2H_5Cl$

The numbers of possible labels for a given position may be indicated by subscripts separated by semicolons added to the atomic symbol(s) in the isotopic descriptor.

Example:

4. $[1\text{-}^{2}H_{1;2}]SiH_3OSiH_2OSiH_3$

IR-4.6 OPTIONAL MODIFIERS OF FORMULAE

IR-4.6.1 **Oxidation state**

The oxidation state of an element in a formula may be indicated by an oxidation number written as a right superscript in Roman numerals. Oxidation state zero may be represented by the numeral 0 but is not usually shown. If an element occurs with more than one oxidation state in the same formula, the element symbol is repeated, each symbol being assigned a numeral, and the symbols cited in order of these numerals.

Examples:

1. $[P^{V}{}_2Mo_{18}O_{62}]^{6-}$
2. $K[Os^{VIII}(N)O_3]$
3. $[Mo^{V}{}_2Mo^{VI}{}_4O_{18}]^{2-}$
4. $Pb^{II}{}_2Pb^{IV}O_4$
5. $[Os^{0}(CO)_5]$
6. $[Mn^{-I}(CO)_5]^{-}$

Where it is not feasible or reasonable to define an oxidation state for each individual member of a group (or cluster), the overall oxidation level of the group should be defined by a formal ionic charge, indicated as in Section IR-4.3. This avoids the use of fractional oxidation states.

Examples:

7. O_2^-
8. $Fe_4S_4^{3+}$

IR-4.6.2 Formulae of radicals

A radical is an atom or molecule with one or more unpaired electrons. It may have positive, negative or zero charge. An unpaired electron may be indicated in a formula by a superscript dot. The dot is placed as a right upper index to the chemical symbol, so as not to interfere with indications of mass number, atomic number or composition. In the case of diradicals, *etc.*, the superscript dot is preceded by the appropriate superscript multiplier. The radical dot with its multiplier, if any, precedes any charge. To avoid confusion, the multiplier and the radical dot can be placed within parentheses.

Metals and their ions or complexes often possess unpaired electrons but, by convention, they are not considered to be radicals, and radical dots are not used in their formulae. However, there may be occasions when a radical ligand is bound to a metal or metal ion where it is desirable to use a radical dot.

Examples:

1. $H^{\bullet}$
2. $HO^{\bullet}$
3. $NO_2^{\bullet}$
4. $O_2^{2\bullet}$
5. $O_2^{\bullet-}$
6. $BH_3^{\bullet+}$
7. $PO_3^{\bullet2-}$
8. $NO^{(2\bullet)-}$
9. $N_2^{(2\bullet)2+}$

IR-4.6.3 Formulae of optically active compounds

The sign of optical rotation is placed in parentheses, the wavelength (in nm) being indicated as a right subscript. The whole symbol is placed before the formula and refers to the sodium D-line unless otherwise stated.

Example:

1. $(+)_{589}$-$[Co(en)_3]Cl_3$

IR-4.6.4 Indication of excited states

Excited electronic states may be indicated by an asterisk as right superscript. This practice does not differentiate between different excited states.

Examples:

1. He*
2. NO*

IR-4.6.5 Structural descriptors

Structural descriptors such as *cis*, *trans*, *etc.*, are listed in Table V. Usually such descriptors are used as italicized prefixes and are connected to the formula by a hyphen.

Examples:

1. *cis*-$[PtCl_2(NH_3)_2]$
2. *trans*-$[PtCl_4(NH_3)_2]$

The descriptor μ designates an atom or group bridging coordination centres.

Example:

3. $[(H_3N)_5Cr(\mu\text{-}OH)Cr(NH_3)_5]^{5+}$

IR-4.7 REFERENCES

1. This is the so-called Hill order. See E.A. Hill, *J. Am. Chem. Soc.*, **22**, 478–494 (1900).
2. For intermetallic compounds, earlier recommendations prescribed alphabetical ordering rather than by electronegativity (see Section I-4.6.6 of *Nomenclature of Inorganic Chemistry, IUPAC Recommendations 1990*, ed. G.J. Leigh, Blackwell Scientific Publications, Oxford, 1990).
3. For example, the ordering of B and C in formulae was inconsistent in *Nomenclature of Inorganic Chemistry, IUPAC Recommendations 1990*, ed. G.J. Leigh, Blackwell Scientific Publications, Oxford, 1990.
4. *Nomenclature of Organic Chemistry, IUPAC Recommendations*, eds. W.H. Powell and H. Favre, Royal Society of Chemistry, in preparation.
5. Chapter II-2 of *Nomenclature of Inorganic Chemistry II, IUPAC Recommendations 2000*, eds. J.A. McCleverty and N.G. Connelly, Royal Society of Chemistry, 2001. (Red Book II.)

IR-5 Compositional Nomenclature, and Overview of Names of Ions and Radicals

CONTENTS

IR-5.1 INTRODUCTION

Compositional nomenclature is formally based on composition, not structure, and may thus be the (only) choice if little or no structural information is available or a minimum of structural information is to be conveyed.

The simplest type of compositional name is a *stoichiometric* name, which is just a reflection of the empirical formula (Section IR-4.2.1) or the molecular formula (Section IR-4.2.2) of the compound. In stoichiometric names, proportions of constituent elements may be indicated in several ways, using multiplicative prefixes, oxidation numbers or charge numbers.

In some cases, a compound may be regarded as composed of constituents that may themselves be given names of any of several types (including stoichiometric names); the overall name of the compound is then assembled from the names of the constituents so as to indicate their proportions. One category of such compositional names is *generalized stoichiometric names* (see Section IR-5.4) in which the various parts may themselves be names of monoatomic and polyatomic ions. For this reason, Section IR-5.3, devoted to the naming of ions, is included. Another category consists of the names devised for addition compounds which have a format of their own, described in Section IR-5.5.

IR-5.2 STOICHIOMETRIC NAMES OF ELEMENTS AND BINARY COMPOUNDS

A purely stoichiometric name carries no information about the structure of the species named.

In the simplest case, the species to be named consists of only one element, and the name is formed by adding the relevant multiplicative prefix to the element name (*e.g.* S_8, octasulfur). This case is exemplified in Section IR-3.4.3.

When constructing a stoichiometric name for a binary compound, one element is designated as the electropositive constituent and the other the electronegative constituent. The electropositive constituent is *by convention* the element that occurs last in the sequence of Table VI* and its name is the unmodified element name (Table I). The name of the electronegative constituent is constructed by modifying the element name with the ending 'ide', as explained in detail for monoatomic anions in Section IR-5.3.3.2. All element names thus modified with the 'ide' ending are given in Table IX.

The stoichiometric name of the compound is then formed by combining the name of the electropositive constituent, cited first, with that of the electronegative constituent, both suitably qualified by any necessary multiplicative prefixes ('mono', 'di', 'tri', 'tetra', 'penta', *etc.*, given in Table IV). The multiplicative prefixes precede the names they multiply, and are joined directly to them without spaces or hyphens. The final vowels of multiplicative prefixes should not be elided (although 'monoxide', rather than 'monooxide', is an allowed exception because of general usage). The two parts of the name are separated by a space in English.

Stoichiometric names may correspond to the empirical formula or to a molecular formula different from the empirical formula (compare Examples 3 and 4 below).

Examples:

1. HCl hydrogen chloride
2. NO nitrogen oxide, or nitrogen monooxide, or nitrogen monoxide
3. NO_2 nitrogen dioxide
4. N_2O_4 dinitrogen tetraoxide
5. OCl_2 oxygen dichloride
6. O_2Cl dioxygen chloride
7. Fe_3O_4 triiron tetraoxide
8. SiC silicon carbide

* Tables numbered with a Roman numeral are collected together at the end of this book.

9. $SiCl_4$ silicon tetrachloride
10. Ca_3P_2 tricalcium diphosphide, or calcium phosphide
11. NiSn nickel stannide
12. Cu_5Zn_8 pentacopper octazincide
13. $Cr_{23}C_6$ tricosachromium hexacarbide

Multiplicative prefixes need not be used in binary names if there is no ambiguity about the stoichiometry of the compound (such as in Example 10 above). The prefix 'mono' is, strictly speaking, superfluous and is only needed for emphasizing stoichiometry when discussing compositionally related substances, such as Examples 2, 3 and 4 above.

Alternatively, proportions of constituents may be indicated by using oxidation numbers or charge numbers (Section IR-5.4.2).

For compounds containing more than two elements, further conventions are required to form a compositional name (see Sections IR-5.4 and IR-5.5).

IR-5.3 NAMES OF IONS AND RADICALS

IR-5.3.1 General

The charges of the atoms need not be specified in a stoichiometric name. In many cases, however, atoms or groups of atoms are known to carry a particular charge. Within compositional nomenclature, the name of a compound can include the names of individual ions constructed as stoichiometric names or according to other principles, as described below.

IR-5.3.2 Cations

IR-5.3.2.1 *General*

A cation is a monoatomic or polyatomic species having one or more positive charges. The charge on a cation can be indicated in names by using the charge number or, in the case of additively named cations, by the oxidation number(s) of the central atom or atoms. Oxidation and charge numbers are discussed in Section IR-5.4.2.2.

IR-5.3.2.2 *Monoatomic cations*

The name of a monoatomic cation is that of the element with an appropriate charge number appended in parentheses. Unpaired electrons in monoatomic cations may be indicated using a radical dot, *i.e.* a centred dot placed in front of the charge, preceded by a number if necessary.

Examples:

1. Na^+ sodium(1+)
2. Cr^{3+} chromium(3+)

3. Cu^{+}	copper(1+)
4. Cu^{2+}	copper(2+)
5. I^{+}	iodine(1+)
6. H^{+}	hydrogen(1+), hydron
7. $^{1}H^{+}$	protium(1+), proton
8. $^{2}H^{+}$	deuterium(1+), deuteron
9. $^{3}H^{+}$	tritium(1+), triton
10. $He^{\bullet +}$	helium(•1+)
11. $O^{\bullet +}$	oxygen(•1+)
12. $N_2^{(2\bullet)2+}$	dinitrogen(2•2+)

The names of the hydrogen isotopes are discussed in Section IR-3.3.2.

IR-5.3.2.3 *Homopolyatomic cations*

Homopolyatomic cations are named by adding the charge number to the stoichiometric name of the corresponding neutral species, *i.e.* the element name with the appropriate multiplicative prefix. Radical dots may be added to indicate the presence of unpaired electrons.

Examples:

1. O_2^{+} or $O_2^{\bullet +}$	dioxygen(1+) or dioxygen(•1+)
2. S_4^{2+}	tetrasulfur(2+)
3. Hg_2^{2+}	dimercury(2+)
4. Bi_5^{4+}	pentabismuth(4+)
5. H_3^{+}	trihydrogen(1+)

IR-5.3.2.4 *Heteropolyatomic cations*

Heteropolyatomic cations are usually named either substitutively (see Section IR-6.4) or additively (see Chapter IR-7). Substitutive names do not require a charge number, because the name itself implies the charge (Examples 2 and 4 below). Radical dots may be added to additive names to indicate the presence of unpaired electrons.

A few cations have established and still acceptable non-systematic names.

Examples:

1. NH_4^{+}	azanium (substitutive), or ammonium (acceptable non-systematic)
2. H_3O^{+}	oxidanium (substitutive), or oxonium (acceptable non-systematic; *not* hydronium)
3. PH_4^{+}	phosphanium (substitutive)
4. H_4O^{2+}	oxidanediium (substitutive)

5. SbF_4^+ tetrafluorostibanium (substitutive), or tetrafluoridoantimony(1+) or tetrafluoridoantimony(V) (both additive)
6. $BH_3^{\bullet+}$ boraniumyl (substitutive) or trihydridoboron(•1+) (additive)

More examples are given in Table IX.

IR-5.3.3 Anions

IR-5.3.3.1 *Overview*

An anion is a monoatomic or polyatomic species having one or more negative charges. The charge on an anion can be indicated in the name by using the charge number or, in the case of an additively named anion, by the oxidation number(s) of the central atom or atoms. Oxidation and charge numbers are discussed in Section IR-5.4.2.2.

The endings in anion names are 'ide' (monoatomic or homopolyatomic species, heteropolyatomic species named from a parent hydride), 'ate' (heteropolyatomic species named additively), and 'ite' (used in a few names which are still acceptable but do not derive from current systematic nomenclature). When there is no ambiguity, the charge number may be omitted, as in Example 1 below. Parent hydride-based names do not carry charge numbers because the name itself implies the charge (Examples 3 and 4 below).

Examples:

1. Cl^- chloride(1−), or chloride
2. S_2^{2-} disulfide(2−)
3. PH_2^- phosphanide
4. PH^{2-} phosphanediide
5. $[CoCl_4]^{2-}$ tetrachloridocobaltate(2−), or tetrachloridocobaltate(II)
6. NO_2^- dioxidonitrate(1−), or nitrite

IR-5.3.3.2 *Monoatomic anions*

The name of a monoatomic anion is the element name (Table I) modified so as to carry the anion designator 'ide', either formed by replacing the ending of the element name ('en', 'ese', 'ic', 'ine', 'ium', 'ogen', 'on', 'orus', 'um', 'ur', 'y' or 'ygen') by 'ide' or by directly adding 'ide' as an ending to the element name.

Examples:

1. chlorine, chloride
2. carbon, carbide
3. xenon, xenonide
4. tungsten, tungstide
5. bismuth, bismuthide

6. sodium, sodide
7. potassium, potasside

In one case, an abbreviated name has to be chosen: germanium, germide. The systematic name ‘germanide’ designates the anion GeH_3^-.

Some names of monoatomic anions are based on the root of the Latin element names. In these the ending ‘um’ or ‘ium’ is replaced by ‘ide’.

Examples:

8. silver, argentum, argentide
9. gold, aurum, auride
10. copper, cuprum, cupride
11. iron, ferrum, ferride
12. lead, plumbum, plumbide
13. tin, stannum, stannide

All element names thus modified are included in Table IX.

Charge numbers and radical dots may be added as appropriate to specify anions fully.

Examples:

14. O^{2-}	oxide(2−), or oxide
15. $O^{\bullet-}$	oxide(•1−)
16. N^{3-}	nitride(3−), or nitride

IR-5.3.3.3 *Homopolyatomic anions*

Homopolyatomic anions are named by adding the charge number to the stoichiometric name of the corresponding neutral species, *i.e.* the element name with the appropriate multiplicative prefix. Again, a radical dot may be added as appropriate.

In a few cases, non-systematic names are still acceptable alternatives.

Examples:

	Systematic name	*Acceptable alternative name*
1. O_2^- or $O_2^{\bullet-}$	dioxide(1−) or dioxide(•1−)	superoxide
2. O_2^{2-}	dioxide(2−)	peroxide
3. O_3^-	trioxide(1−)	ozonide
4. I_3^-	triiodide(1−)	
5. $Cl_2^{\bullet-}$	dichloride(•1−)	
6. C_2^{2-}	dicarbide(2−)	acetylide
7. N_3^-	trinitride(1−)	azide

8. S_2^{2-} disulfide(2−)
9. Sn_5^{2-} pentastannide(2−)
10. Pb_9^{4-} nonaplumbide(4−)

In some cases, homopolyatomic anions may be considered as derived from a parent hydride by removal of hydrons (see Section IR-6.4).

Examples:

11. O_2^{2-} dioxidanediide
12. S_2^{2-} disulfanediide

IR-5.3.3.4 *Heteropolyatomic anions*

Heteropolyatomic anions are usually named either substitutively (see Section IR-6.4.4) or additively (see Chapter IR-7 and Section IR-9.2.2). Radical dots may be added to additive names to indicate the presence of unpaired electron(s).

A few heteropolyatomic anions have established and still acceptable non-systematic names.

Examples:

1. NH_2^- azanide (substitutive), dihydridonitrate(1−) (additive), or amide (acceptable non-systematic)
2. GeH_3^- germanide (substitutive), or trihydridogermanate(1−) (additive)
3. HS^- sulfanide (substitutive), or hydridosulfate(1−) (additive)
4. H_3S^- sulfanuide or λ^4-sulfanide (both substitutive), or trihydridosulfate(1−) (additive)
5. $H_2S^{\bullet-}$ sulfanuidyl or λ^4-sulfanidyl (both substitutive), or dihydridosulfate(•1−) (additive)
6. SO_3^{2-} trioxidosulfate(2−) (additive), or sulfite (acceptable non-systematic)
7. OCl^- chloridooxygenate(1−) (additive), or hypochlorite (acceptable non-systematic)
8. ClO_3^- trioxidochlorate(1−) (additive), or chlorate (acceptable non-systematic)
9. $[PF_6]^-$ hexafluoro-λ^5-phosphanuide (substitutive), or hexafluoridophosphate(1−) (additive)
10. $[CuCl_4]^{2-}$ tetrachloridocuprate(II) (additive)
11. $[Fe(CO)_4]^{2-}$ tetracarbonylferrate(−II) (additive)

All acceptable, but not fully systematic, anion names are given in Table IX.

Note that in Ref. 1, radical anions consisting of only hydrogen and one other element were named additively using the ending 'ide' rather than the ending 'ate' (*e.g.* Example 5 above). Making this exception to the general system of additive nomenclature for these particular cases is now discouraged.

When one or more hydron(s) are attached to an anion at (an) unknown position(s), or at (a) position(s) which one cannot or does not wish to specify, a 'hydrogen name' (see Section IR-8.4) may be used. Such names may also be used for simpler compounds, such as partially dehydronated oxoacids. Certain of these names have accepted abbreviated forms, such as hydrogencarbonate, dihydrogenphosphate, *etc.* All such accepted abbreviated names are given in Section IR-8.5.

Examples:

12. $HMo_6O_{19}^-$ hydrogen(nonadecaoxidohexamolybdate)(1−)
13. HCO_3^- hydrogen(trioxidocarbonate)(1−), or hydrogencarbonate
14. $H_2PO_4^-$ dihydrogen(tetraoxidophosphate)(1−), or dihydrogenphosphate

IR-5.4 GENERALIZED STOICHIOMETRIC NAMES

IR-5.4.1 Order of citation of electropositive and electronegative constituents

The constituents of the compound to be named are divided into formally electropositive and formally electronegative constituents. There must be at least one electropositive and one electronegative constituent. Cations are electropositive and anions electronegative, by definition. Electropositive elements occur later in Table VI than electronegative elements by convention.

In principle, the division into electropositive and electronegative constituents is arbitrary if the compound contains more than two elements. In practice, however, there is often no problem in deciding where the division lies.

The names of the electropositive constituents precede those of the electronegative constituents in the overall name. The order of citation is alphabetical within each class of constituents (multiplicative prefixes being ignored), except that hydrogen is always cited last among electropositive constituents if actually classified as an electropositive constituent.

This principle for constructing generalized stoichiometric names parallels the principle for constructing 'generalized salt formulae' in Section IR-4.4.3.4. However, the order of citation in a generalized stochiometric name is not necessarily the same as the order of symbols in the corresponding generalized salt formula, as is seen from Examples 4, 5 and 7 below.

The following generalized stoichiometric names, based only on single-element constituents, do not carry information about the structure.

Examples:

1. IBr iodine bromide
2. PBrClI phosphorus bromide chloride iodide
3. ArHF or ArFH argon hydrogen fluoride, or argon fluoride hydride

4. ClOF or OClF	chlorine oxygen fluoride or oxygen chloride fluoride
5. CuK_5Sb_2 or K_5CuSb_2	copper pentapotassium diantimonide, or pentapotassium cupride diantimonide

Note from these examples that the order of any two elements in the name depends on the arbitrary division of elements into electropositive and electronegative constituents. (The same applies to the order of the element symbols in the formulae as illustrated in Section IR-4.4.3.4.) Additive names representing the actual structure of the compounds in Examples 3 and 4 (FArH and FClO, respectively) are given in Section IR-7.2.

In some cases, the use of substitutive or additive nomenclature for naming an ion is not possible or desirable because of the lack of structural information. In such cases, it may be best to give a stoichiometric name and add the charge number. Parentheses are needed to make it clear that the charge number denotes the overall charge of the ion.

Example:

6. $O_2Cl_2^+$	(dioxygen dichloride)(1+)

When names of polyatomic ions occur as constituents in a generalized stoichiometric name, a certain amount of structural information is often implied by the name.

Example:

7. $NaNH_4[HPO_4]$	ammonium sodium hydrogenphosphate

IR-5.4.2 Indication of proportions of constituents

IR-5.4.2.1 *Use of multiplicative prefixes*

The proportions of the constituents, be they monoatomic or polyatomic, may be indicated in generalized stoichiometric names by multiplicative prefixes, as was the case for the constituents of binary compounds (Section IR-5.2).

Examples:

1. Na_2CO_3	disodium trioxidocarbonate, or sodium carbonate
2. $K_4[Fe(CN)_6]$	tetrapotassium hexacyanidoferrate
3. PCl_3O	phosphorus trichloride oxide
4. $KMgCl_3$	magnesium potassium trichloride

When the name of the constituent itself starts with a multiplicative prefix (as in disulfate, dichromate, triphosphate, tetraborate, *etc.*), or when ambiguity could otherwise arise, the alternative multiplicative prefixes 'bis', 'tris', 'tetrakis', 'pentakis', *etc.* (Table IV) are used and the name of the group acted upon by the alternative prefix is placed in parentheses.

Examples:

5. $Ca(NO_3)_2$	calcium bis(trioxidonitrate), or calcium nitrate
6. $(UO_2)_2SO_4$	bis(dioxidouranium) tetraoxidosulfate
7. $Ba(BrF_4)_2$	barium bis(tetrafluoridobromate)
8. $U(S_2O_7)_2$	uranium bis(disulfate)
9. $Ca_3(PO_4)_2$	tricalcium bis(phosphate)
10. $Ca_2P_2O_7$	calcium diphosphate
11. $Ca(HCO_3)_2$	calcium bis(hydrogencarbonate)

IR-5.4.2.2 *Use of charge and oxidation numbers*

It is possible to provide information on the proportions of the constituents in names by using one of two other devices: the charge number, which designates ionic charge, and the oxidation number, which designates oxidation state. In nomenclature, the use of the charge number is preferred as the determination of the oxidation number is sometimes ambiguous and subjective. It is advisable to use oxidation numbers only when there is no uncertainty about their assignment.

The *charge number* is a number whose magnitude is the ionic charge. It is written in parentheses immediately after the name of an ion, without a space. The charge is written in arabic numerals, followed by the sign of the charge. Note that unity is always indicated, unlike in superscript charge designations (which are used in formulae). No charge number is used after the name of a neutral species.

Examples:

1. $FeSO_4$	iron(2+) sulfate
2. $Fe_2(SO_4)_3$	iron(3+) sulfate
3. $(UO_2)_2SO_4$	dioxidouranium(1+) sulfate
4. UO_2SO_4	dioxidouranium(2+) sulfate
5. $K_4[Fe(CN)_6]$	potassium hexacyanidoferrate(4−)
6. $[Co(NH_3)_6]Cl(SO_4)$	hexaamminecobalt(3+) chloride sulfate

The *oxidation number* (see Sections IR-4.6.1 and IR-9.1.2.8) of an element is indicated by a Roman numeral placed in parentheses immediately following the name (modified by the ending 'ate' if necessary) of the element to which it refers. The oxidation number may be positive, negative or zero (represented by the numeral 0). An oxidation number is always non-negative unless the minus sign is explicitly used (the positive sign is never used). Non-integral oxidation numbers are not used for nomenclature purposes.

Examples:

7. PCl_5	phosphorus(V) chloride
8. $Na[Mn(CO)_5]$	sodium pentacarbonylmanganate(−I)
9. $[Fe(CO)_5]$	pentacarbonyliron(0)

Several conventions are observed for inferring oxidation numbers, the use of which is particularly common in the names of compounds of transition elements. Hydrogen is considered positive (oxidation number I) in combination with non-metallic elements and negative (oxidation number −I) in combination with metallic elements. Organic groups combined with metal atoms are treated sometimes as anions (for example, a methyl ligand is usually considered to be a methanide ion, CH_3^-), sometimes as neutral (*e.g.* carbon monooxide). Bonds between atoms of the same species make no contribution to oxidation number.

Examples:

10. N_2O	nitrogen(I) oxide
11. NO_2	nitrogen(IV) oxide
12. Fe_3O_4	iron(II) diiron(III) oxide
13. MnO_2	manganese(IV) oxide
14. CO	carbon(II) oxide
15. $FeSO_4$	iron(II) sulfate
16. $Fe_2(SO_4)_3$	iron(III) sulfate
17. SF_6	sulfur(VI) fluoride
18. $(UO_2)_2SO_4$	dioxidouranium(V) sulfate
19. UO_2SO_4	dioxidouranium(VI) sulfate
20. $K_4[Fe(CN)_6]$	potassium hexacyanidoferrate(II), or potassium hexacyanidoferrate(4−)
21. $K_4[Ni(CN)_4]$	potassium tetracyanidonickelate(0), or potassium tetracyanidonickelate(4−)
22. $Na_2[Fe(CO)_4]$	sodium tetracarbonylferrate(−II), or sodium tetracarbonylferrate(2−)
23. $[Co(NH_3)_6]Cl(SO_4)$	hexaamminecobalt(III) chloride sulfate, or hexaamminecobalt(3+) chloride sulfate
24. $Fe_4[Fe(CN)_6]_3$	iron(III) hexacyanidoferrate(II), or iron(3+) hexacyanidoferrate(4−)

Note that oxidation numbers are no longer recommended when naming homopolyatomic ions. This is to avoid ambiguity. Oxidation numbers refer to the individual atoms of the element in question, even if they are appended to a name containing a multiplicative prefix, *cf.* Example 12 above. To conform to this practice, dimercury(2+) (see Section IR-5.3.2.3) would have to be named dimercury(I); dioxide(2−) (see Section IR-5.3.3.3) would be dioxide(−I); and ions such as pentabismuth(4+) (see Section IR-5.3.2.3) and dioxide(1−) (see Section IR-5.3.3.3), with fractional formal oxidation numbers, could not be named at all.

IR-5.4.2.3 *Multiple monoatomic constituents vs. homopolyatomic constituents*

Care should be taken to distinguish between multiple monoatomic constituents and polyatomic constituents. This distinction is often not apparent from the formula, but is tacitly implied.

Examples:

1. TlI_3 thallium tris(iodide), or thallium(III) iodide, or thallium(3+) iodide
2. $Tl(I_3)$ thallium triiodide(1−), or thallium(I) (triiodide), or thallium(1+) (triiodide)

Both compounds in Examples 1 and 2 have the overall formula TlI_3 and both could be named by the simple stoichiometric name thallium triiodide. However, it is possible, and usually desirable, to convey more information in the name.

The compound in Example 1 consists of iodide, I^-, and thallium, in the proportion 3:1, whereas the compound in Example 2 consists of triiodide(1−), I_3^-, and thallium in the proportion 1:1. In the first name for the first compound, then, the multiplicative prefix 'tris' is used to make it completely clear that three iodide ions are involved rather than one triiodide ion. The alternative names use the oxidation number III for thallium and the charge number 3+, respectively, to convey indirectly the proportions of the constituents.

In the first name in Example 2, it is clear that the electronegative constituent is a homopolyatomic entity with charge −1. The next two names convey this indirectly by adding the oxidation number or the charge number to the name thallium; including the parentheses around the name of the electronegative part reinforces that it is a homopolyatomic entity.

For both compounds, fully explicit names including the charge number for the thallium ion, although partly redundant, are also acceptable. Thus, thallium(3+) tris(iodide) and thallium(1+) triiodide(1−), for Examples 1 and 2 respectively, may be preferable in systematic contexts such as indexes and registries.

Examples:

3. $HgCl_2$ mercury dichloride, or mercury(II) chloride, or mercury(2+) chloride
4. Hg_2Cl_2 dimercury dichloride, or (dimercury) dichloride, or dimercury(2+) chloride

In Example 4, the first name is purely stoichiometric, whereas the second name contains more information in indicating that the compound contains a homodiatomic cation. In the last name, where the charge of the dication is specified, the prefix 'di' for 'chloride' is not necessary.

Examples:

5. Na_2S_3 disodium (trisulfide) (this indicates the presence of the polyatomic anion), or sodium trisulfide(2−) (with the charge on the anion indicated, the multiplicative prefix on the cation name is not necessary)
6. Fe_2S_3 diiron tris(sulfide), or iron(III) sulfide

Salts which contain anions that are S_n^{2-} chains, as well as those containing several S^{2-} anions, are both referred to as 'polysulfides' but, as demonstrated, names may be given that provide a distinction between these cases.

Examples:

7. K_2O dipotassium oxide
8. K_2O_2 dipotassium (dioxide), or potassium dioxide(2−)
9. KO_2 monopotassium (dioxide), or potassium dioxide(1−)
10. KO_3 potassium (trioxide), or potassium trioxide(1−)

Clearly, a simple stoichiometric name like 'potassium dioxide', although strictly speaking unambiguous (referring to the compound in Example 9), could easily be misinterpreted. In other cases, based on chemical knowledge, there is no chance of misinterpretation in practice, and the simple stoichiometric name will most often be used, as in Examples 11 and 12 below.

Examples:

11. BaO_2 barium dioxide (simple stoichiometric name), or barium (dioxide) or barium dioxide(2−) (specifying the diatomic anion), or barium peroxide (using the acceptable alternative name for the anion)
12. MnO_2 manganese dioxide (simple stoichiometric name), or manganese bis(oxide) (specifies two oxide ions rather than a diatomic anion), or manganese(IV) oxide

IR-5.5 NAMES OF (FORMAL) ADDITION COMPOUNDS

The term *addition compounds* covers donor-acceptor complexes (adducts) and a variety of lattice compounds. The method described here, however, is relevant not just to such compounds, but also to multiple salts and to certain compounds of uncertain structure or compounds for which the full structure need not be communicated.

The names of the individual components of such a generalized addition compound are each constructed by using an appropriate nomenclature system, whether compositional, substitutive or additive. The overall name of the compound is then formed by connecting the names of the components by 'em' dashes; the proportions of the components are indicated after the name by a stoichiometric descriptor consisting of arabic numerals separated by a solidus or solidi. The descriptor, in parentheses, is separated from the compound name by a space. The order of names of the individual components is, firstly, according to the increasing number of the components and, secondly, alphabetical. As the only exception, the component name 'water' is always cited last. (Note that this represents a change from the rule in Ref. 2 according to which the component names must follow the order given by the formula.) The numerals in the descriptor appear in the same order as the corresponding component names.

For addition compounds containing water as a component, the class name 'hydrates' is acceptable because of well established use, even though the ending 'ate' might seem to indicate an anionic component. For hydrates with a simple stoichiometry, names of the classical 'hydrate' type are acceptable, but rules have not been formulated for non-integer stoichiometries such as that in Example 12 below. Also, because of their ambiguity, the terms 'deuterate' and tritiate' are not acceptable for addition compounds of $^{2}H_2O$ and $^{3}H_2O$ or other isotope-modified water species. Example 3 shows a formula and a name for a compound of the present type with isotope modification. In this case the modified component formula and name are presented according to the rules of Section II-2.3.3 of Ref. 3.

Examples:

1. $BF_3{\cdot}2H_2O$ — boron trifluoride—water (1/2)
2. $8Kr{\cdot}46H_2O$ — krypton—water (8/46)
3. $8Kr{\cdot}46{}^{3}H_2O$ — krypton—($^{3}H_2$)water (8/46)
4. $CaCl_2{\cdot}8NH_3$ — calcium chloride—ammonia (1/8)
5. $AlCl_3{\cdot}4EtOH$ — aluminium chloride—ethanol (1/4)
6. $BiCl_3{\cdot}3PCl_5$ — bismuth(III) chloride—phosphorus(V) chloride (1/3)
7. $2Na_2CO_3{\cdot}3H_2O_2$ — sodium carbonate—hydrogen peroxide (2/3)
8. $Co_2O_3{\cdot}nH_2O$ — cobalt(III) oxide—water (1/*n*)
9. $Na_2SO_4{\cdot}10H_2O$ — sodium sulfate—water (1/10), or sodium sulfate decahydrate
10. $Al_2(SO_4)_3{\cdot}K_2SO_4{\cdot}24H_2O$ — aluminium sulfate—potassium sulfate—water (1/1/24)
11. $AlK(SO_4)_2{\cdot}12H_2O$ — aluminium potassium bis(sulfate) dodecahydrate
12. $3CdSO_4{\cdot}8H_2O$ — cadmium sulfate—water (3/8)

There is no difference between donor-acceptor complexes and coordination compounds from a nomenclature point of view. Thus, for such systems an additive name such as described in Sections IR-7.1 to IR-7.3 and in Chapter IR-9 may be given.

Example:

13. $BH_3{\cdot}(C_2H_5)_2O$ or $[B\{(C_2H_5)_2O\}H_3]$ — borane—ethoxyethane (1/1), or (ethoxyethane)trihydridoboron

In Section P-68.1 of Ref. 4, a slightly different nomenclature is presented for organic donor-acceptor complexes.

IR-5.6 SUMMARY

Compositional names are either of the stoichiometric type (which, furthermore, are of the binary type except in the case of homoatomic species) or of the addition compound type. Compositional nomenclature is used if little or no structural information is to be conveyed by the name. However, substitutive or additive nomenclature may be used to indicate the structure of constituents of a compound that is named overall by compositional nomenclature. Substitutive nomenclature is described in Chapter IR-6 and additive nomenclature in Chapters IR-7, IR-8 and IR-9.

IR-5.7 REFERENCES

1. Names for Inorganic Radicals, W.H. Koppenol, *Pure Appl. Chem.*, **72**, 437–446 (2000).
2. *Nomenclature of Inorganic Chemistry, IUPAC Recommendations 1990*, ed. G.J. Leigh, Blackwell Scientific Publications, Oxford, 1990.
3. *Nomenclature of Inorganic Chemistry II, IUPAC Recommendations 2000*, eds. J.A. McCleverty and N.G. Connelly, Royal Society of Chemistry, 2001.
4. *Nomenclature of Organic Chemistry, IUPAC Recommendations*, eds. W.H. Powell and H. Favre, Royal Society of Chemistry, in preparation.

IR-6 Parent Hydride Names and Substitutive Nomenclature

CONTENTS

IR-6.1 INTRODUCTION

Substitutive nomenclature is a system in which names are based on the names of *parent hydrides*, which define a standard population of hydrogen atoms attached to a skeletal structure. Names of derivatives of the parent hydrides are formed by citing prefixes or suffixes appropriate to the *substituent groups* (or substituents) replacing the hydrogen atoms (preceded by locants when required), joined without a break to the name of the unsubstituted parent hydride.

Substitutive nomenclature is recommended only for derivatives of the parent hydrides named in Table IR-6.1 (in Section IR-6.2.1), and derivatives of polynuclear hydrides containing only these elements (see Sections IR-6.2.2 to IR-6.2.4). The bonding numbers of the skeletal atoms are understood to be as in Table IR-6.1 (these bonding numbers, *e.g.* 4 for Si and 2 for Se, are termed *standard bonding numbers*). Other bonding numbers must be indicated by an appropriate designator (the 'λ convention', see Section IR-6.2.2.2 and Section P-14.1 of Ref. 1).

In general, relevant practices and conventions of substitutive nomenclature as applied to organic compounds[1] are also followed here.

Constructing a substitutive name generally involves the replacement of hydrogen atoms in a parent structure with other atoms or atom groups. Related operations, often considered to be part of substitutive nomenclature, are *skeletal replacement* (Section IR-6.2.4.1) and *functional replacement* in oxoacid parents (Section IR-8.6). Note that some operations in parent hydride-based nomenclature are not substitutive operations (*e.g.* formation of cations and anions by addition of H^+ and H^-, respectively, *cf.* Sections IR-6.4.1 and IR-6.4.5). Names formed by the modifications of parent hydride names described in those sections are still considered part of substitutive nomenclature.

In most cases, the compounds named substitutively in the present chapter may alternatively and equally systematically be named additively (Chapter IR-7), but it is important to note that for the parent hydrides presented here such additive names cannot be used as parent names in substitutive nomenclature.

Neutral boron hydrides are called boranes. The basic aspects of borane nomenclature are provided in Section IR-6.2.3; more advanced aspects will be treated in a future IUPAC publication.

IR-6.2 PARENT HYDRIDE NAMES

IR-6.2.1 Mononuclear parent hydrides with standard and non-standard bonding numbers

The mononuclear hydrides of elements of groups 13–17 of the periodic table play a central role in substitutive nomenclature. They are used as parent hydrides as indicated above with the parent names given in Table IR-6.1.

In cases where the bonding number deviates from the standard number defined above, it must be indicated in the hydride name by means of an appropriate superscript appended to the Greek letter λ, this symbol being separated from the name in Table IR-6.1 by a hyphen.

Table IR-6.1 *Parent names of mononuclear hydrides*

BH_3	borane	CH_4	methane[a]	NH_3	azane[b]	H_2O	oxidane[b,c]	HF	fluorane[d]
AlH_3	alumane[c]	SiH_4	silane	PH_3	phosphane[e]	H_2S	sulfane[c,f]	HCl	chlorane[d]
GaH_3	gallane	GeH_4	germane	AsH_3	arsane[e]	H_2Se	selane[c,f]	HBr	bromane[d]
InH_3	indigane[g]	SnH_4	stannane	SbH_3	stibane[e]	H_2Te	tellane[c,f]	HI	iodane[d]
TlH_3	thallane	PbH_4	plumbane	BiH_3	bismuthane[c]	H_2Po	polane[c,f]	HAt	astatane[d]

[a] The systematic analogue is 'carbane'. Because of the universal use of the name 'methane' for CH_4, 'carbane' is not recommended.

[b] The names 'azane' and 'oxidane' are only intended for use in naming derivatives of ammonia and water, respectively, by substitutive nomenclature, and they form the basis for naming polynuclear entities (*e.g.* triazane, dioxidane). Examples of such use may be found in Section IR-6.4 and Table IX. In Section P-62 of Ref. 1 many organic derivatives of ammonia are named on the basis of the substituent group suffixes 'amine' and 'imine'.

[c] The names aluminane, bismane, oxane, thiane, selenane, tellurane and polonane cannot be used since they are the names of saturated six-membered heteromonocycles in the Hantzsch–Widman system (see Section IR-6.2.4.3). The name 'alane' has been used for AlH_3, but must be discarded because the systematically derived name of the substituent group $-AlH_2$ would be 'alanyl' which is the well-established name of the acyl group derived from the amino acid alanine.

[d] The names 'fluorane', 'chlorane', 'bromane', 'iodane' and 'astatane' are included here because they are the basis for the formation of substitutive names of ions, radicals and substituent groups (see IR-6.4.7 and Table IX for examples). The unsubstituted hydrides may also be named 'hydrogen fluoride', 'hydrogen bromide', *etc.* (compositional nomenclature, Chapter IR-5). However, these compositional names cannot be used as parent names.

[e] The systematic names 'phosphane', 'arsane' and 'stibane' are used throughout this book. The names 'phosphine', 'arsine' and 'stibine' are no longer acceptable.

[f] Sulfane, when unsubstituted, may also be named 'hydrogen sulfide' or, better, 'dihydrogen sulfide' (compositional nomenclature, Chapter IR-5). However, a compositional name cannot be used as a parent name. Corresponding remarks apply to selane, tellane and polane.

[g] The analogous systematic name for InH_3 would be 'indane' which is, however, well established as the name of the hydrocarbon 2,3-dihydroindene. The name 'indiane' would lead to confusion when naming unsaturated derivatives, *e.g.* 'triindiene' could mean a compound with two double bonds (a diene) as well as the mono-unsaturated derivative of triindiane. The parent name 'indigane' derives from the etymological source 'indigo' (from the flame colour of indium).

Examples:

1. PH_5 — λ^5-phosphane
2. PH — λ^1-phosphane
3. SH_6 — λ^6-sulfane
4. SnH_2 — λ^2-stannane

IR-6.2.2 Homopolynuclear parent hydrides (other than boron and carbon hydrides)

IR-6.2.2.1 *Homonuclear acyclic parent hydrides in which all atoms have their standard bonding number*

Names are constructed by prefixing the 'ane' name of the corresponding mononuclear hydride from Table IR-6.1 with the appropriate multiplicative prefix ('di', 'tri', 'tetra', *etc.*; see Table IV*) corresponding to the number of atoms of the chain bonded in series.

Examples:

1. HOOH — dioxidane, or hydrogen peroxide
2. H_2NNH_2 — diazane, or hydrazine
3. H_2PPH_2 — diphosphane
4. H_3SnSnH_3 — distannane
5. HSeSeSeH — triselane
6. $H_3SiSiH_2SiH_2SiH_3$ — tetrasilane

The compositional name 'hydrogen peroxide' (*cf.* Chapter IR-5) is an alternative to 'dioxidane' for H_2O_2 itself, but is not applicable as a parent hydride name in substitutive nomenclature.

In Section P-68.3 of Ref. 1 organic derivatives of H_2NNH_2 are named on the basis of 'hydrazine' as a parent name.

IR-6.2.2.2 *Homonuclear acyclic parent hydrides with elements exhibiting non-standard bonding numbers*

In cases where the skeletal atoms of a hydride chain are the same but one or more has a bonding number different from the standard values defined by Table IR-6.1, the name of the hydride is formed as if all the atoms showed standard bonding numbers, but is preceded by locants, one for each non-standard atom, each locant qualified without a space by λ^n, where n is the appropriate bonding number.

When a choice is needed between the same skeletal atom in different valence states, the one in a non-standard valence state is preferred for assignment of the lower locant. If a further choice is needed between the same skeletal atom in two or more non-standard valence states, preference for the lower locant or locants is given in order of decreasing numerical value of the bonding number, *e.g.* λ^6 is preferred to λ^4.

* Tables numbered with a Roman numeral are collected together at the end of this book.

Examples:

1.
$\overset{1}{H_5S}\overset{2}{S}\overset{3}{S}H_4\overset{4}{S}H$
$1\lambda^6,3\lambda^6$-tetrasulfane (*not* $2\lambda^6,4\lambda^6$)

2.
$H\overset{1}{S}\overset{2}{S}H_4\overset{3}{S}H_4\overset{4}{S}H_2\overset{5}{S}H$
$2\lambda^6,3\lambda^6,4\lambda^4$-pentasulfane (*not* $2\lambda^4,3\lambda^6,4\lambda^6$)

3.
$H_4PPH_3PH_3PH_4$
$1\lambda^5,2\lambda^5,3\lambda^5,4\lambda^5$-tetraphosphane

4.
HPbPbPbH
$1\lambda^2,2\lambda^2,3\lambda^2$-triplumbane

IR-6.2.2.3 *Unsaturated homonuclear acyclic hydrides*

Chains containing unsaturation are accommodated in substitutive nomenclature by the methods used with alkenes and alkynes (see Section P-31.1 of Ref. 1), *i.e.* the name of the corresponding saturated chain hydride is modified by replacing the 'ane' ending with 'ene' in the case of a double bond and 'yne' in the case of a triple bond. If there is one of each, the ending becomes 'en' ...'yne' with appropriate locants; 'diene' is used when there are two double bonds, and so on. In each case the position(s) of unsaturation is (are) indicated by (a) numerical locant(s) immediately preceding the suffix(es). Locants are chosen to be as low as possible.

Examples:

1.
HN=NH diazene

2.
HSb=SbH distibene

3.
$H_2\overset{1}{N}\overset{2}{N}=\overset{3}{N}\overset{4}{N}H\overset{5}{N}H_2$ pentaaz-2-ene (*not* pentaaz-3-ene)

Unsaturated acyclic hydrides are not classified as parent hydrides. Because of the hierarchical rules of substitutive nomenclature, the numbering of the double and triple bonds may not be fixed until various groups and modifications have been numbered. (See Section IR-6.4.9 for an example.)

IR-6.2.2.4 *Homonuclear monocyclic parent hydrides*

There are three main ways of giving parent names to homonuclear monocyclic hydrides:

(i) by using the Hantzsch–Widman (H–W) name (see Section IR-6.2.4.3 and Section P-22.2 of Ref. 1);
(ii) by using the relevant replacement prefix ('a' term) from Table X together with the appropriate multiplicative prefix to indicate replacement of carbon atoms in the corresponding carbocyclic compound name (see Section P-22.2 of Ref. 1);
(iii) by adding the prefix 'cyclo' to the name of the corresponding unbranched, unsubstituted chain (see Sections IR-6.2.2.1 to IR-6.2.2.3 and Section P-22.2 of Ref. 1).

Each method is used in Examples 1–4 below. When naming organic derivatives of non-carbon homonuclear monocyclic parent hydrides, the Hantzsch–Widman name is preferred for rings with 3 to 10 members. For larger rings, the names given by the second method are used. For more detailed rules on large-ring parent hydrides, see Section P-22.2 of Ref. 1.

Examples:

1.

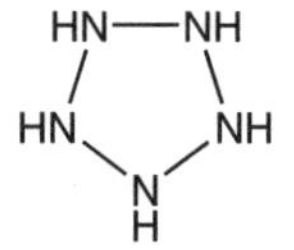

(i) H–W name: pentaazolidine
(ii) pentaazacyclopentane
(iii) cyclopentaazane

2.

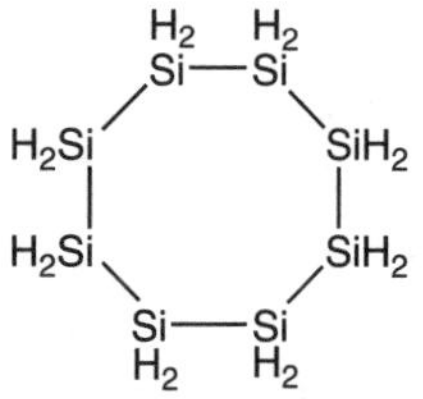

(i) H–W name: octasilocane
(ii) octasilacyclooctane
(iii) cyclooctasilane

3.

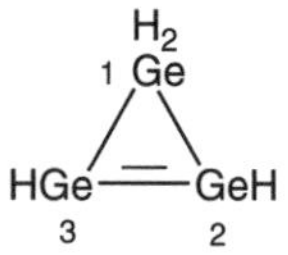

(i) H–W name: 1*H*-trigermirene
(ii) trigermacyclopropene
(iii) cyclotrigermene

4.

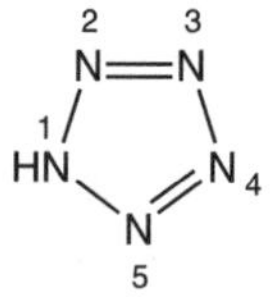

(i) H–W name: 1*H*-pentaazole

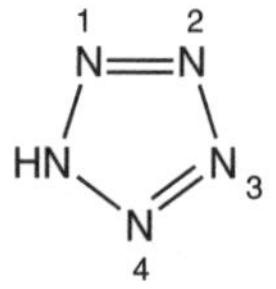

(ii) pentaazacyclopenta-1,3-diene
(iii) cyclopentaaza-1,3-diene

Note that in Example 4 the numbering for the H–W name differs from that for the other two methods; H–W priorities depend on the H-atom position, and those in (ii) and (iii) on the locations of the double bonds.

IR-6.2.2.5 *Homonuclear polycyclic parent hydrides*

Parent names of homonuclear polycycles may be constructed by three methods:

(i) by specifying the fusion of relevant monocycles (see Section P-25.3 of Ref. 1), each named by the Hantzsch–Widman system (see Section IR-6.2.4.3);
(ii) by using a skeletal replacement prefix ('a' term) from Table X together with the appropriate multiplicative prefix to indicate replacement of the carbon atoms in the corresponding carbocyclic compound;
(iii) by specifying the ring structure using the von Baeyer notation (see Section P-23.4 of Ref. 1) in combination with the name of the corresponding linear hydride as derived in Section IR-6.2.2.1.

Examples:

1.

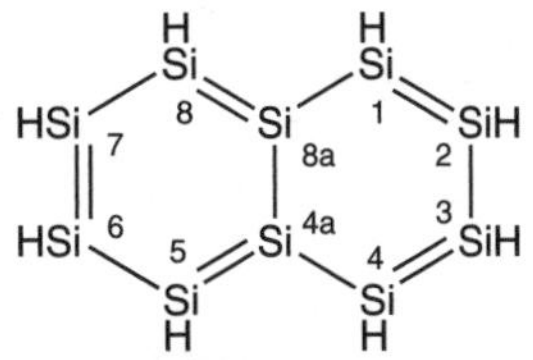

(i) hexasilinohexasiline
(ii) decasilanaphthalene

2.

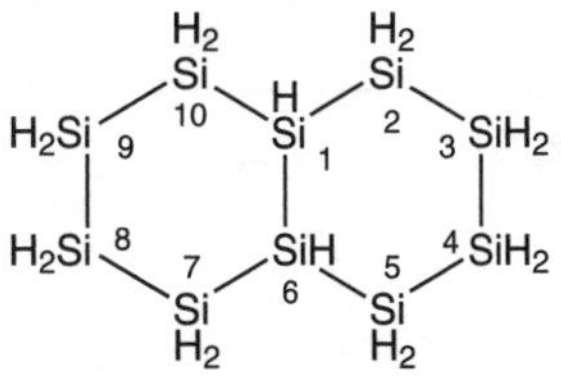

(ii) and (iii) decasilabicyclo[4.4.0]decane
(iii) bicyclo[4.4.0]decasilane

IR-6.2.3 **Boron hydrides**

IR-6.2.3.1 *Stoichiometric names*

Neutral polyboron hydrides are called boranes and the simplest possible parent structure, BH_3, is given the name 'borane'. The number of boron atoms in a boron hydride molecule is indicated by a multiplicative prefix. The principal difference between this system of naming and hydrocarbon nomenclature is that the number of hydrogen atoms must be defined; it cannot be inferred from simple bonding considerations. The number of hydrogen atoms is indicated by the appropriate arabic numeral in parentheses directly following the name. Such names convey only compositional information.

Examples:

1. B_2H_6 diborane(6)
2. $B_{20}H_{16}$ icosaborane(16)

IR-6.2.3.2 *Structural descriptor names*

More structural information is obtained by augmenting the stoichiometric name by a structural descriptor. The descriptor is based on electron-counting relationships[2] and is presented in Table IR-6.2.

Table IR-6.2 *Summary of common polyboron hydride structure types according to stoichiometry and electron-counting relationships*[a]

Descriptor	*Skeletal electron pairs*	*Parent hydride*	*Description of structure*
closo	$n+1$	B_nH_{n+2}	Closed polyhedral structure with triangular faces only.
nido	$n+2$	B_nH_{n+4}	Nest-like non-closed polyhedral structure; n vertices of the parent ($n+1$)-atom *closo* polyhedron occupied.
arachno	$n+3$	B_nH_{n+6}	Web-like non-closed polyhedral structure; n vertices of the parent ($n+2$)-atom *closo* polyhedron occupied.
hypho	$n+4$	B_nH_{n+8}	Net-like non-closed polyhedral structure; n vertices of the parent ($n+3$)-atom *closo* polyhedron occupied.
klado	$n+5$	B_nH_{n+10}	Open branch-like polyhedral structure; n vertices of the parent ($n+4$)-atom *closo* polyhedron occupied.

[a] The structural relationships are often represented by a Rudolph diagram.[3]

Examples:

1.

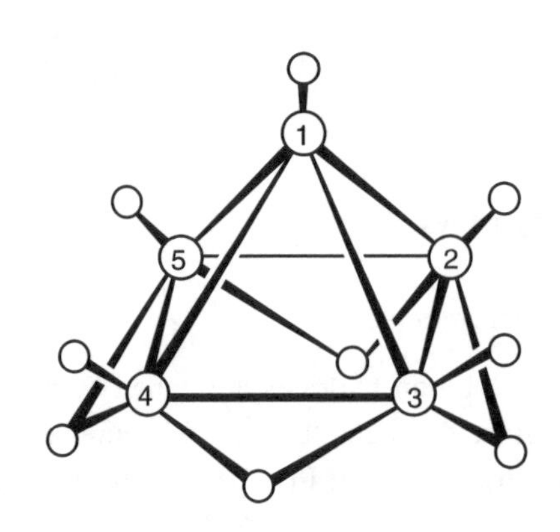

nido-pentaborane(9), B_5H_9

2.

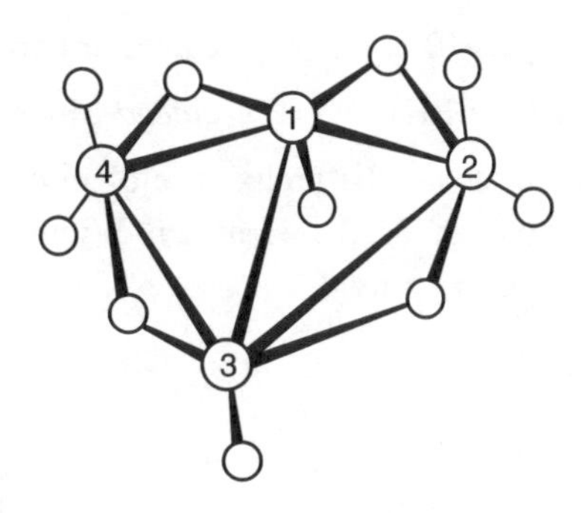

arachno-tetraborane(10), B_4H_{10}

The two structures in Examples 1 and 2 can be thought of as related to that of *closo*-$B_6H_6^{2-}$ as follows:

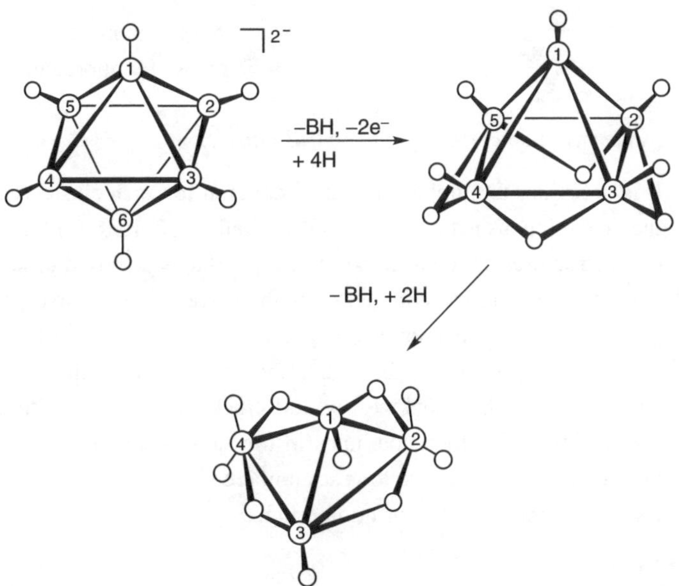

The structures are obtained formally by removal of one (Example 1) or two (Example 2) BH groups and the addition of the appropriate number of hydrogen atoms.

It should be noted that the prefixes *nido*, *arachno*, *etc.* are not used for the simplest boranes for which formal derivation from *closo* parent structures by successive subtractions might seem to be far-fetched.

Chain compounds may be explicitly specified as such by using the prefix '*catena*'.

Examples:

3.

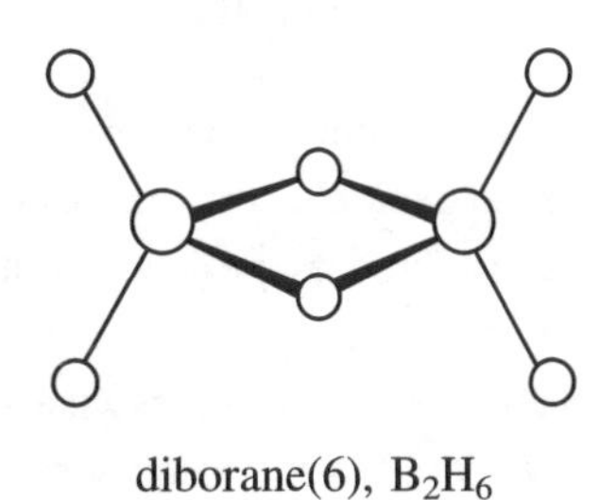

diborane(6), B_2H_6

4. H_2BBHBH_2 *catena*-triborane(5)
5. $HB{=}BBH_2$ *catena*-triborene(3)

For cyclic systems, the prefix 'cyclo' in connection with the name of the corresponding chain compound, or the Hantzsch–Widman (H–W) nomenclature system (see Section IR-6.2.4.3), may be used.

Example:

6.

cyclotetraborane
H–W name: tetraboretane

IR-6.2.3.3 *Systematic numbering of polyhedral clusters*

It is necessary to number the boron skeleton for each cluster systematically, so as to permit the unambiguous naming of the substituted derivatives. For this purpose, the boron atoms of *closo* structures are considered to occupy planes disposed sequentially, perpendicular to the axis of highest order symmetry. (If there are two such axes, the 'longer', in terms of the greater number of perpendicular planes crossed, is chosen.)

Numbering begins at the nearest boron atom when the cluster is viewed along this axis and proceeds either clockwise or anti-clockwise, dealing with all skeletal atoms of the first plane. Numbering then continues in the same sense in the next plane, beginning with the boron atom nearest to the lowest numbered boron atom in the preceding plane when going forward in the direction of numbering.

Example:

1.

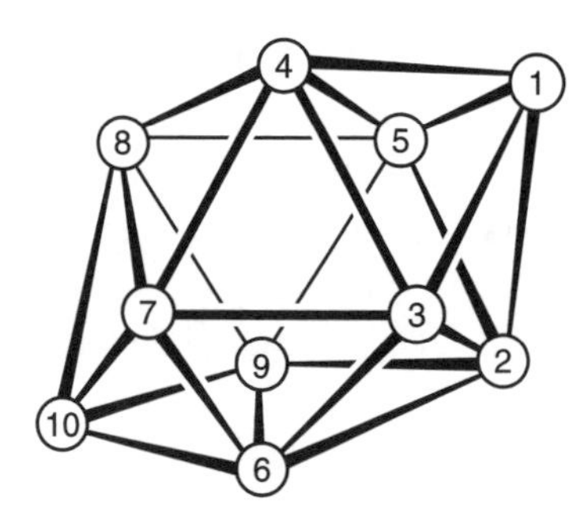

closo-$B_{10}H_{10}{}^{2-}$ (hydrogen atoms omitted for clarity)

The numbering in *nido* clusters is derived from that of the related *closo* cluster. In the case of *arachno* and more open clusters, the opened side is presented towards the observer and the boron atoms considered as projected onto a plane at the rear. They are then numbered sequentially in zones, commencing at the central boron atom of highest connectivity and proceeding clockwise or anti-clockwise until the innermost zone is completed. The next zone is then numbered in the same sense starting from the 12 o'clock position, and so on until the outermost zone is completed. This treatment means that the numbering of the *closo* parent is unlikely to carry over into the corresponding *arachno* system.

Example:

2.

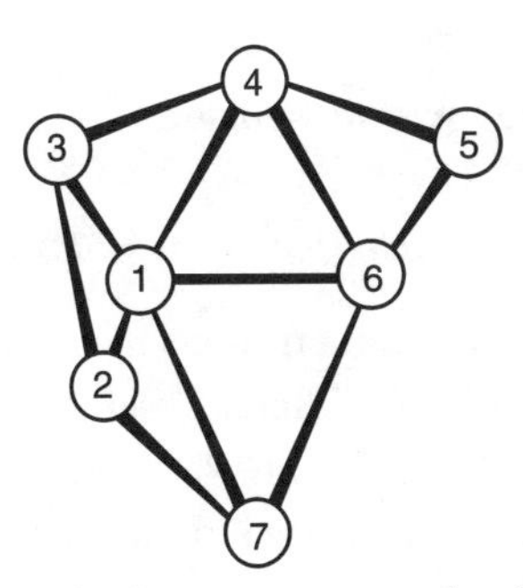

arachno-B_7H_{13} (hydrogen atoms omitted for clarity)

When there is a choice, the molecule is so oriented that the 12 o'clock position is decided by sequential application of the following criteria:

(i) the 12 o'clock position lies in a symmetry plane, that contains as few boron atoms as possible;
(ii) the 12 o'clock position lies in that portion of the symmetry plane which contains the greatest number of skeletal atoms;
(iii) the 12 o'clock position lies opposite the greater number of bridging atoms.

The use of criteria (i)–(iii) may fail to effect a decision, and where a symmetry plane is lacking they are inapplicable. In such cases the general principles of organic numbering are used, such as choosing a numbering scheme which gives substituted atoms the lowest locants.

IR-6.2.3.4 *Systematic naming giving hydrogen atom distribution*

In open boranes each boron atom can be assumed to carry at least one terminal hydrogen atom. However, it is necessary to specify the positions of the bridging hydrogen atoms by using the symbol μ, preceded by the locants for the skeletal positions so bridged in ascending numerical order. The designator *H* is used for the bridging hydrogen atoms in the name.

Example:

1.

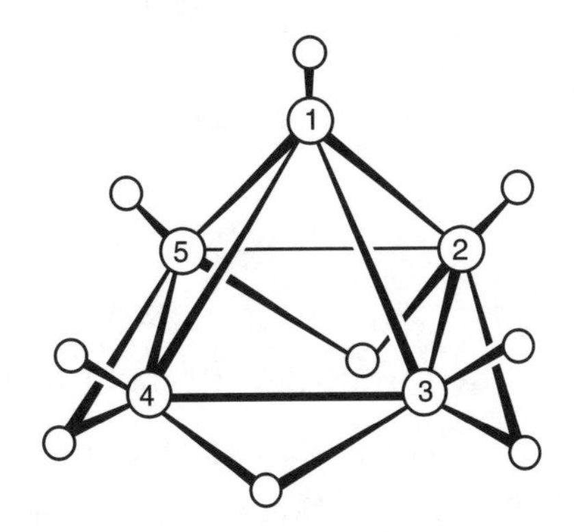

2,3:2,5:3,4:4,5-tetra-μ*H*-*nido*-pentaborane(9)

This method of locating bridging hydrogen atoms is adapted from the 'indicated hydrogen' method in organic nomenclature (see Section P-14.6 of Ref. 1). The 'indicated hydrogen' method would yield the name (2,3-μ*H*),(2,5-μ*H*),(3,4-μ*H*),(4,5-μ*H*)-*nido*-pentaborane(9).

IR-6.2.4 Heteronuclear parent hydrides

IR-6.2.4.1 *Heteronuclear acyclic parent hydrides in general*

When at least four carbon atoms in an unbranched-chain parent hydrocarbon are replaced by heteroatoms, alike or different, and the terminal carbon atoms either remain or are replaced by P, As, Sb, Bi, Si, Ge, Sn, Pb, B, Al, Ga, In, or Tl, *skeletal replacement nomenclature* ('a' nomenclature) may be used to indicate the heteroatoms (see Sections P-15.4 and P-21.2 of Ref. 1).

In this method, the chain is named first as if it were composed entirely of carbon atoms. Any heteroatoms in the chain are then designated by appropriate replacement prefixes ('a' terms) from Table X cited in the order given by Table VI, each preceded by its appropriate locant. The locants are assigned by numbering the chain from that end which gives lower locants to the heteroatom set as a whole and, if these are equal, from that end which gives the lower locant or locant set to the replacement prefix first cited. If there is still a choice, lower locants are assigned to the sites of unsaturation.

Only chains with four or more heteroatoms (or strictly speaking, four or more heterounits) are given parent names constructed in this way. A heterounit is a sequence of heteroatoms which is in itself the skeleton of a parent hydride, *e.g.* Se and SS and SiOSi (see Section IR-6.2.4.2), but not OSiO. Heteroatoms must not belong to the principal characteristic group (see Section IR-6.3.1) (if there is one) when counting them for this purpose. Heteronuclear chains with fewer heterounits, and heteronuclear chains not terminating in any of the atoms listed above, are named substitutively as derivatives of homonuclear parent hydrides and are not themselves used as parents.

Examples:

1.

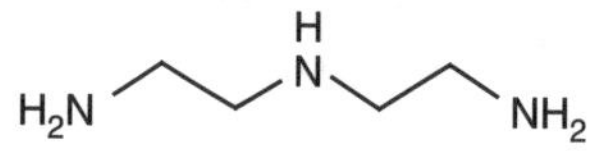

H_2N ... N(H) ... NH_2

N-(2-aminoethyl)ethane-1,2-diamine

2.

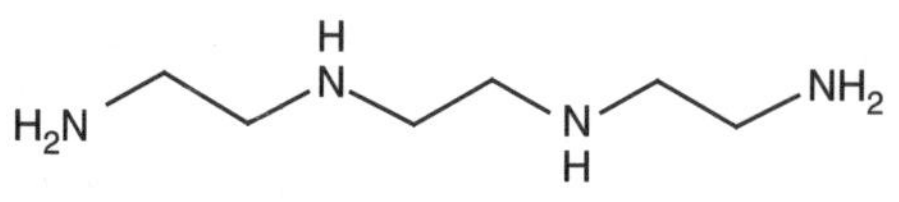

H_2N ... N(H) ... N(H) ... NH_2

N,*N*′-bis(2-aminoethyl)ethane-1,2-diamine

3.

$\overset{11}{C}H_3\overset{10}{O}\,\overset{9}{C}H_2\overset{8}{C}H_2\overset{7}{O}\overset{6}{C}H_2\overset{5}{C}H_2\overset{4}{Si}H_2\overset{3}{C}H_2\overset{2}{S}\overset{1}{C}H_3$

7,10-dioxa-2-thia-4-silaundecane

(Parent name. Note the name is not 2,5-dioxa-10-thia-8-silaundecane because the locant set 2,4,7,10 takes precedence over the locant set 2,5,8,10.)

Unambiguous parent names for non-carbon-containing heteronuclear chains can be derived from a hydrocarbon parent or a non-carbon homonuclear chain parent (*cf.* Section IR-6.2.2.1). Alternatively, heteronuclear chains may be named additively by the method described in Section IR-7.4. However, such names cannot be used as parent names in substitutive nomenclature.

Example:

4. $\overset{1}{\mathrm{Si}}\mathrm{H_3}\overset{2}{\mathrm{Si}}\mathrm{H_2}\overset{3}{\mathrm{Si}}\mathrm{H_2}\overset{4}{\mathrm{Ge}}\mathrm{H_2}\overset{5}{\mathrm{Si}}\mathrm{H_3}$

   1,2,3,5-tetrasila-4-germapentane (*not* 1,3,4,5-tetrasila-2-germapentane), or
   2-germapentasilane (note: based on different numbering), or
   1,1,1,2,2,3,3,4,4,5,5,5-dodecahydrido-4-germy-1,2,3,5-tetrasily-[5]catena

IR-6.2.4.2 *Hydrides consisting of chains of alternating skeletal atoms*

Chain hydrides with a backbone of alternating atoms of two elements A and E, neither of which is carbon, *i.e.* of sequences $(AE)_nA$, where element A occurs later in the sequence of Table VI, can be named by successive citation of the following name parts:

(i) a multiplicative prefix (Table IV) denoting the number of atoms of element A, with no elision of a terminal vowel of this prefix;
(ii) replacement prefixes ending in 'a' (Table X) denoting elements A and E in that order (with elision of the terminal 'a' of the replacement prefix before another 'a' or an 'o');
(iii) the ending 'ne'.

Examples:

1. $SnH_3OSnH_2OSnH_2OSnH_3$	tetrastannoxane
2. $SiH_3SSiH_2SSiH_2SSiH_3$	tetrasilathiane
3. $PH_2NHPHNHPH_2$	triphosphazane
4. SiH_3NHSiH_3	disilazane
5. $\overset{1}{\mathrm{P}}\mathrm{H_2}\overset{2}{\mathrm{N}}{=}\overset{3}{\mathrm{P}}\overset{4}{\mathrm{N}}\mathrm{H}\overset{5}{\mathrm{P}}\mathrm{H}\overset{6}{\mathrm{N}}\mathrm{H}\overset{7}{\mathrm{P}}\mathrm{H_2}$	tetraphosphaz-2-ene

The first four structures are parent hydrides, but not the unsaturated compound (see remarks in Section IR-6.2.2.3).

IR-6.2.4.3 *Heteronuclear monocyclic parent hydrides; Hantzsch–Widman nomenclature*

For heteronuclear monocyclic parent hydrides there are two general naming systems and, in certain cases, a third possibility.

(i) In the (extended) Hantzsch–Widman (H–W) system (Section P-22.2 of Ref. 1), names are constructed so as to convey the ring size, the presence of heteroatoms (*i.e.* non-carbon atoms) and the degree of hydrogenation (either *mancude*, *i.e.* with the maximum number of non-cumulative double bonds, or saturated) by means of characteristic prefixes and endings. The endings are given in Table IR-6.3. (Hydrides with intermediate degrees of hydrogenation are named by the use of the prefix 'hydro' together with an appropriate multiplicative prefix. However, such hydrides are not parents.)

The order of citation of the heteroatoms follows Table VI, *i.e.* F > Cl > Br > I > O > ... *etc.*, where ' > ' means 'is cited before'. Locants are assigned to the heteroatoms so as to ensure first that the locant '1' is given to the atom cited first and then that the total set of locants is as low as possible consistent with sequential numbering of the ring positions (ordering locant sets alphanumerically). The heteroatoms are cited by the replacement prefixes ('a' terms) given in Table X together with appropriate multiplicative prefixes. (As exceptions, the 'a' terms for aluminium and indium in the Hantzsch–Widman system are 'aluma' and 'indiga'.) In the case of six-membered rings, the ring heteroatom which is cited last decides which of the alternative endings in Table IR-6.3 is chosen.

Tautomers may be distinguished using *indicated hydrogen* to specify the location of the hydrogen atom(s) which can be placed in more than one way [and thus, indirectly, the location of the double bond(s)], as in Example 2 below.

Table IR-6.3 *Endings in the Hantzsch–Widman system*

Number of atoms in ring	*Mancude*[a]	*Saturated*
3	irene ('irine' for rings with N as only heteroatom)	irane ('iridine' for rings containing N)
4	ete	etane ('etidine' for rings containing N)
5	ole	olane ('olidine' for rings containing N)
6(A)[b]	ine	ane
6(B)[b]	ine	inane
6(C)[b]	inine	inane
7	epine	epane
8	ocine	ocane
9	onine	onane
10	ecine	ecane

[a] Maximum number of non-cumulative double bonds.
[b] 6(A) is used when the last-cited heteroatom is O, S, Se, Te, Po or Bi; 6(B) is used when the last-cited heteroatom is N, Si, Ge, Sn or Pb; and 6(C) is used when the last-cited heteroatom is F, Cl, Br, I, P, As, Sb, B, Al, Ga, In or Tl.

(ii) Alternatively, the name is based on the name of the corresponding carbocycle, and the heteroatoms are indicated by the replacement prefixes ('a' terms) from Table X together with appropriate multiplicative prefixes. The order of citation is again given by Table VI.

(iii) For the special case of saturated rings of two alternating skeletal atoms (as in Examples 3–6 below), the name may be constructed using the prefix 'cyclo' followed by the replacement prefixes (Table X) cited in the *reverse* of the order in which the corresponding elements appear in Table VI. The name ends with 'ane'.

The Hantzsch–Widman names are preferred for rings with up to 10 members, in organic nomenclature. For saturated rings and mancude rings (rings with the maximum number of non-cumulative double bonds) with more than 10 members method (ii) is used. For more detailed rules on large-ring parent hydrides, see Section P-22.2 of Ref. 1.

Examples:

1.

(i) H–W name: disilagermirane
(ii) disilagermacyclopropane

2.

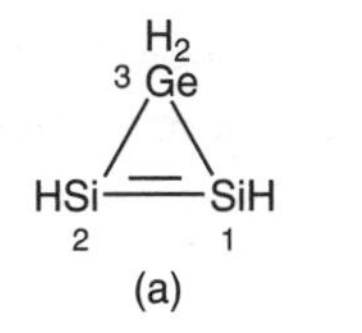

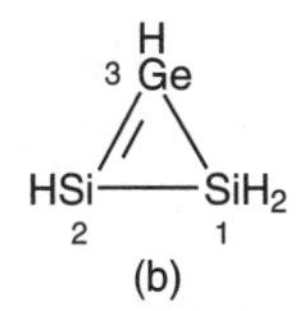

H–W names: 3*H*-1,2,3-disilagermirene (a), and 1*H*-1,2,3-disilagermirene (b)

3.

(i) H–W name: 1,3,2,4-dioxadistibetane
(ii) 1,3-dioxa-2,4-distibacyclobutane
(iii) cyclodistiboxane

4.

(i) H–W name: 1,3,5,2,4,6-triazatriborinane
(ii) 1,3,5-triaza-2,4,6-triboracyclohexane
(iii) cyclotriborazane

5.

(i) H–W name: 1,3,5,2,4,6-trioxatriborinane
(ii) 1,3,5-trioxa-2,4,6-triboracyclohexane
(iii) cyclotriboroxane

6.

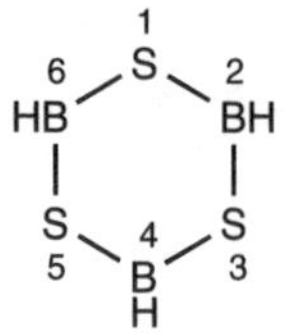

(i) H–W name: 1,3,5,2,4,6-trithiatriborinane
(ii) 1,3,5-trithia-2,4,6-triboracyclohexane
(iii) cyclotriborathiane

The names borazole, boroxole and borthiole, respectively, for the three compounds in Examples 4, 5 and 6 have been abandoned long ago as they imply five-membered rings in the Hantzsch–Widman system. The names borazin(e), boroxin and borthiin indicate six-membered rings with unsaturation and only one boron atom and one other heteroatom (although the order of the element name stems is wrong) and are also not acceptable.

Example:

7.

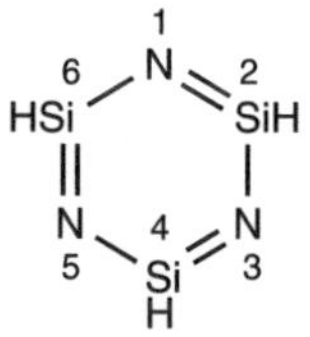

(i) H–W name: 1,3,5,2,4,6-triazatrisiline
(ii) 1,3,5-triaza-2,4,6-trisilacyclohexa-1,3,5-triene

Where ring atoms have a connectivity different from their standard bonding number (see Section IR-6.2.1), their actual bonding number is expressed as an arabic superscript to the Greek letter lambda following immediately after an appropriate locant.

Example:

8.

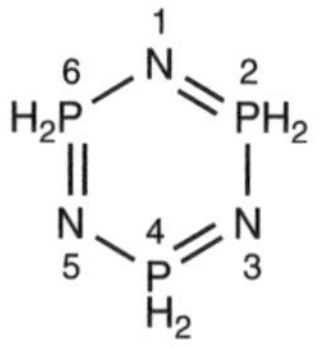

(i) H–W name: 1,3,5,$2\lambda^5$,$4\lambda^5$,$6\lambda^5$-triazatriphosphinine
(ii) 1,3,5-triaza-$2\lambda^5$,$4\lambda^5$,$6\lambda^5$-triphosphacyclohexa-1,3,5-triene

IR-6.2.4.4 *Skeletal replacement in boron hydrides*

It is possible that the essential skeletal structure of the boron hydrides is preserved in derivatives in which one or more of the boron atoms are replaced by other atoms. The names of such species are formed by an adaptation of replacement nomenclature, giving carbaboranes, azaboranes, phosphaboranes, thiaboranes, *etc.*

In the heteroboranes, the number of nearest neighbours to the heteroatom is variable and can be 5, 6, 7, *etc.* Therefore, in the adaptation of replacement nomenclature to polyborane compounds, the replacement of a boron atom by another atom is indicated in the name along with the number of hydrogen atoms in the resulting polyhedral structure. The prefixes *closo*, *nido*, *arachno*, *etc.*, are retained as described for boron hydrides (Section IR-6.2.3.2). The positions of the supplanting heteroatoms in the polyhedral framework are indicated by locants which are the lowest possible numbers taken as a set consistent with the numbering of the parent polyborane. If a choice remains for locant assignment within a given set, then lower numbering should be assigned to the element encountered first using Table VI.

The hydrogen atom population of the actual compound concerned (and not that of the parent all-boron skeletal compound) is added as an arabic numeral in parentheses at the end of the name. The numeral is retained upon hydrogen substitution.

Examples:

1. $B_{10}C_2H_{12}$ dicarba-*closo*-dodecaborane(12)
2. $B_3C_2H_5$

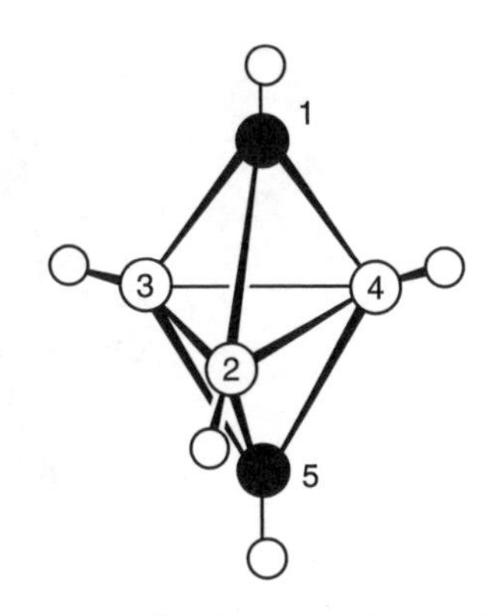

1,5-dicarba-*closo*-pentaborane(5)

3. $B_4C_2H_8$

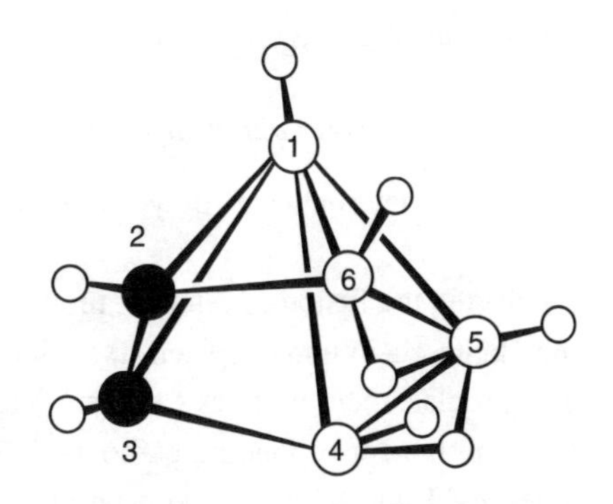

4,5:5,6-di-μ*H*-2,3-dicarba-*nido*-hexaborane(8)

Note that locants for skeletal replacement take precedence over those for bridging hydrogen atoms. The number of bridging hydrogen atoms is usually different for heteroboranes compared with parent polyboranes, and for numbering purposes only the parent boron skeleton is considered.

Examples:

4.

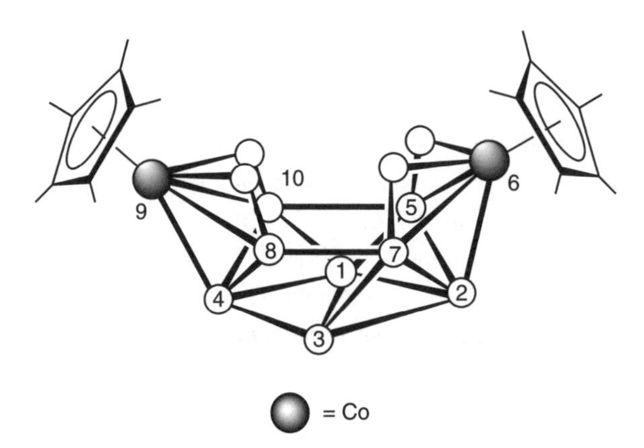

6,9-bis(pentamethyl-η^5-cyclopentadienyl)-5,6:6,7:8,9:9,10-tetra-μH-6,9-dicobalta-*nido*-decaborane(12) (one terminal hydrogen atom on each boron atom omitted for clarity)

5.

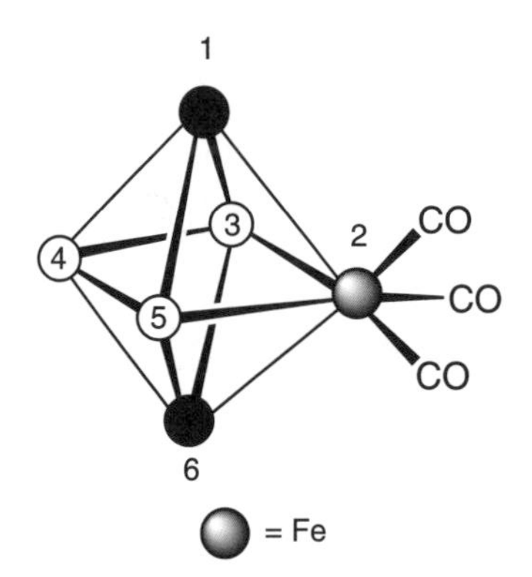

2,2,2-tricarbonyl-1,6-dicarba-2-ferra-*closo*-hexaborane(5) (one terminal hydrogen atom on each boron and carbon atom omitted for clarity)

IR-6.2.4.5 *Heteronuclear polycyclic parent hydrides*

Parent names of heteronuclear polycycles may be constructed by three methods:

(i) specifying the fusion of relevant monocycles (see Section P-25.3 of Ref. 1), named by the Hantzsch–Widman system (see Section IR-6.2.4.3);

(ii) using replacement prefixes ('a' terms) from Table X to specify replacement of carbon atoms in the corresponding carbocyclic compound. Heteroatoms are cited in the order given by Table VI and appropriate multiplicative prefixes are added;

(iii) for ring systems consisting of repeating units, using the von Baeyer notation to specify the ring structure combined with the appropriate multiplicative prefix and the replacement prefixes from Table X appropriate to the repeating unit, *cf.* the names discussed in Section IR-6.2.4.2.

Example:

1.

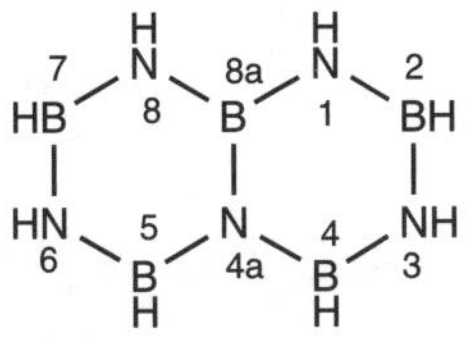

{Numbering is only for method (ii)}

(i) octahydro[1,3,5,2,4,6]triazatriborinino[1,3,5,2,4,6]triazatriborinine
(ii) octahydro-1,3,4a,6,8-pentaaza-2,4,5,7,8a-pentaboranaphthalene
(iii) bicyclo[4.4.0]pentaborazane

In this example, names (i) and (ii) need the additional 'octahydro' prefix because the available parent hydrides for these constructions (triazatriborinine and naphthalene, respectively) are mancude (*i.e.* have the maximum number of non-cumulative double bonds).

IR-6.3 SUBSTITUTIVE NAMES OF DERIVATIVES OF PARENT HYDRIDES

IR-6.3.1 **Use of suffixes and prefixes**

Substituent groups (or substituents), considered as replacing hydrogen atoms in parent hydrides, are named using appropriate suffixes ('ol', 'thiol', 'peroxol', 'carboxylic acid', *etc.*) and prefixes ('hydroxy', 'phosphanyl', 'bromo', 'nitro', *etc.*). Substituent suffixes are ranked in Section P-43 of Ref. 1. Prefixes are extensively listed in Appendix 2 of Ref. 1. The case of substituents formed by removal of one or more hydrogen atoms from a parent hydride is explained briefly, with examples, in Section IR-6.4.7, and prefixes for many common inorganic substituents are included in Table IX.

Some substituents are always cited as prefixes, most notably halogen atoms. Otherwise, the highest-ranking substituent (the principal characteristic group) is cited as a suffix and the rest of the substituents as prefixes. Except for 'hydro', prefixes are cited in alphabetical order before the name of the parent hydride, parentheses being used to avoid ambiguity.

Multiplicative prefixes indicate the presence of two or more identical substituents; if the substituents themselves are substituted, the prefixes 'bis', 'tris', 'tetrakis', *etc.* are used. In the case of a multiplicative prefix ending in 'a' and a suffix starting with a vowel, the 'a' is elided (see Example 2 below). The final 'e' of a parent hydride name is elided in front of a suffix starting with a vowel (see Examples 1 and 5 below).

Where there is a choice of parent hydride among those listed in Table IR-6.1 (or corresponding hydrides with non-standard bonding numbers, *cf.* Section IR-6.2.2.2), the name is based on the parent hydride of the element occurring first in the sequence: N, P, As, Sb, Bi, Si, Ge, Sn, Pb, B, Al, Ga, In, Tl, O, S, Se, Te, C, F, Cl, Br, I.

The above exposition is only a very brief overview of the most important principles of substitutive nomenclature. In Ref. 1, an extensive system of rules is developed for choosing one name among the many unambiguous substitutive names that may often be constructed for organic compounds. A corresponding extensive set of rules has not been developed for non-carbon-containing compounds, partly because many such compounds can just as well be given additive names (Chapter IR-7), and often are.

The following names exemplify the above principles. In some cases, additive names are given for comparison.

Examples:

1.	SiH_3OH	silanol
2.	$Si(OH)_4$	silanetetrol (substitutive), or tetrahydroxidosilicon (additive)
3.	SF_6	hexafluoro-λ^6-sulfane (substitutive), or hexafluoridosulfur (additive)
4.	TlH_2CN	thallanecarbonitrile (substitutive), or cyanidodihydridothallium (additive)
5.	SiH_3NH_2	silanamine (substitutive), or amidotrihydridosilicon (additive)
6.	PH_2Cl	chlorophosphane
7.	PH_2Et	ethylphosphane
8.	$TlH_2OOOTlH_2$	trioxidanediylbis(thallane)
9.	$PbEt_4$	tetraethylplumbane (substitutive), or tetraethyllead (additive)
10.	$GeH(SMe)_3$	tris(methylsulfanyl)germane
11.	$PhGeCl_2SiCl_3$	trichloro[dichloro(phenyl)germyl]silane, *not* dichloro(phenyl)(trichlorosilyl)germane
12.	$MePHSiH_3$	methyl(silyl)phosphane, *not* (methylphosphanyl)silane or (silylphosphanyl)methane

For polynuclear parent hydrides, numerical locants are often needed to specify the positions of substituent groups. If there are several equivalent numberings of the parent hydride skeletal atoms relative to the substituents after relevant rules from Section IR-6.2 have been applied, the numbering is chosen which leads to the lowest set of locants for the compound as a whole. If there is still a choice, lowest locants are assigned to the substituent cited first in the name. If all substitutable hydrogen atoms are replaced by the same substituent, the locants can be omitted, as in Example 20 below.

Examples:

13.	$H_3GeGeGeH_2GeBr_3$	4,4,4-tribromo-2λ^2-tetragermane (numbering of parent fixed by λ designator)
14.	$HOOC\overset{1}{Si}H_2\overset{2}{Si}H_2\overset{3}{Si}H_3$	trisilane-1-carboxylic acid
15.	$H\overset{1}{N}=\overset{2}{N}\overset{3}{N}HMe$	3-methyltriaz-1-ene (*not* 1-methyltriaz-3-ene) (numbering of skeletal atoms fixed by position of unsaturation)

16. $\overset{1}{Cl}\overset{}{Si}H_2\overset{2}{Si}HCl\overset{3}{Si}H_2\overset{4}{Si}H_2\overset{5}{Si}H_2Cl$ — 1,2,5-trichloropentasilane (*not* 1,4,5-)

17. $Br\overset{1}{Sn}H_2\overset{2}{Sn}Cl_2\overset{3}{Sn}H_2C_3H_7$ — 1-bromo-2,2-dichloro-3-propyltristannane (bromo preferred to propyl for lowest locant)

18. $H\overset{1}{Sn}Cl_2\overset{2}{O}\overset{3}{Sn}H_2\overset{4}{O}\overset{5}{Sn}H_2\overset{6}{O}\overset{7}{Sn}H_2Cl$ — 1,1,7-trichlorotetrastannoxane

19. Ring structure: N(H) (1), Me_2Si (2), N(H) (3), SiHEt (4)

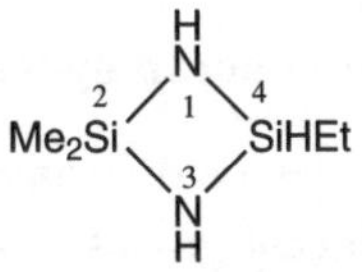

4-ethyl-2,2-dimethylcyclodisilazane
H–W name: 4-ethyl-2,2-dimethyl-1,3,2,4-diazadisiletane
(locant set 2,2,4 preferred to 2,4,4 in both names)

20. $Et_3PbPbEt_3$ — 1,1,1,2,2,2-hexaethyldiplumbane, or hexaethyldiplumbane (substitutive), or bis(triethyllead)(*Pb—Pb*) (additive)

21. MeNHN=NMe — 1,3-dimethyltriaz-1-ene

The names of branched structures are based on the longest available unbranched chain, which is regarded as defining the parent hydride, and the names of the shorter chains, which are treated as substituents and appropriately cited. Once the longest chain has been chosen, it is numbered so as to give the lowest set of locants to the substituents.

Examples:

22. Structure: $(H_2B)_2B—BH_2$

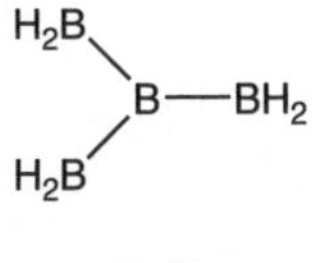

2-boranyltriborane(5)

23. Structure: H_3SiSiH_2 (1 2) and H_3Si on HSi (3); SiH (4) bearing $SiH_2SiH_2SiH_3$ (5 6 7) and SiH_2SiH_3

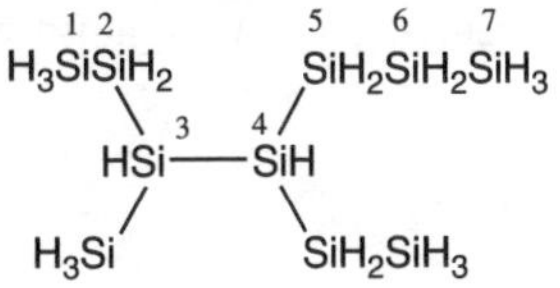

4-disilanyl-3-silylheptasilane
(*not* 4-disilanyl-5-silylheptasilane)

If a choice of principal chain cannot be made on the basis of chain length alone, unsaturation is the next selection criterion and then the greatest number of substituents.

Example:

24.

ClH_2Si, SiH_2Cl, HSi—Si—SiH, Cl_3Si, $SiHCl_2$ (Si atoms numbered 1–5)

1,1,1,5,5-pentachloro-2,4-bis(chlorosilyl)pentasilane
(all other five-silicon chains have fewer substituents)

IR-6.3.2 Hydrogen substitution in boron hydrides

The construction of names of derivatives of boron hydrides where hydrogen atoms have been replaced by substituent groups follows the procedures given in Section IR-6.3.1. The only special feature is the need for specifying replacement of a bridging hydrogen atom, in which case the designator 'μ-' is used in front of the substituent group name, as in Example 4 below.

Examples:

1.

F_2B, B—BF_2, F_2B

2-(difluoroboranyl)-1,1,3,3-tetrafluorotriborane(5)

2.

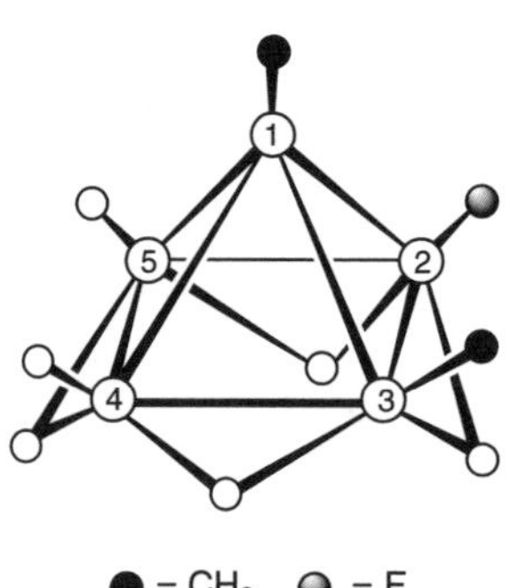

● = CH_3 ● = F

2-fluoro-1,3-dimethylpentaborane(9), or
2-fluoro-1,3-dimethyl-2,3:2,5:3,4:4,5-tetra-μ*H*-*nido*-pentaborane(9)

3.

● = NH_2

diboran(6)amine

4.

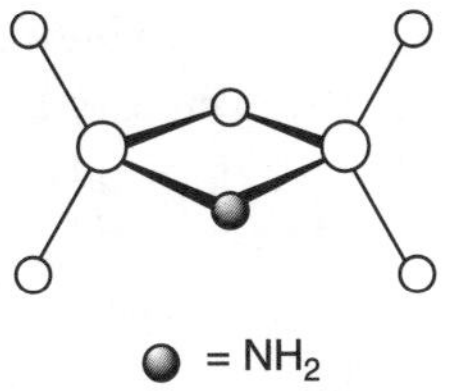

= NH_2

diboran(6)-μ-amine

IR-6.4 NAMES OF IONS AND RADICALS DERIVED FROM PARENT HYDRIDES

This section presents names of ions and radicals that can be formally derived from hydrides by the operations of removal or addition of hydrogen atoms, hydride ions or hydrons. A great many ions and radicals can also be named by additive methods, as described in Chapter IR-7. Many simple ions and radicals are named in Table IX, often by both nomenclature types.

IR-6.4.1 Cations derived from parent hydrides by addition of one or more hydrons

The name of an ion formally derived by adding a hydron to a parent hydride is obtained by adding the suffix 'ium' to the name of the parent hydride, with elision of a final 'e'. For polycations formed in this way, the suffixes 'diium', 'triium', *etc.*, are used without elision of any final 'e' on the parent hydride name. Any necessary locants are placed immediately preceding the suffix. Locants for added hydrons take precedence over locants for unsaturation, as in Example 8 below.

The alternative names ammonium, hydrazinium, hydrazinediium and oxonium are used for naming organic derivatives, see Section IR-6.4.3 and Section P-73.1 of Ref. 1.

Examples:

1. NH_4^+ azanium, or ammonium
2. $N_2H_5^+$ diazanium, or hydrazinium
3. $N_2H_6^{2+}$ diazanediium, or hydrazinediium
4. H_3O^+ oxidanium, or oxonium (*not* hydronium)
5. H_4O^{2+} oxidanediium
6. $H_3O_2^+$ dioxidanium
7. $^+H_3PPHPH_3^+$ triphosphane-1,3-diium
8. $^+H_3NN{=}NH$ triaz-2-en-1-ium

IR-6.4.2 Cations derived from parent hydrides by loss of one or more hydride ions

A cation produced by formal loss of a hydride ion from a parent hydride is named by adding the suffix 'ylium' to the parent name, with elision of a final 'e'. For polycations formed in this way, the suffixes 'diylium', 'triylium', *etc.*, are used without elision of any final 'e' on the parent hydride name. Any necessary locants are placed immediately preceding the suffix. Locants for removed hydride ions take precedence over locants for unsaturation, as in Example 5 below.

For the names silane, germane, stannane and plumbane, as well as a number of hydrocarbon names, 'ylium' *replaces* the ending 'ane' of the parent hydride (*cf.* Section P-73.2 of Ref. 1).

Examples:

1. PH_2^+ phosphanylium
2. $Si_2H_5^+$ disilanylium
3. SiH_3^+ silylium
4. BH_2^+ boranylium
5. $^+HNN{=}NH$ triaz-2-en-1-ylium

IR-6.4.3 Substituted cations

Names of substituted derivatives of cations are formed from the modified parent hydride names (as described in IR-6.4.1 and IR-6.4.2) by adding appropriate substituent prefixes. When numbering derivatives of polynuclear parents, the locants for added hydrons or removed hydride ions take precedence over locants for substituents, as in Example 6 below.

Examples:

1. $[NF_4]^+$ tetrafluoroazanium, or tetrafluoroammonium
2. $[PCl_4]^+$ tetrachlorophosphanium
3. $[NMe_4]^+$ tetramethylazanium, or tetramethylammonium
4. $[SEtMePh]^+$ ethyl(methyl)phenylsulfanium
5. $[MeOH_2]^+$ methyloxidanium, or methyloxonium
6. $[ClPHPH_3]^+$ 2-chlorodiphosphan-1-ium

IR-6.4.4 Anions derived from parent hydrides by loss of one or more hydrons

An anion formally obtained by removal of one or more hydrons from a parent hydride is named by adding 'ide', 'diide', *etc.*, to the parent name, with elision of a terminal 'e' before 'ide' but not in any other cases. Any necessary locants are placed immediately preceding the suffix. Locants for removed hydrons take precedence over locants for unsaturation, as in Example 10 below.

Examples:

1. NH_2^- azanide, or amide
2. NH^{2-} azanediide, or imide
3. H_2NNH^- diazanide, or hydrazinide

4. H_2NN^{2-} diazane-1,1-diide, or hydrazine-1,1-diide
5. $^{-}HNNH^{-}$ diazane-1,2-diide, or hydrazine-1,2-diide
6. SiH_3^{-} silanide
7. GeH_3^{-} germanide
8. SnH_3^{-} stannanide
9. SH^{-} sulfanide
10. $^{-}HNN{=}NH$ triaz-2-en-1-ide

Names of anions derived by formal loss of one or more hydrons from hydroxy groups and their chalcogen analogues (characterized by suffixes such as 'ol' and 'thiol') are formed by adding the ending 'ate' to the appropriate name.

Examples:

11. SiH_3O^{-} silanolate
12. PH_2S^{-} phosphanethiolate

The anion in Example 12 may also be named as a derivative of phosphinothious acid, H_2PSH, thus giving the name 'phosphinothioite'. This type of name is used as the basis for naming organic derivatives of H_2PSH. (See discussion of inorganic acids in Chapter IR-8.)

IR-6.4.5 Anions derived from parent hydrides by addition of one or more hydride ions

The addition of a hydride ion to a parent hydride is designated by the ending 'uide' (see Section P-72.3 of Ref. 1). Rules regarding locants are analogous to the rules for the 'ide' suffix (see Section IR-6.4.4). For compounds of this kind, additive names (Chapter IR-7) are common and acceptable alternatives.

Example:

1. $[BH_4]^{-}$ boranuide (from borane), or tetrahydridoborate(1−) (additive)

IR-6.4.6 Substituted anions

Names of substituted derivatives of anions are formed from parent hydride names modified as above (see Sections IR-6.4.4 and IR-6.4.5) by further adding appropriate prefixes for the substituents. When numbering the structure, the position where a hydron was removed or a hydride ion was added takes precedence over the positions with substituents, as in Example 4 below. In many cases, additive names are common and acceptable alternatives.

Examples:

1. $SnCl_3^{-}$ trichlorostannanide (from stannane), or trichloridostannate(1−) (additive)
2. $MePH^{-}$ methylphosphanide

3. $MeNH^-$ — methylazanide, or methylamide, or methanaminide (all substitutive, see Section P-72.2 of Ref. 1)

4. $\overset{1}{^-Sn}H_2\overset{2}{O}\overset{3}{Sn}H_2\overset{4}{O}\overset{5}{Sn}H_2\overset{6}{O}\overset{7}{Sn}H_2\overset{8}{O}\overset{9}{Sn}Cl_3$

   9,9,9-trichloropentastannoxan-1-ide

5. $[BH_3CN]^-$ — cyanoboranuide (from borane), or cyanidotrihydridoborate(1−) (additive)

6. $[PF_6]^-$ — hexafluoro-λ^5-phosphanuide (from phosphane), or hexafluoridophosphate(1−) (additive)

IR-6.4.7 Radicals and substituent groups

Radicals and substituent groups derived from parent hydrides by removal of one or more hydrogen atoms are named by modifying the parent hydride name as follows:

(i) removal of one hydrogen atom: add suffix 'yl' (eliding final 'e' of parent hydride name);
(ii) removal of two or more hydrogen atoms: add suffix 'yl' with appropriate multiplicative prefix (no vowel elision).

The suffix 'ylidene' is used on a substituent group if a double bond is implied when a skeletal atom has formally lost two hydrogen atoms. If a triple bond is implied, the ending 'ylidyne' is used. With these endings, the ending 'e' of the parent hydride name is again elided.

For radicals, if two hydrogens are removed from the same atom the suffix 'ylidene' is used.

Locants may be needed to indicate the skeletal atoms from which hydrogen atoms have been removed. Such locants are placed immediately before the suffix. When numbering the structure, the positions where hydrogen atoms were removed take precedence over unsaturation, as in Example 9 below.

Radicals may also be named using additive nomenclature, see Section IR-7.1.4 and examples in subsequent sections of Chapter IR-7.

Examples:

1. $NH^{2\bullet}$ — azanylidene
2. $PH_2^\bullet$ and H_2P- — phosphanyl
3. $PH^{2\bullet}$ and $HP=$ — phosphanylidene
4. HP< — phosphanediyl
5. $P\equiv$ — phosphanylidyne
6. $H_2Br^\bullet$ and H_2Br- — λ^3-bromanyl
7. $H_2NNH^\bullet$ and H_2NNH- — diazanyl or hydrazinyl
8. $^\bullet HNNH^\bullet$ and $-HNNH-$ — diazane-1,2-diyl or hydrazine-1,2-diyl
9. $HP{=}NP^\bullet NHPH^\bullet$ and $HP{=}N\overset{|}{P}NHPH-$ — triphosphaz-4-ene-1,3-diyl

In a number of cases, the established name of a substituent group or radical is non-systematic or is a shorter version obtained by *replacing* the ending 'ane' of the parent name by the suffix 'yl' or the suffix 'ylidene':

Examples:

10. $OH^{\bullet}$	hydroxyl (for oxidanyl)
11. $OH-$	hydroxy (for oxidanyl)
12. $NH_2^{\bullet}$	aminyl (for azanyl)
13. NH_2-	amino (for azanyl)
14. $CH_2^{2\bullet}$	methylidene (for methanylidene), or λ^2-methane, or carbene
15. $SiH_3^{\bullet}$ and SiH_3-	silyl (for silanyl)
16. $GeH_3^{\bullet}$ and GeH_3-	germyl (for germanyl)
17. $SnH_3^{\bullet}$ and SnH_3-	stannyl (for stannanyl)
18. $PbH_3^{\bullet}$ and PbH_3-	plumbyl (for plumbanyl)
19. $SiH_2^{2\bullet}$	silylidene

This list is exhaustive as far as non-carbon parent hydrides are concerned. A number of established shortened or entirely non-systematic names are also used for carbon-based hydrides: methyl, ethyl, propyl, butyl, pentyl, hexyl, cyclohexyl, phenyl, naphthyl, *etc.*

IR-6.4.8 Substituted radicals or substituent groups

Radicals or substituent groups formally derived by removing one or more hydrogen atoms and introducing substituents in parent hydrides are named using prefixes for the substituents as explained in Section IR-6.3.1. The positions from which the hydrogen atoms were removed take priority over the positions with substituents. Several simple such radicals and substituent groups are named in Table IX. In a few cases the name of a radical and the corresponding substituent group as used in organic nomenclature may differ (see Example 2 below).

Examples:

1. $NH_2O^{\bullet}$ and NH_2O-	aminooxidanyl
2. $HONH^{\bullet}$	hydroxyazanyl
$HONH-$	hydroxyamino
3. $Me_3PbPbMe_2^{\bullet}$ and $Me_3PbPbMe_2-$	1,1,2,2,2-pentamethyldiplumban-1-yl (*not* 1,1,1,2,2-pentamethyldiplumban-2-yl)

IR-6.4.9 Anionic and cationic centres and radicals in a single molecule or ion

If several of the above features [cationic moiety, anionic moiety, radical formed by removal of hydrogen atom(s)] are present in a molecule or ion, a rule is needed to decide in which order to cite the various modifications of the parent hydride name.

The order is:

cation < anion < radical

in the sense that:

(i) the suffixes indicating these modifications are cited in that order;
(ii) the lowest locants are given to positions where hydrogen atoms have been removed, if any; anion sites, if any, are numbered using the next lowest locants; finally, any cationic sites are numbered. All these take precedence over unsaturation and over substituents cited by prefixes.

Examples:

1. $H_2Te^{\bullet +}$ tellaniumyl
2. $H_2Te^{\bullet -}$ tellanuidyl
3. $Me_3\overset{2}{N}^{+}-\overset{1}{N}^{-}-Me$ 1,2,2,2-tetramethyldiazan-2-ium-1-ide
4. $Me\overset{3}{N}=\overset{2}{N}^{\bullet +}-\overset{1}{N}^{-}-SiMe_3$ 3-methyl-1-(trimethylsilyl)triaz-2-en-2-ium-1-id-2-yl

Further complications arise if one wishes to name a substituent group containing a radical centre (see Section P-71.5 of Ref. 1).

IR-6.5 REFERENCES

1. *Nomenclature of Organic Chemistry, IUPAC Recommendations*, eds. W.H. Powell and H. Favre, Royal Society of Chemistry, in preparation.
2. K. Wade, *Adv. Inorg. Chem. Radiochem.*, **18**, 1–66 (1976); R.E. Williams, *Adv. Inorg. Chem. Radiochem.*, **18**, 67–142 (1976); D.M.P. Mingos, *Acc. Chem. Res.*, **17**, 311–319 (1984).
3. R.W. Rudolph and W.R. Pretzer, *Inorg. Chem.*, **11**, 1974–1978 (1972); R.W. Rudolph, *Acc. Chem. Res.*, **9**, 446–452 (1976).

IR-7 Additive Nomenclature

CONTENTS

IR-7.1 INTRODUCTION

IR-7.1.1 General

Additive nomenclature was originally developed for Werner-type coordination compounds, which were regarded as composed of a central atom (or atoms) surrounded by added groups known as ligands, but many other types of compound may also be conveniently given additive names. Such names are constructed by placing the names of the ligands (sometimes modified) as prefixes to the name(s) of the central atom(s).

This Chapter deals with the general characteristics of additive nomenclature and provides examples of additive names for simple mononuclear and polynuclear compounds. Chain and ring compounds are then treated using additive principles supplemented by further conventions. Additive names for inorganic acids are discussed in Chapter IR-8. Additive nomenclature as applied to metal coordination compounds is described in further detail in Chapter IR-9 (where a flowchart, Figure IR-9.1, provides a general procedure for naming coordination compounds). Additive names for a large number of simple compounds are given in Table IX*.

* Tables numbered with a Roman numeral are collected together at the end of this book.

Note that in some cases, a compound named additively may alternatively and equally systematically be named substitutively on the basis of a suitably chosen parent structure (Chapter IR-6). It is important to note, however, that additive names for parent hydrides cannot be used as parent names in substitutive nomenclature.

IR-7.1.2 Choosing a central atom or atoms, or a chain or ring structure

Making a choice of central atom or atoms is a key step in the process of naming a compound using additive nomenclature. If there are (one or more) metal atoms in the compound, these should be chosen as the central atom(s). Such atom(s) should also be relatively central in the structure and, where possible, should be chosen to make use of molecular symmetry (thereby shortening the name). Usually hydrogen atoms are disregarded when choosing central atoms.

For some compounds, a choice of central atom or atoms will remain. The atom(s) that occur(s) latest when following the arrow in Table VI should be chosen as the central atom(s).

If there is more than one central atom in a structure according to the above criteria then the compound can be named as a dinuclear or polynuclear compound.

As an alternative to the procedure above, a group of atoms forming a chain or ring sub-structure within a compound may be chosen in order to give the compound an additive name using the chains and rings nomenclature outlined in Section IR-7.4.

IR-7.1.3 Representing ligands in additive names

Additive names are constructed by placing (sometimes modified) ligand names as prefixes to the name of the central atom. For anionic ligands, the anion endings 'ide', 'ate' and 'ite' (see Section IR-5.3.3) are changed to 'ido', 'ato' and 'ito', respectively, when generating these prefixes. Names of neutral and cationic ligands are used unchanged, except in a few special cases, most notably water (prefix 'aqua'), ammonia (prefix 'ammine'), carbon monoxide bound through carbon (prefix 'carbonyl'), and nitrogen monoxide bound through nitrogen (prefix 'nitrosyl') (*cf.* Section IR-9.2.4.1).

In principle, it is a matter of convention whether a ligand is considered to be anionic, neutral or cationic. The default is to consider ligands as anionic, so that OH is 'hydroxido', Cl 'chlorido', SO_4 'sulfato', *etc.* Some ligands are conventionally regarded as neutral, *e.g.* amines and phosphanes and ligands derived from hydrocarbons by removal of a hydrogen atom, such as methyl, benzyl, *etc.*

Appropriate prefixes to represent many simple ligands within names are given in Table IX. For further details, see Section IR-9.2.2.3.

IR-7.1.4 Ions and radicals

Anionic species take the ending 'ate' in additive nomenclature, whereas no distinguishing termination is used for cationic or neutral species. Additive names of ions end with the charge number (see Section IR-5.4.2.2). In additive names of radicals, the radical character of the compound may be indicated by a radical dot, •, added in parentheses and centred, after the name of the compound. Polyradicals are indicated by the appropriate numeral placed before the dot. For example, a diradical is indicated by '(2•)'.

IR-7.2 MONONUCLEAR ENTITIES

Names of mononuclear compounds and ions, *i.e.* of species with a single central atom, are formed by citing the appropriate prefixes for the ligands alphabetically before the name of the central atom. Ligands occurring more than once are collected in the name by means of multiplicative prefixes (Table IV), *i.e.* 'di', 'tri', 'tetra', *etc.*, for simple ligands such as chlorido, benzyl, aqua, ammine and hydroxido, and 'bis', 'tris', 'tetrakis', *etc.*, for more complex ligands, *e.g.* 2,3,4,5,6-pentachlorobenzyl and triphenylphosphane. The latter prefixes are also used to avoid any ambiguity which might attend the use of 'di', 'tri', *etc.* Multiplicative prefixes which are not inherent parts of the ligand name do not affect the alphabetical ordering.

Prefixes representing ligands can be separated using enclosing marks (see also Section IR-9.2.2.3), and this should be done for all but the simplest ligands, including organic ligands. In some cases the use of enclosing marks is essential in order to avoid ambiguity, as in Examples 10 and 11 below.

In several of the examples below, substitutive names (see Chapter IR-6) are also given. In some cases, however, there is no parent hydride available for the construction of a substitutive name (see Examples 9 and 11). Note also that the formulae given below in square brackets are coordination compound-type formulae with the central atom listed first.

Examples:

1. $Si(OH)_4$ — tetrahydroxidosilicon (additive),
   or silanetetrol (substitutive)
2. $B(OMe)_3$ — trimethoxidoboron or
   tris(methanolato)boron (both additive),
   or trimethoxyborane (substitutive)
3. FClO or [ClFO] — fluoridooxidochlorine (additive),
   or fluoro-λ^3-chloranone (substitutive)
4. ClOCl or $[OCl_2]$ — dichloridooxygen (additive),
   or dichlorooxidane (substitutive)
5. $[Ga\{OS(O)Me\}_3]$ — tris(methanesulfinato)gallium (additive),
   or tris(methanesulfinyloxy)gallane (substitutive)
6. $MeP(H)SiH_3$ or $[SiH_3\{P(H)Me\}]$ — trihydrido(methylphosphanido)silicon (additive),
   or methyl(silyl)phosphane (substitutive)
7. $NH^{2\bullet}$ — hydridonitrogen(2•) (additive),
   or azanylidene (substitutive)
8. $HOC(O)^{\bullet}$ — hydroxidooxidocarbon(•) (additive),
   or hydroxyoxomethyl (substitutive)
9. FArH or [ArFH] — fluoridohydridoargon
10. [HgMePh] — methyl(phenyl)mercury (additive)
11. $[Hg(CHCl_2)Ph]$ — (dichloromethyl)(phenyl)mercury

12. $[Te(C_5H_9)Me(NCO)_2]$

bis(cyanato-*N*)(cyclopentyl)(methyl)tellurium (additive),
or cyclopentyldiisocyanato(methyl)-λ^4-tellane (substitutive)

13. $[Al(POCl_3)_6]^{3+}$ — hexakis(trichloridooxidophosphorus)aluminium(3+)
14. $[Al(OH_2)_6]^{3+}$ — hexaaquaaluminium(3+)
15. $[H(py)_2]^+$ — bis(pyridine)hydrogen(1+)
16. $[H(OH_2)_2]^+$ — diaquahydrogen(1+)
17. $[BH_2(py)_2]^+$ — dihydridobis(pyridine)boron(1+)
18. $[PFO_3]^{2-}$ — fluoridotrioxidophosphate(2−)
19. $[Sb(OH)_6]^-$ — hexahydroxidoantimonate(1−) (additive), or hexahydroxy-λ^5-stibanuide (substitutive)
20. $[HF_2]^-$ — difluoridohydrogenate(1−)
21. $[BH_2Cl_2]^-$ — dichloridodihydridoborate(1−) (additive), or dichloroboranuide (substitutive)
22. $OCO^{\bullet-}$ — dioxidocarbonate(•1−)
23. $NO^{(2\bullet)-}$ — oxidonitrate(2•1−)
24. $PO_3^{\bullet 2-}$ — trioxidophosphate(•2−)
25. $[ICl_2]^+$ — dichloridoiodine(1+) (additive), or dichloroiodanium (substitutive)
26. $[BH_4]^-$ — tetrahydridoborate(1−) (additive), or boranuide (substitutive)
27. CH_5^- — pentahydridocarbonate(1−) (additive), or methanuide (substitutive)
28. $[PH_6]^-$ — hexahydridophosphate(1−) (additive), or λ^5-phosphanuide (substitutive)
29. $[PF_6]^-$ — hexafluoridophosphate(1−) (additive), or hexafluoro-λ^5-phosphanuide (substitutive)

IR-7.3 POLYNUCLEAR ENTITIES

IR-7.3.1 Symmetrical dinuclear entities

In symmetrical dinuclear entities, each of the central atoms is of the same element and they are identically ligated. Below, additive names of several formats are given to a number of such species. Again, in some cases substitutive names are also easily constructed, as exemplified below.

The general procedure for naming a symmetrical dinuclear entity is as follows.

The ligands are represented in the usual way and the multiplicative affix 'di' is added immediately before the name of the central atom. The name of the central element is modified to the 'ate' form if the compound is an anion.

A bond between the two central atoms, if there is one, is indicated by adding to the name the italicized symbols for those two atoms, separated by an 'em' dash and enclosed in parentheses.

In bridged dinuclear species, bridging ligands are indicated by the Greek letter μ, placed before the ligand name and separated from it by a hyphen. The whole term, *e.g.* 'μ-chlorido', is separated from the rest of the name by hyphens. If the bridging ligand occurs more than once, multiplicative prefixes are employed (see also Sections IR-9.1.2.10 and IR-9.2.5.2).

Examples:

1. $[Et_3PbPbEt_3]$ hexaethyldilead(*Pb*—*Pb*) (additive), or hexaethyldiplumbane (substitutive)
2. $HSSH^{\bullet -}$ dihydridodisulfate(*S*—*S*)(•1−) (additive), or disulfanuidyl (substitutive)
3. NCCN dinitridodicarbon(*C*—*C*)
4. $NCCN^{\bullet -}$ dinitridodicarbonate(*C*—*C*)(•1−)
5. (NC)SS(CN) bis(nitridocarbonato)disulfur(*S*—*S*), or dicyanidodisulfur(*S*—*S*)
6. $(NC)SS(CN)^{\bullet -}$ bis(nitridocarbonato)disulfate(*S*—*S*)(•1−), or dicyanidodisulfate(*S*—*S*)(•1−)
7. OClO μ-chlorido-dioxygen, or dioxidochlorine
8.

$Al_2Cl_4(\mu\text{-}Cl)_2$

di-μ-chlorido-tetrachloridodialuminium

A variant of the format in the above additive names involves starting with 'bis' and then citing the name of the half-molecule or ion in parentheses. Thus, Examples 1–6 and 8 become:

Examples:

9. $[Et_3PbPbEt_3]$ bis(triethyllead)(*Pb*—*Pb*)
10. $HSSH^{\bullet -}$ bis(hydridosulfate)(*S*—*S*)(•1−)
11. NCCN bis(nitridocarbon)(*C*—*C*)
12. $NCCN^{\bullet -}$ bis(nitridocarbonate)(*C*—*C*)(•1−)

13. (NC)SS(CN) bis[(nitridocarbonato)sulfur](*S*—*S*), or bis(cyanidosulfur)(*S*—*S*)

14. $(NC)SS(CN)^{\bullet-}$ bis[(nitridocarbonato)sulfate](*S*—*S*)(•1−), or bis(cyanidosulfate)(*S*—*S*)(•1−)

15. $Cl_2Al(\mu\text{-}Cl)_2AlCl_2$ di-μ-chlorido-bis(dichloridoaluminium)

Note that the five compounds in Examples 10–14 may also easily be named as chain compounds, as shown in Section IR-7.4. The name in Example 14 differs from that given in Ref. 1 (in which the sulfur–sulfur bond was indicated as above, but the carbon atoms were treated as the central atoms).

The species in Examples 13 and 14 may also be regarded as containing a bridging ligand, as demonstrated in Examples 16 and 17.

Examples:

16. [NCSSCN] μ-disulfanediido-bis(nitridocarbon)

17. $[NCSSCN]^{\bullet-}$ μ-disulfanediido-bis(nitridocarbonate)(•1−)

IR-7.3.2 Non-symmetrical dinuclear compounds

There are two types of non-symmetrical dinuclear compounds: (i) those with identical central atoms differently ligated, and (ii) those with different central atoms. In both cases names are formed by means of the procedure described in Section IR-9.2.5, which also deals with bridging groups.

Priority is assigned to the central atoms as follows. For cases of type (i) the central atom carrying the greater number of alphabetically preferred ligands is numbered 1. For cases of type (ii) the number 1 is assigned to the higher priority central element of Table VI, whatever the ligand distribution.

In both types of compound, names are constructed in the usual way, by first citing the prefixes representing the ligands in alphabetical order. Each prefix representing a ligand is followed by a hyphen, the number(s) assigned to the central atom(s) to which the ligand is attached (see below), the Greek letter κ (kappa) (see Section IR-9.2.4.2) with a right superscript denoting the number of such ligands bound to the central atom(s) (the number 1 being omitted for a single ligand), and the italic element symbol for the ligating atom by which the ligand is attached to the central atom(s). This describes the ligands and their mode of attachment. The κ construction can be omitted in very simple cases (see Examples 1–3 below) or when the distribution of the ligands on the central atoms is obvious (see Example 4 below).

The central atom names are listed after the ligand names. The multiplicative prefix 'di' is used where the central atoms are the same element. Otherwise, the order of the central atom names is obtained using Table VI. The order of the central atom names is reflected in the numbering employed with the κ symbols. The ending 'ate' is added if the dinuclear compound is an anion, and a radical dot may be added for radicals. In the case of two different central atoms, the two names are cited inside parentheses and 'ate' is added outside the parentheses.

Examples:

1. ClClO oxido-1κ*O*-dichlorine(*Cl*—*Cl*),
   or oxidodichlorine(*Cl*—*Cl*)
2. $ClOO^{\bullet}$ chlorido-1κ*Cl*-dioxygen(*O*—*O*)(•),
   or chloridodioxygen(*O*—*O*)(•)
3. $ClClF^{+}$ fluorido-1κ*F*-dichlorine(*Cl*—*Cl*)(1+),
   or fluoridodichlorine(*Cl*—*Cl*)(1+)
4. $[O_3POSO_3]^{2-}$ μ-oxido-hexaoxido-1κ^3*O*,2κ^3*O*-(phosphorussulfur)ate(2−),
   or μ-oxido-hexaoxido(phosphorussulfur)ate(2−)
5.

   Me_3Sn (1) — S (thiophene ring) — BiMeEt (2)

   ethyl-2κ*C*-tetramethyl-1κ^3*C*,2κ*C*-μ-thiophene-2,5-diyl-tinbismuth
6. $[Cl(PhNH)_2\overset{2}{Ge}\overset{1}{Ge}Cl_3]$ tetrachlorido-1κ^3*Cl*,2κ*Cl*-bis(phenylamido-2κ*N*)-
   digermanium(*Ge*—*Ge*)
7. $\overset{1}{Li}\overset{2}{Pb}Ph_3$ triphenyl-2κ^3*C*-lithiumlead(*Li*—*Pb*)

Where the precise positions of ligation are unknown, the kappa convention cannot be used.

Examples:

8. $[Pb_2(CH_2Ph)_2F_4]$ dibenzyltetrafluoridodilead
9. $[Ge_2(CH_2Ph)Cl_3(NHPh)_2]$ (benzyl)trichloridobis(phenylamido)digermanium

IR-7.3.3 Oligonuclear compounds

In simple cases, the principles of the preceding sections may be generalized for the naming of oligonuclear compounds. Again, there are compounds which are also easily named by substitutive nomenclature because of the availability of obvious parent hydrides.

Examples:

1. $HO_3^{\bullet}$ hydridotrioxygen(•)
2. $HON_3^{\bullet-}$ hydroxido-1κ*O*-trinitrate(2 *N*—*N*)(•1−)
3. $Cl_3SiSiCl_2SiCl_3$ octachloridotrisilicon(2 *Si*—*Si*) (additive),
   or octachlorotrisilane (substitutive)
4. $FMe_2SiSiMe_2SiMe_3$ fluorido-1κ*F*-heptamethyltrisilicon(2 *Si*—*Si*) (additive),
   or 1-fluoro-1,1,2,2,3,3,3-heptamethyltrisilane (substitutive)

(An alternative additive name for the compound in Example 3, based on the longest chain in the molecule, can also be constructed by the method described in Section IR-7.4.3; see Example 6 therein.)

For heterooligonuclear systems, more conventions are needed to identify and name the collection of central atoms, and to number the central atoms so as to provide locants for the ligands.

Example:

5. $Me_3SiSeSiMe_3$
μ-selenido-bis(trimethylsilicon) (additive), or
hexamethyl-1κ^3*C*,2κ^3*C*-disiliconselenium(2 *Si—Se*) (additive), or
1,1,1,3,3,3-hexamethyldisilaselenane (substitutive)

Note that in the last example one can choose to name the compound as dinuclear or trinuclear. The complexities deriving from the structural variations which may occur with homonuclear and heteronuclear central atom clusters and bridging groups are dealt with in more detail in Sections IR-9.2.5.6 to IR-9.2.5.7.

IR-7.4 INORGANIC CHAINS AND RINGS

IR-7.4.1 General

Inorganic chain and ring structures may be named with no implications about the nature of bonds, except for the connectivity of the molecule or ion, using a particular system of additive nomenclature. The method can be applied to all chain and ring compounds although it is principally intended for species composed mainly of atoms other than carbon. While small molecules can be named more conveniently by using several alternative methods, the advantage of this nomenclature system lies in the simplicity with which complicated structures can be derived from the name and *vice versa*. Details of this system are given in Ref. 2; a simplified treatment is provided here.

The overall topology of the structure is specified as follows. A neutral chain compound is called 'catena' preceded by a multiplicative prefix, 'di', 'tri', *etc.*, to indicate the number of branches in the molecule. Likewise, cyclic compounds are called 'cycle' preceded by the appropriate multiplicative prefix. A mixed chain and ring compound is classified as an *assembly* composed of acyclic and cyclic *modules* and, if neutral, is named as 'catenacycle', with appropriate multiplicative prefixes inserted as in Example 3 below.

Examples:

1.

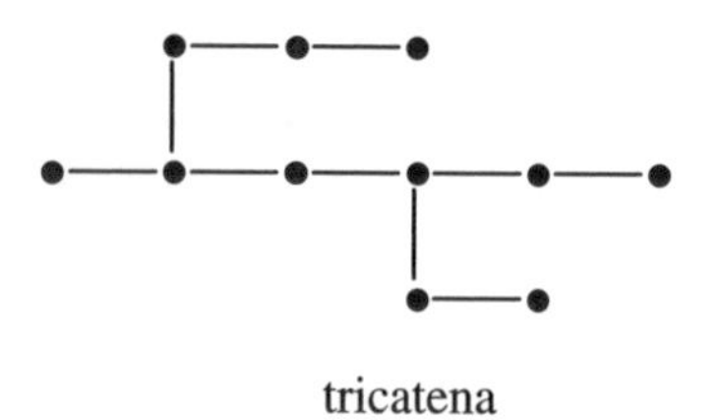

tricatena

2.

dicycle

3.

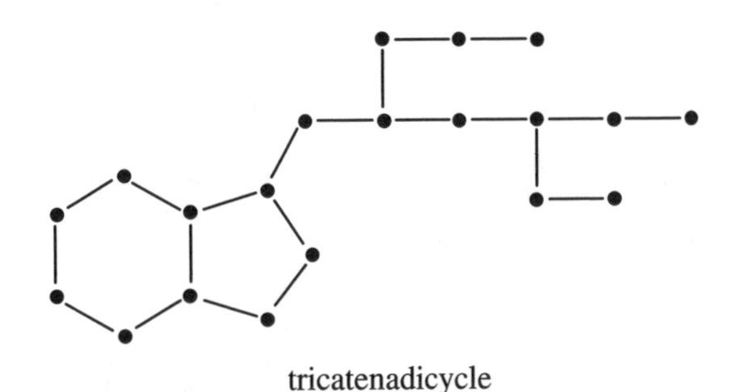

tricatenadicycle

IR-7.4.2 Nodal descriptor

The connectivity in the molecular framework is indicated by a *nodal descriptor*, which is placed in square brackets immediately before the terms 'catena', 'cycle' or 'catenacycle'. The atoms are numbered according to the general nodal nomenclature regardless of their identity. Only in the case of ambiguity are the identities of the atoms taken into consideration.

The first part of the descriptor indicates the number of atoms in the main chain. The arabic numerals after the full stop indicate the lengths of the branches cited in priority order. A superscript locant for each branch denotes the atom in the part of the molecule already numbered to which the branch is attached.

A zero in the descriptor indicates a ring and is followed by an arabic numeral indicating the number of atoms in the main ring. For polycyclic systems, the numbering begins from one of the bridgeheads and proceeds in the direction which gives the lowest possible locant for the other bridgehead. In this case, the number of atoms in the bridge is cited after the full stop. A pair of superscript locants is inserted for each such bridge numeral, separated by a comma and cited in increasing numerical order.

An assembly descriptor consists of square brackets enclosing, in the order of their seniority (see Ref. 2 for rules), the nodal descriptors of each module in parentheses. Between the descriptors of the modules, the locants of the nodes linking the modules are indicated. These locants, separated by a colon, are the atom numbers in the *final* numbering of the entire assembly (compare Example 7 below with Examples 5 and 6).

Examples:

1.

1 7

descriptor: [7]

2.

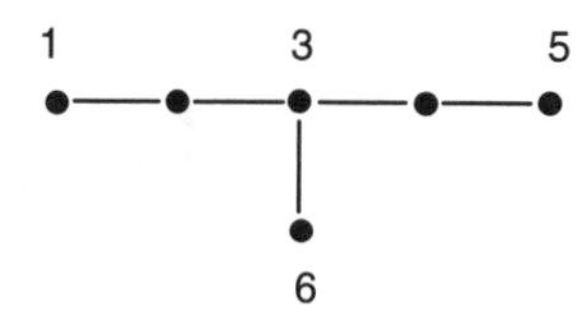

descriptor: $[5.1^3]$

3.

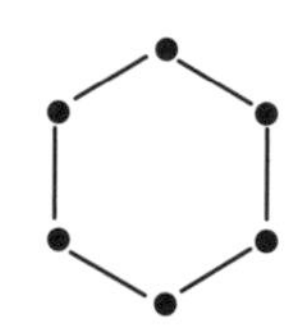

descriptor: [06]

4.

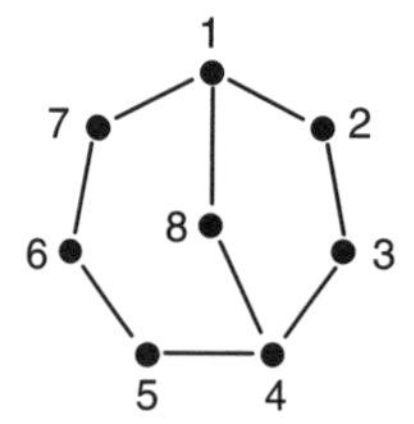

descriptor: $[07.1^{1,4}]$

5.

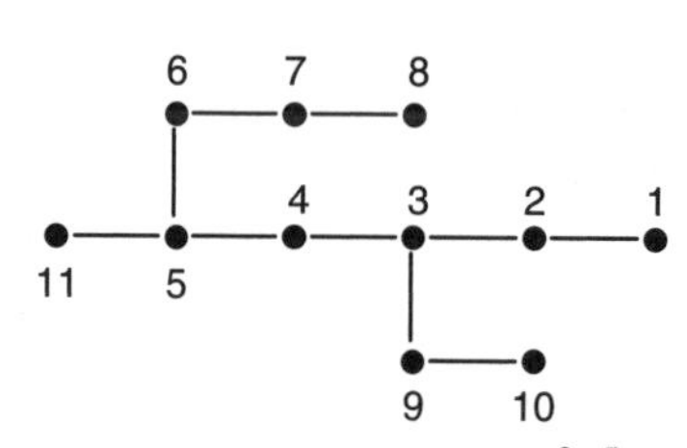

descriptor: $[8.2^3 1^5]$

6.

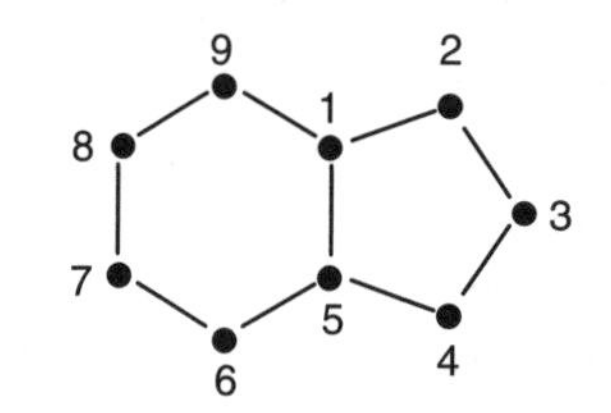

descriptor: $[09.0^{1,5}]$

7.

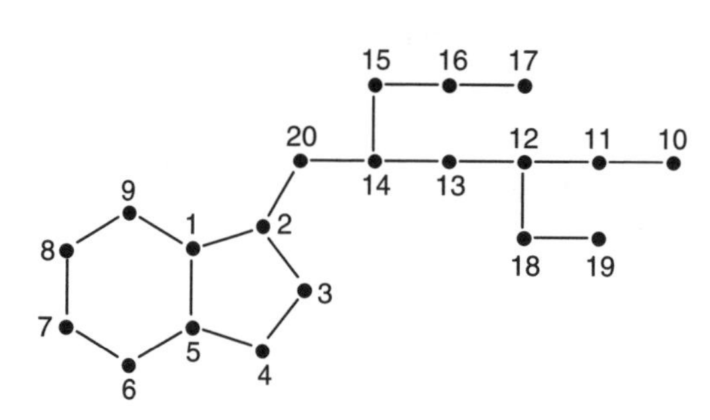

descriptor: $[(09.0^{1,5})2{:}20(8.2^3 1^5)]$

IR-7.4.3 Name construction

The atoms forming the nodal skeleton are listed in alphabetical order complete with their locants and are named using 'y' terms, examples of which are given in Table IR-7.1; a full list is given in Table X.

Table IR-7.1 Some 'y' terms for naming elements in the nodal framework

H	hydrony	C	carby	N	azy	O	oxy
B	bory	Si	sily	P	phosphy	S	sulfy
		Ge	germy	As	arsy	Se	seleny
		Sn	stanny	Sb	stiby	Te	tellury

Atoms and groups of atoms which are not part of the nodal framework are named as ligands (Section IR-7.1.3) and are cited in alphabetical order, together with their locants, before the sequence of names of the atoms constituting the nodal framework. The nodal descriptor is given next. The 'catena', 'cycle' or 'catenacycle' term is added at the end, *cf.* Section IR-7.4.1. (Note that bridging ligands are not employed in this system.)

In the case of anionic and cationic species, these terms are modified by the endings 'ate' and 'ium' respectively, to yield the terms 'catenate', 'catenium', 'cyclate', 'cyclium', 'catenadicyclium', 'catenacyclate', *etc.*, and a charge number is added at the end of the name. Radical species may be indicated analogously by using the radical dot (see Section IR-7.1.4).

Examples 1–6, which demonstrate the use of the system described here, were also named in Section IR-7.3.1. Examples 7–13 cannot be named so easily by other methods.

Examples:

1. NCCN 1,4-diazy-2,3-dicarby-[4]catena
2. $NCCN^{\bullet -}$ 1,4-diazy-2,3-dicarby-[4]catenate(•1−)
3. NCSSCN 1,6-diazy-2,5-dicarby-3,4-disulfy-[6]catena
4. $NCSSCN^{\bullet -}$ 1,6-diazy-2,5-dicarby-3,4-disulfy-[6]catenate(•1−)
5. $HSSH^{\bullet -}$ 1,2-dihydrido-1,2-disulfy-[2]catenate(•1−)
6. $Cl_3SiSiCl_2SiCl_3$ 2,2,3,3,4,4-hexachlorido-1,5-dichlory-2,3,4-trisily-[5]catena
7. $ClSiH_2SiH(Me)NSO$ 2,2,3-trihydrido-3-methyl-4-azy-1-chlory-6-oxy-2,3-disily-5-sulfy-[6]catena
8.

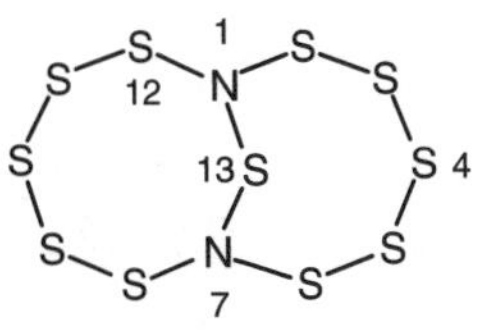

1,7-diazyundecasulfy-[012.1^{1,7}]dicycle

Because this compound contains only nitrogen and sulfur it is not necessary to indicate the locants of all sulfur atoms. Only the locants of the two nitrogen atoms are needed. The same applies to several of the following examples.

9.

3,6-diiodido-1,3,4,6-tetraphosphy-2,5,7-trisulfy-$[06.1^{1,4}]$dicycle

10.

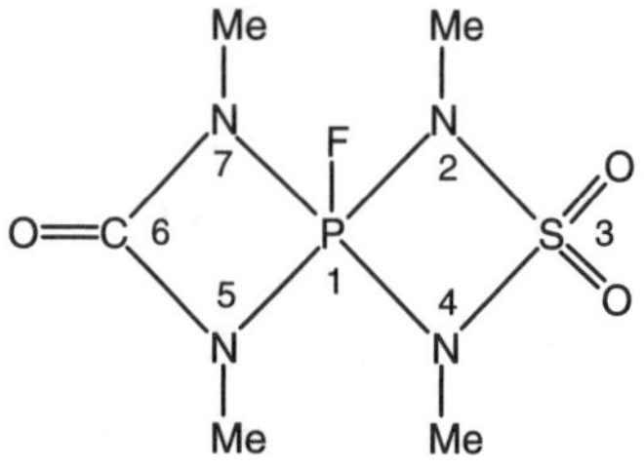

1-fluorido-2,4,5,7-tetramethyl-3,3,6-trioxido-2,4,5,7-tetraazy-6-carby-1-phosphy-3-sulfy-$[04.3^{1,1}]$dicycle

11.

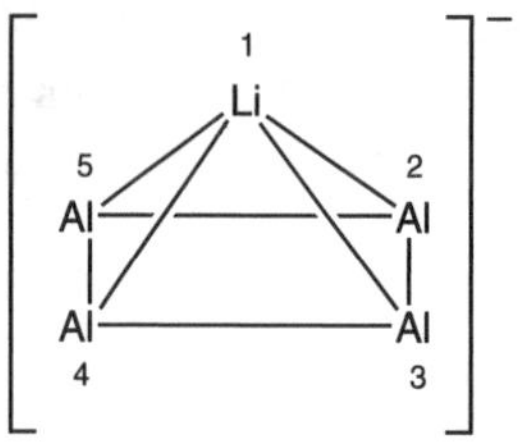

tetraaluminy-1-lithy-$[05.0^{1,3}0^{1,4}0^{2,5}]$tetracyclate(1−)

12.

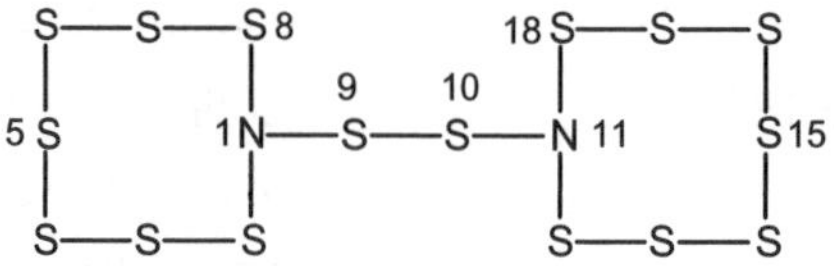

1,11-diazyhexadecasulfy-[(08)1:9(2)10:11(08)]catenadicycle

13.

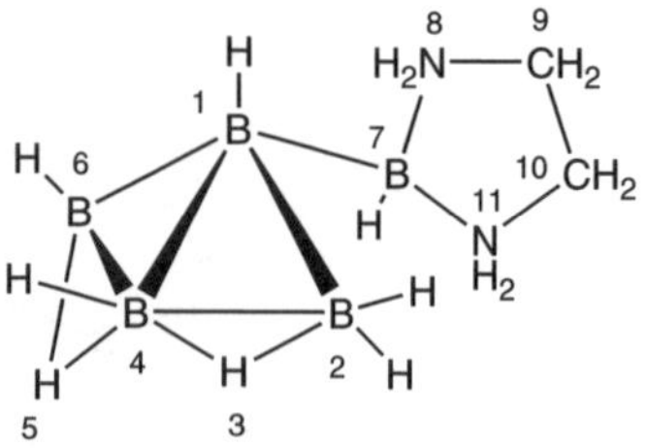

1,2,2,4,6,7,8,8,9,9,10,10,11,11-tetradecahydrido-8,11-diazy-1,2,4,6,7-pentabory-9,10-dicarby-3,5-dihydrony-$[(06.0^{1,4}0^{2,4}0^{4,6})1{:}7(05)]$pentacycle

IR-7.5 REFERENCES

1. Names for Inorganic Radicals, W.H. Koppenol, *Pure Appl. Chem.*, **72**, 437–446 (2000).
2. Nomenclature of Inorganic Chains and Ring Compounds, E.O. Fluck and R.S. Laitinen, *Pure Appl. Chem.*, **69**, 1659–1692 (1997); Chapter II-5 in *Nomenclature of Inorganic Chemistry II, IUPAC Recommendations 2000*, eds. J.A. McCleverty and N.G. Connelly, Royal Society of Chemistry, 2001.

IR-8 Inorganic Acids and Derivatives

CONTENTS

IR-8.1 INTRODUCTION AND OVERVIEW

Certain inorganic and simple carbon-containing compounds are commonly given non-systematic or semi-systematic names containing the word 'acid'. Examples are boric acid or orthoboric acid, metaboric acid, phosphoric acid, diphosphoric acid, *cyclo*-triphosphoric acid, *catena*-triphosphoric acid, dithionous acid, peroxodisulfuric acid or peroxydisulfuric acid, *etc*. These names are unique in modern nomenclature in that, interpreted literally, they describe a particular chemical property of the compounds in question. Systematic names are otherwise based solely on composition and structure.

All such acids may also be given structure-based systematic names using principles already described in preceding chapters on substitutive and additive nomenclature, so in that respect the 'acid'-containing names are superfluous. Furthermore, many species which would be classified as acids based on their chemical properties are never named as such, *e.g.* aqua ions such as hexaaquaaluminium(3+), and hydrides and derivatives such as ammonium, hydrogen sulfide (sulfane), *etc*. The term 'acid' is thus not used consistently.

Based on these considerations, the use of the word 'acid' in any new name in inorganic nomenclature is discouraged. However, a number of the existing 'acid' names are so commonly used (sulfuric acid, perchloric acid, *etc.*) that it would be unrealistic to suggest replacing them altogether by systematic alternatives. Another reason to include them in the present recommendations is that the acids in question are used as parent structures in the nomenclature of certain organic (*i.e.* carbon-containing) derivatives so that the derivative names are directly or indirectly based on the names containing the word 'acid'. See examples below and Section IR-8.6.

The main purposes of this chapter are:

(a) to show how the inorganic species commonly named as acids may be given systematic additive names (Section IR-8.3 and Tables IR-8.1 and IR-8.2);

(b) to list the 'acid' names that are still acceptable due to common usage and/or because they are needed in organic nomenclature (see Tables IR-8.1 and IR-8.2).

In addition, Sections IR-8.4 and IR-8.5 deal with a further type of names, denoted here as *hydrogen names*. These names can be viewed as generalizations of common anion names such as 'hydrogencarbonate', but they are not necessary for naming completely specified molecular structures and can be regarded as a special topic.

Thus, this Chapter provides several acceptable names for many inorganic acids; it is left to practitioners to choose the name most suitable for a particular application. In the future, IUPAC aims to select *preferred* names for inorganic species, including the acids dealt with here, just as Ref. 1 does for organic species.

Finally, names which do not denote compounds of a definite composition, such as hydrochloric acid, stannic acid, tungstic acid, *etc.*, are outside the scope of the systematic nomenclature presented here. However, the chemical systems involved can always be discussed using systematic names such as hydrogen chloride, tin(IV) oxide, tungsten(VI) oxide, *etc.*

A few examples are given now in order to illustrate some of the general remarks above. In these examples, and in the remainder of this chapter, alternative formulae are sometimes provided for clarity in connection with the discussion of additive names. These are based on a perception of the structures in question as generalized coordination entities. For mononuclear entities, this means that the central atom symbol is listed first and then the ligand symbols in alphabetical order, as prescribed in Section IR-4.4.3.2.

Example:

1. phosphoric acid = H_3PO_4 or $[PO(OH)_3]$

Based on the structure, the compound can be named substitutively (Chapter IR-6) as a derivative of the parent hydride λ^5-phosphane (PH_5), leading to the name trihydroxy-λ^5-phosphanone, or additively (Chapter IR-7) as trihydroxidooxidophosphorus.

As opposed to the two last names, the name phosphoric acid does not convey the structure, but does fit into a general pattern whereby the ending 'ic' denotes a higher or the highest possible oxidation state (compare nitric acid, sulfuric acid). Examples 2 and 3 show organic derivatives named on the basis of phosphoric acid as the parent.

Examples:

2. $PO(OMe)_3$ trimethyl phosphate
3. $PO(NMe_2)_3$ hexamethylphosphoric triamide

Each of these two compounds could also be named substitutively, on the basis of the above parent hydride, or additively but the names given here are preferred IUPAC names (see Section P-67.1 of Ref. 1).

Some organic derivative names still contain the word 'acid', as in the following derivatives of arsonic acid (H_2AsHO_3 or $[AsHO(OH)_2]$).

Examples:

4. $PhAsO(OH)_2$ phenylarsonic acid
5. EtAsCl(OH)S ethylarsonochloridothioic *O*-acid

The name in Example 4 regards the compound as derived from arsonic acid, by substitution of a phenyl group for the hydrogen atom bound directly to arsenic. The name in Example 5, in addition to the hydrogen substitution, involves functional replacement nomenclature (Section IR-8.6).

Note that there is one general case where the word 'acid' may appear in a fully systematic name of an inorganic compound, namely when substitutive nomenclature is used and prescribes a suffix for the highest ranking substituent group which ends with the word 'acid'.

Consider the polythionic acids, $H_2S_nO_6 = [(HO)(O)_2SS_{n-2}S(O)_2(OH)]$ ($n \geq 2$), which have the common names dithionic acid, trithionic acid, tetrathionic acid, *etc.* They may be named systematically using additive nomenclature, as shown in Table IR-8.1. For $n \geq 3$, they may also be named substitutively on the basis of the central (poly)sulfane skeleton, as exemplified below.

Examples:

6. $H_2S_3O_6 = [(HO)(O)_2SSS(O)_2(OH)]$ sulfanedisulfonic acid
7. $H_2S_4O_6 = [(HO)(O)_2SSSS(O)_2(OH)]$ disulfanedisulfonic acid

IR-8.2 GENERAL PRINCIPLES FOR SYSTEMATIC NAMING OF ACIDS

Molecular compounds and ions commonly regarded as inorganic acids are treated no differently than other molecular species when constructing systematic names.

The most easily applied general principle for systematic naming is that of additive nomenclature, exemplified in Section IR-8.3. As mentioned in Section IR-8.1, substitutive nomenclature could also be generally applied. However, this is not further elaborated here.

Sections IR-8.4 and IR-8.5 describe hydrogen names, which are related to additive names and only needed in special cases.

The method called 'acid nomenclature' in Section I-9.6 of Ref. 2 is little used and not needed. Its use is therefore no longer recommended.

Table IR-8.1 *Acceptable common names and systematic (additive) names for oxoacid and related structures*

This Table includes compounds containing oxygen and hydrogen and at least one other element and with at least one OH group; certain isomers; and examples of corresponding partially and fully dehydronated species. Formulae are given in the classical oxoacid format with the 'acid' (oxygen-bound) hydrogens listed first, followed by the central atom(s), then the hydrogen atoms bound directly to the central atom, and then the oxygen atoms (*e.g.* HBH_2O, $H_2P_2H_2O_5$), *except* for chain compounds such as *e.g.* HOCN. In most cases formulae are also written as for coordination entities, assembled according to the principles of Chapter IR-7 (*e.g.* the Table gives '$HBH_2O = [BH_2(OH)]$' and '$H_2SO_4 = [SO_2(OH)_2]$'). More names of oxoanions are given in Table IX.
Note that Section P-42 of Ref. 1 lists a great many inorganic oxoacid species for use as parent structures in the naming of organic derivatives. (See the discussion in Section IR-8.1.) Most of those species, but not all, are included here. In particular, several dinuclear and polynuclear acids are not explicitly included.

Formula	*Acceptable common name (unless otherwise stated)*	*Systematic additive name(s)*
$H_3BO_3 = [B(OH)_3]$	boric acid[a]	trihydroxidoboron
$H_2BO_3^- = [BO(OH)_2]^-$	dihydrogenborate	dihydroxidooxidoborate(1−)
$HBO_3^{2-} = [BO_2(OH)]^{2-}$	hydrogenborate	hydroxidodioxidoborate(2−)
$[BO_3]^{3-}$	borate	trioxidoborate(3−)
$(HBO_2)_n = -\!(B(OH)O)\!-_n$	metaboric acid	*catena*-poly[hydroxidoboron-μ-oxido]
$(BO_2^-)_n = -\!(OBO)\!-_n^{n-}$	metaborate	*catena*-poly[(oxidoborate-μ-oxido)(1−)]
$H_2BHO_2 = [BH(OH)_2]$	boronic acid	hydridodihydroxidoboron
$HBH_2O = [BH_2(OH)]$	borinic acid	dihydridohydroxidoboron
$H_2CO_3 = [CO(OH)_2]$	carbonic acid	dihydroxidooxidocarbon
$HCO_3^- = [CO_2(OH)]^-$	hydrogencarbonate	hydroxidodioxidocarbonate(1−)
$[CO_3]^{2-}$	carbonate	trioxidocarbonate(2−)
$HOCN = [C(N)OH]$	cyanic acid	hydroxidonitridocarbon
$HNCO = [C(NH)O]$	isocyanic acid	azanediidooxidocarbon, (hydridonitrato)oxidocarbon
$OCN^- = [C(N)O]^-$	cyanate	nitridooxidocarbonate(1−)
$HONC = [N(C)OH]$	[b]	carbidohydroxidonitrogen
$HCNO = [N(CH)O]$	[b]	(hydridocarbonato)oxidonitrogen
$ONC^- = [N(C)O]^-$	[b]	carbidooxidonitrate(1−)
$H_4SiO_4 = [Si(OH)_4]$	silicic acid[a]	tetrahydroxidosilicon
$[SiO_4]^{4-}$	silicate	tetraoxidosilicate(4−)
$(H_2SiO_3)_n = -\!(Si(OH)_2O)\!-_n$	metasilicic acid	*catena*-poly[dihydroxidosilicon-μ-oxido]

(*continued*)

Table IR-8.1 *Continued*

Formula	*Acceptable common name (unless otherwise stated)*	*Systematic additive name(s)*
$-(SiO_3)-_n{}^{2n-}$	metasilicate	*catena*-poly[(dioxidosilicate-μ-oxido)(1−)]
$H_6Si_2O_7 = [(HO)_3SiOSi(OH)_3]$	disilicic acid[c]	μ-oxido-bis(trihydroxidosilicon)
$[Si_2O_7]^{6-} = [O_3SiOSiO_3]^{6-}$	disilicate	μ-oxido-bis(trioxidosilicate)(6−)
$H_2NO_3{}^{+} = [NO(OH)_2]^{+}$	[d]	dihydroxidooxidonitrogen(1+), dihydrogen(trioxidonitrate)(1+)
$HNO_3 = [NO_2(OH)]$	nitric acid	hydroxidodioxidonitrogen
$[NO_3]^{-}$	nitrate	trioxidonitrate(1−)
$H_2NHO = [NH_2OH]$	hydroxylamine[e]	dihydridohydroxidonitrogen
$H_2NHO_3 = [NHO(OH)_2]$	azonic acid	hydridodihydroxidooxidonitrogen
$HNO_2 = [NO(OH)]$	nitrous acid	hydroxidooxidonitrogen
$[NO_2]^{-}$	nitrite	dioxidonitrate(1−)
$HNH_2O_2 = [NH_2O(OH)]$	azinic acid	dihydridohydroxidooxidonitrogen
$H_2N_2O_2 = [HON{=}NOH]$	diazenediol[f]	bis(hydroxidonitrogen)(*N*—*N*), or 1,4-dihydrido-2,3-diazy-1,4-dioxy-[4]catena
$HN_2O_2{}^{-} = [HON{=}NO]^{-}$	2-hydroxydiazene-1-olate[f]	hydroxido-1κ*O*-oxido-2κ*O*-dinitrate(*N*—*N*)(1−), or 1-hydrido-2,3-diazy-1,4-dioxy-[4]catenate(1−)
$[N_2O_2]^{2-} = [ON{=}NO]^{2-}$	diazenediolate[f]	bis(oxidonitrate)(*N*—*N*)(2−), or 2,3-diazy-1,4-dioxy-[4]catenate(2−)
$H_3PO_4 = [PO(OH)_3]$	phosphoric acid[a]	trihydroxidooxidophosphorus
$H_2PO_4{}^{-} = [PO_2(OH)_2]^{-}$	dihydrogenphosphate	dihydroxidodioxidophosphate(1−)
$HPO_4{}^{2-} = [PO_3(OH)]^{2-}$	hydrogenphosphate	hydroxidotrioxidophosphate(2−)
$[PO_4]^{3-}$	phosphate	tetraoxidophosphate(3−)
$H_2PHO_3 = [PHO(OH)_2]$	phosphonic acid[g]	hydridodihydroxidooxidophosphorus
$[PHO_2(OH)]^{-}$	hydrogenphosphonate	hydridohydroxidodioxidophosphate(1−)
$[PHO_3]^{2-}$	phosphonate	hydridotrioxidophosphate(2−)
$H_3PO_3 = [P(OH)_3]$	phosphorous acid[g]	trihydroxidophosphorus
$H_2PO_3{}^{-} = [PO(OH)_2]^{-}$	dihydrogenphosphite	dihydroxidooxidophosphate(1−)
$HPO_3{}^{2-} = [PO_2(OH)]^{2-}$	hydrogenphosphite	hydroxidodioxidophosphate(2−)
$[PO_3]^{3-}$	phosphite	trioxidophosphate(3−)
$HPO_2 = [P(O)OH]$	hydroxyphosphanone[h]	hydroxidooxidophosphorus
$HPO_2 = [P(H)O_2]$	λ^5-phosphanedione[h]	hydridodioxidophosphorus
$H_2PHO_2 = [PH(OH)_2]$	phosphonous acid	hydridodihydroxidophosphorus

$HPH_2O_2 = [PH_2O(OH)]$	phosphinic acid	dihydridohydroxidooxidophosphorus
$HPH_2O = [PH_2(OH)]$	phosphinous acid	dihydridohydroxidophosphorus
$H_4P_2O_7 = [(HO)_2P(O)OP(O)(OH)_2]$	diphosphoric acid[c]	μ-oxido-bis(dihydroxidooxidophosphorus)
$(HPO_3)_n = \text{-(}P(O)(OH)O\text{)-}_n$	metaphosphoric acid	*catena*-poly[hydroxidooxidophosphorus-μ-oxido]
$H_4P_2O_6 = [(HO)_2P(O)P(O)(OH)_2]$	hypodiphosphoric acid	bis(dihydroxidooxidophosphorus)(*P*—*P*)
$H_2P_2H_2O_5 = [(HO)P(H)(O)OP(H)(O)(OH)]$	diphosphonic acid	μ-oxido-bis(hydridohydroxidooxidophosphorus)
$P_2H_2O_5^{2-} = [O_2P(H)OP(H)(O)_2]^{2-}$	diphosphonate	μ-oxido-bis(hydridodioxidophosphate)(2−)
$H_3P_3O_9$	*cyclo*-triphosphoric acid	tri-μ-oxido-tris(hydroxidooxidophosphorus), or 2,4,6-trihydroxido-2,4,6-trioxido-1,3,5-trioxy-2,4,6-triphosphy-[6]cycle
$H_5P_3O_{10}$	*catena*-triphosphoric acid triphosphoric acid[c]	pentahydroxido-1κ²*O*,2κ*O*,3κ²*O*-di-μ-oxido-trioxido-1κ*O*,2κ*O*,3κ*O*-triphosphorus, or μ-(hydroxidotrioxidophosphato-1κ*O*,2κ*O*)-bis(dihydroxidooxidophosphorus), or 1,7-dihydrido-2,4,6-trihydroxido-2,4,6-trioxido-1,3,5,7-tetraoxy-2,4,6-triphosphy-[7]catena
$H_3AsO_4 = [AsO(OH)_3]$	arsenic acid, arsoric acid[i]	trihydroxidooxidoarsenic
$H_3AsO_3 = [As(OH)_3]$	arsenous acid, arsorous acid[i]	trihydroxidoarsenic
$H_2AsHO_3 = [AsHO(OH)_2]$	arsonic acid	hydridodihydroxidooxidoarsenic
$H_2AsHO_2 = [AsH(OH)_2]$	arsonous acid	hydridodihydroxidoarsenic
$HAsH_2O_2 = [AsH_2O(OH)]$	arsinic acid	dihydridohydroxidooxidoarsenic
$HAsH_2O = [AsH_2(OH)]$	arsinous acid	dihydridohydroxidoarsenic
$H_3SbO_4 = [SbO(OH)_3]$	antimonic acid, stiboric acid[i]	trihydroxidooxidoantimony
$H_3SbO_3 = [Sb(OH)_3]$	antimonous acid, stiborous acid[i]	trihydroxidoantimony
$H_2SbHO_3 = [SbHO(OH)_2]$	stibonic acid	hydridodihydroxidooxidoantimony
$H_2SbHO_2 = [SbH(OH)_2]$	stibonous acid	hydridodihydroxidoantimony
$HSbH_2O_2 = [SbH_2O(OH)]$	stibinic acid	dihydridohydroxidooxidoantimony
$HSbH_2O = [SbH_2(OH)]$	stibinous acid	dihydridohydroxidoantimony
$H_3SO_4^+ = [SO(OH)_3]^+$	[d]	trihydroxidooxidosulfur(1+), trihydrogen(tetraoxidosulfate)(1+)
$H_2SO_4 = [SO_2(OH)_2]$	sulfuric acid	dihydroxidodioxidosulfur
$HSO_4^- = [SO_3(OH)]^-$	hydrogensulfate	hydroxidotrioxidosulfate(1−)
$[SO_4]^{2-}$	sulfate	tetraoxidosulfate(2−)
$HSHO_3 = [SHO_2(OH)]$	sulfonic acid[j]	hydridohydroxidodioxidosulfur
$H_2SO_3 = [SO(OH)_2]$	sulfurous acid	dihydroxidooxidosulfur

(*continued*)

Table IR-8.1 *Continued*

Formula	*Acceptable common name (unless otherwise stated)*	*Systematic additive name(s)*
$HSO_3^- = [SO_2(OH)]^-$	hydrogensulfite	hydroxidodioxidosulfate(1−)
$[SO_3]^{2-}$	sulfite	trioxidosulfate(2−)
$HSHO_2 = [SHO(OH)]$	sulfinic acid[j]	hydridohydroxidooxidosulfur
$H_2SO_2 = [S(OH)_2]$	sulfanediol[k]	dihydroxidosulfur
$[SO_2]^{2-}$	sulfanediolate[k]	dioxidosulfate(2−)
$HSOH = [SH(OH)]$	sulfanol[k]	hydridohydroxidosulfur
$HSO^- = [SHO]^-$	sulfanolate[k]	hydridooxidosulfate(1−)
$H_2S_2O_7 = [(HO)S(O)_2OS(O)_2(OH)]$	disulfuric acid[c]	μ-oxido-bis(hydroxidodioxidosulfur)
$[S_2O_7]^{2-} = [(O)_3SOS(O)_3]^{2-}$	disulfate	μ-oxido-bis(trioxidosulfate)(2−)
$H_2S_2O_6 = [(HO)(O)_2SS(O)_2(OH)]$	dithionic acid[c,l]	bis(hydroxidodioxidosulfur)(*S*—*S*), or 1,4-dihydrido-2,2,3,3-tetraoxido-1,4-dioxy-2,3-disulfy-[4]catena
$[S_2O_6]^{2-} = [O_3SSO_3]^{2-}$	dithionate	bis(trioxidosulfate)(*S*—*S*)(2−), or 2,2,3,3-tetraoxido-1,4-dioxy-2,3-disulfy-[4]catenate(2−)
$H_2S_3O_6 = [(HO)(O)_2SSS(O)_2(OH)]$	trithionic acid [c,m]	1,5-dihydrido-2,2,4,4-tetraoxido-1,5-dioxy-2,3,4-trisulfy-[5]catena
$H_2S_4O_6 = [(HO)(O)_2SSSS(O)_2(OH)]$	tetrathionic acid [c,m]	1,6-dihydrido-2,2,5,5-tetraoxido-1,6-dioxy-2,3,4,5-tetrasulfy-[6]catena
$H_2S_2O_5 = [(HO)(O)_2SS(O)OH]$	disulfurous acid[n]	dihydroxido-1κO,2κO-trioxido-1$\kappa^2 O$,2κO-disulfur(*S*—*S*)
$[S_2O_5]^{2-} = [O(O)_2SS(O)O]^{2-}$	disulfite[n]	pentaoxido-1$\kappa^3 O$,2$\kappa^2 O$-disulfate(*S*—*S*)(2−)
$H_2S_2O_4 = [(HO)(O)SS(O)(OH)]$	dithionous acid[c,l]	bis(hydroxidooxidosulfur)(*S*—*S*), or 1,4-dihydrido-2,3-dioxido-1,4-dioxy-2,3-disulfy-[4]catena
$[S_2O_4]^{2-} = [O_2SSO_2]^{2-}$	dithionite	bis(dioxidosulfate)(*S*—*S*)(2−), or 2,3-dioxido-1,4-dioxy-2,3-disulfy-[4] catenate(2−)
$H_2SeO_4 = [SeO_2(OH)_2]$	selenic acid	dihydroxidodioxidoselenium
$[SeO_4]^{2-}$	selenate	tetraoxidoselenate(2−)
$H_2SeO_3 = [SeHO_2(OH)]$	selenonic acid [j,o]	hydridohydroxidodioxidoselenium
$H_2SeO_3 = [SeO(OH)_2]$	selenous acid[o]	dihydroxidooxidoselenium
$[SeO_3]^{2-}$	selenite	trioxidoselenate(2−)
$HSeHO_2 = [SeHO(OH)]$	seleninic acid[j]	hydridohydroxidooxidoselenium
$H_6TeO_6 = [Te(OH)_6]$	orthotelluric acid[a]	hexahydroxidotellurium
$[TeO_6]^{6-}$	orthotellurate[a]	hexaoxidotellurate(6−)

$H_2TeO_4 = [TeO_2(OH)_2]$	telluric acid[a]	dihydroxidodioxidotellurium
$[TeO_4]^{2-}$	tellurate[a]	tetraoxidotellurate(2−)
$H_2TeO_3 = [TeO(OH)_2]$	tellurous acid	dihydroxidooxidotellurium
$HTeHO_3 = [TeHO_2(OH)]$	telluronic acid[j]	hydridohydroxidodioxidotellurium
$HTeHO_2 = [TeHO(OH)]$	tellurinic acid[j]	hydridohydroxidooxidotellurium
$HClO_4 = [ClO_3(OH)]$	perchloric acid	hydroxidotrioxidochlorine
$[ClO_4]^-$	perchlorate	tetraoxidochlorate(1−)
$HClO_3 = [ClO_2(OH)]$	chloric acid	hydroxidodioxidochlorine
$[ClO_3]^-$	chlorate	trioxidochlorate(1−)
$HClO_2 = [ClO(OH)]$	chlorous acid	hydroxidooxidochlorine
$[ClO_2]^-$	chlorite	dioxidochlorate(1−)
$HClO = [O(H)Cl]$	hypochlorous acid	chloridohydridooxygen
$[OCl]^-$	hypochlorite	chloridooxygenate(1–)
$HBrO_4 = [BrO_3(OH)]$	perbromic acid	hydroxidotrioxidobromine
$[BrO_4]^-$	perbromate	tetraoxidobromate(1−)
$HBrO_3 = [BrO_2(OH)]$	bromic acid	hydroxidodioxidobromine
$[BrO_3]^-$	bromate	trioxidobromate(1−)
$HBrO_2 = [BrO(OH)]$	bromous acid	hydroxidooxidobromine
$[BrO_2]^-$	bromite	dioxidobromate(1−)
$HBrO = [O(H)Br]$	hypobromous acid	bromidohydridooxygen
$[OBr]^-$	hypobromite	bromidooxygenate(1–)
$H_5IO_6 = [IO(OH)_5]$	orthoperiodic acid[a]	pentahydroxidooxidoiodine
$[IO_6]^{5-}$	orthoperiodate[a]	hexaoxidoiodate(5−)
$HIO_4 = [IO_3(OH)]$	periodic acid[a]	hydroxidotrioxidoiodine
$[IO_4]^-$	periodate[a]	tetraoxidoiodate(1−)
$HIO_3 = [IO_2(OH)]$	iodic acid	hydroxidodioxidoiodine
$[IO_3]^-$	iodate	trioxidoiodate(1−)
$HIO_2 = [IO(OH)]$	iodous acid	hydroxidooxidoiodine
$[IO_2]^-$	iodite	dioxidoiodate(1−)
$HIO = [O(H)I]$	hypoiodous acid	hydridoiodidooxygen
$[OI]^-$	hypoiodite	iodidooxygenate(1−)

(*continued*)

[a] The prefix 'ortho' has not been used consistently in the past (including in Chapter I-9 of Ref. 2). Here, it has been removed in the cases of boric acid, silicic acid and phosphoric acid where there is no ambiguity in the names without 'ortho'. The only cases where 'ortho' distinguishes between two different compounds are the telluric and periodic acids (and corresponding anions).

[b] The names fulminic acid and isofulminic acid have been used inconsistently in the past. The compound originally named fulminic acid is HCNO, which is not an oxoacid, while the esters usually called fulminates in organic chemistry are RONC, corresponding to the oxoacid HONC. The additive names in the right hand column specify the structures unambiguously. The preferred organic names are formonitrile oxide for HCNO and λ^2-methylidenehydroxylamine for HONC. (See Section P-61.9 of Ref. 1. See also Table IX under entries CHNO and CNO).

[c] The oligomeric series can be continued, *e.g.* diphosphoric acid, triphosphoric acid, *etc.*; dithionic acid, trithionic acid, tetrathionic acid, *etc.*; dithionous, trithionous, *etc.*

[d] The names nitric acidium, sulfuric acidium, *etc.* for the hydronated acids represent a hybrid of several nomenclatures and are difficult to translate into certain languages. They are no longer acceptable.

[e] The substitutive name would be azanol. However, for preferred names for certain organic derivatives, NH_2OH itself is regarded as a parent with the name hydroxylamine. See Ref. 1, Section P-68.3.

[f] These are systematic substitutive names. The traditional names hyponitrous acid and hyponitrite are not acceptable; the systematics otherwise adhered to for use of the prefix 'hypo' would have prescribed hypodinitrous and hypodinitrite.

[g] The name phosphorous acid and the formula H_3PO_3 have been used in the literature for both $[P(OH)_3]$ and $[PHO(OH)_2]$. The present choice of names for these two species is in accord with the parent names given in Sections P-42.3 and P-42.4 of Ref. 1.

[h] These are substitutive names. No 'acid' names are commonly used for the two isomers of HPO_2.

[i] The names arsoric, arsorous, stiboric and stiborous are included because they are used as parent names in Section P-42.4 of Ref. 1.

[j] Caution is needed if using the names sulfonic acid, sulfinic acid, selenonic acid, *etc.* for these compounds. Substitutive nomenclature prescribes using substitution into parent hydrides rather than into the acids when naming corresponding functional derivatives, *e.g.* trisulfanedisulfonic acid (*not* trisulfanediyl...), see footnote m; methaneseleninic acid (*not* methyl...); *etc.* Note that the substituent groups 'sulfonyl', 'sulfinyl', *etc.*, are $-S(O)_2-$,$-S(O)-$, *etc.*, *not* HS(O)<, HS(O)−, *etc.*

[k] These are systematic substitutive names. Names based on the traditional names sulfoxylic acid for $S(OH)_2$ and sulfenic acid for HSOH, and indeed these names themselves, are no longer acceptable.

[l] Systematic use of the prefix 'hypo' would give the names hypodisulfuric acid for dithionic acid and hypodisulfurous acid for dithionous acid.

[m] The homologues trithionic acid, tetrathionic acid, *etc.*, may be alternatively named by substitutive nomenclature as sulfanedisulfonic acid, disulfanedisulfonic acid, *etc.*

[n] This common name presents a problem because the unsymmetrical structure is not the structure which would otherwise be associated with the 'diacid' construction (disulfurous acid would systematically be [HO(O)SOS(O)OH]). The use of an additive name eliminates this potential confusion, but the problem with the use of disulfurous acid as a parent name persists in the naming of organic derivatives.

[o] The formula H_2SeO_3 has been used in the literature for both selenonic acid and selenous acid. The present choice of names for the two structures shown is in accord with the parent names given in Sections P-42.1 and P-42.4 of Ref. 1.

IR-8.3 ADDITIVE NAMES

Molecules or ions that can *formally* be regarded as mononuclear coordination entities may be named additively, applying the rules described in Chapter IR-7.

Examples:

1. $H_3SO_4^+ = [SO(OH)_3]^+$ trihydroxidooxidosulfur(1+)
2. $H_2SO_4 = [SO_2(OH)_2]$ dihydroxidodioxidosulfur
3. $HSO_4^- = [SO_3(OH)]^-$ hydroxidotrioxidosulfate(1−)

Structures which can be regarded as oligonuclear coordination entities may be named as such (Section IR-7.3) or may be named using the system for inorganic chains and rings (Section IR-7.4).

In principle, the choice of method in the latter case is arbitrary. However, the machinery of coordination compound nomenclature was developed to enable the handling of complex structures involving polyatomic, and particularly polydentate, ligands and sometimes multiply bridging ligands. Furthermore, the separation into ligands and central atoms, obvious in compounds most usually classified as coordination compounds, may be less obvious in the polyoxoacids. Thus, additive nomenclature of the coordination type tends to be more intricate than necessary when naming polyoxoacids forming relatively simple chains and rings. Here the chains and rings system is easily applied, and the names so derived are easy to decipher. However, this system can lead to long names with many locants.

Both types of additive names are exemplified below for oligonuclear systems.

Examples:

4. The compound commonly named diphosphoric acid, $H_4P_2O_7 = [(HO)_2P(O)OP(O)(OH)_2]$, is named according to the coordination-type additive nomenclature as:

   μ-oxido-bis[dihydroxidooxidophosphorus]
   or as a five-membered chain with ligands:

   1,5-dihydrido-2,4-dihydroxido-2,4-dioxido-1,3,5-trioxy-2,4-diphosphy-[5]catena

5. The compound commonly named *cyclo*-triphosphoric acid:

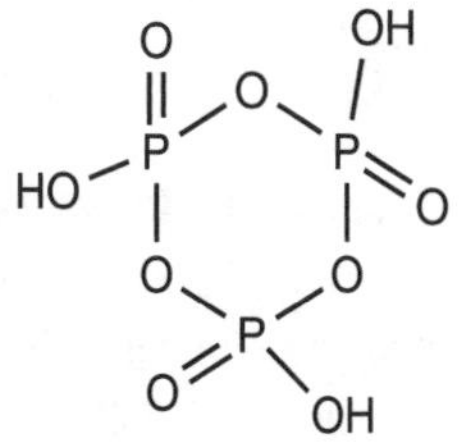

   $H_3P_3O_9$

   may be named according to coordination-type additive nomenclature as:
   tri-μ-oxido-tris(hydroxidooxidophosphorus),
   or as a six-membered ring with ligands:
   2,4,6-trihydroxido-2,4,6-trioxido-1,3,5-trioxy-2,4,6-triphosphy-[6]cycle

6. The related compound, *catena*-triphosphoric acid

$$\mathrm{HO{-}P(=O)(OH){-}O{-}P(=O)(OH){-}O{-}P(=O)(OH){-}OH}$$

$H_5P_3O_{10}$

may be named as a trinuclear coordination entity:

pentahydroxido-1κ^2O,2κ^2O,3κO-di-μ-oxido-1:3κ^2O;2:3κ^2O-trioxido-1κO,2κO,3κO-triphosphorus,

or as a symmetrical dinuclear coordination entity with a bridging phosphate ligand:

μ-(hydroxidotrioxido-1κO,2$\kappa O'$-phosphato)-bis(dihydroxidooxidophosphorus),

or as a mononuclear coordination entity with two phosphate ligands:

bis(dihydroxidodioxidophosphato)hydroxidooxidophosphorus,

or as a seven-membered chain with ligands:

1,7-dihydrido-2,4,6-trihydroxido-2,4,6-trioxido-1,3,5,7-tetraoxy-2,4,6-triphosphy-[7]catena.

All inorganic oxoacids for which a common name containing the word 'acid' is still acceptable according to the present recommendations are listed in Table IR-8.1 together with additive names to illustrate how systematic names may be given.

Several names omitted from Ref. 2, *e.g.* selenic acid and hypobromous acid, are reinstated because they are unambiguous and remain in common use (including their use as parent names in functional replacement nomenclature, see Section IR-8.6).

Table IR-8.1 also includes anions derived from the neutral oxoacids by successive dehydronation. Many of these anions also have common names that are still acceptable, in some cases in spite of the fact that they are based on nomenclature principles that are now otherwise abandoned (*e.g.* nitrate/nitrite and perchlorate/chlorate/chlorite/hypochlorite). For names involving the prefix 'hydrogen', see Sections IR-8.4 and IR-8.5.

It is important to note that the presence of a species in Table IR-8.1 does not imply that it has been described in the literature or that there has been a need to name it in the past. Several names are included only for completeness and to make parent names available for naming organic derivatives.

IR-8.4 HYDROGEN NAMES

An alternative nomenclature for hydrogen-containing compounds and ions is described here. The word 'hydrogen', with a multiplicative prefix if relevant, is joined (with no space) to an anion name formed by additive nomenclature and placed within appropriate enclosing marks (see Section IR-2.2). This construction is followed (again with no space) by a charge number indicating the total charge of the species or structural unit being named (except for neutral species/units).

Hydrogen names are useful when the connectivity (the positions of attachment of the hydrons) in a hydron-containing compound or ion is unknown or not specified (*i.e.* when which of two or more tautomers is not specified, or when one does not wish to specify a complex connectivity, such as in network compounds).

Some of the following examples are discussed in detail below.

Examples:

1. $H_2P_2O_7^{2-}$
   dihydrogen(diphosphate), or
   dihydrogen[μ-oxidobis(trioxidophosphate)](2−)
2. $H_2B_2(O_2)_2(OH)_4$
   dihydrogen(tetrahydroxidodi-μ-peroxido-diborate)
3. $H_2Mo_6O_{19} = H_2[Mo_6O_{19}]$
   dihydrogen(nonadecaoxidohexamolybdate)
4. $H_4[SiW_{12}O_{40}] = H_4[W_{12}O_{36}(SiO_4)]$
   tetrahydrogen[(tetracontaoxidosilicondodecatungsten)ate], or
   tetrahydrogen[hexatriacontaoxido(tetraoxidosilicato)dodecatungstate], or
   tetrahydrogen(silicododecatungstate)
5. $H_4[PMo_{12}O_{40}] = H_4[Mo_{12}O_{36}(PO_4)]$
   tetrahydrogen[tetracontaoxido(phosphorusdodecamolybdenum)ate], or
   tetrahydrogen[hexatriacontaoxido(tetraoxidophosphato)dodecamolybdate], or
   tetrahydrogen(phosphododecamolybdate)
6. $H_6[P_2W_{18}O_{62}] = H_6[W_{18}O_{54}(PO_4)_2]$
   hexahydrogen[dohexacontaoxido(diphosphorusoctadecatungsten)ate], or
   hexahydrogen[tetrapentacontaoxidobis(tetraoxidophosphato)octadecatungstate],
   or hexahydrogen(diphosphooctadecatungstate)
7. $H_4[Fe(CN)_6]$
   tetrahydrogen(hexacyanidoferrate)
8. $H_2[PtCl_6]\cdot 2H_2O$
   dihydrogen(hexachloridoplatinate)—water (1/2)
9. HCN
   hydrogen(nitridocarbonate)

In Example 1, the two hydrons could be located either on two oxygen atoms on the same phosphorus atom or one on each of the phosphorus atoms. Thus, as already indicated, hydrogen names do not necessarily fully specify the structure.

In the same way, the hydrogen name in Example 9 covers, in principle, two tautomers. This also applies to the common compositional name 'hydrogen cyanide'. The names 'hydridonitridocarbon' (additive nomenclature), 'methylidyneazane' (substitutive nomenclature) and 'formonitrile' (functional organic nomenclature) all specify the tautomer HCN.

Hydrogen names may also be used for molecular compounds and ions with no tautomerism problems if one wishes to emphasize the conception of the structure as hydrons attached to the anion in question:

Examples:

10. $HMnO_4$ hydrogen(tetraoxidomanganate)
11. H_2MnO_4 dihydrogen(tetraoxidomanganate)

12. H_2CrO_4 dihydrogen(tetraoxidochromate)
13. $HCrO_4^-$ hydrogen(tetraoxidochromate)(1−)
14. $H_2Cr_2O_7$ dihydrogen(heptaoxidodichromate)
15. H_2O_2 dihydrogen(peroxide)
16. HO_2^- hydrogen(peroxide)(1−)
17. H_2S dihydrogen(sulfide)
18. $H_2NO_3^+$ dihydrogen(trioxidonitrate)(1+)

Note the difference from *compositional names* such as 'hydrogen peroxide' for H_2O_2 and 'hydrogen sulfide' for H_2S (Chapter IR-5) in which (in English) there is a space between the electropositive and electronegative component(s) of the name.

Compositional names of the above type, containing the word 'hydrogen', were classified as 'hydrogen nomenclature' in the discussion of oxoacids in Section I-9.5 of Ref. 2, and such names were extensively exemplified. However, in order to avoid ambiguity, their general use is not encouraged here. Consider, for example, that the compositional names 'hydrogen sulfide' and 'hydrogen sulfide(2−)' can both be interpreted as H_2S as well as HS^-. The situation with H_2S is completely analogous to that with Na_2S which may be named sodium sulfide, disodium sulfide, sodium sulfide(2−) and disodium sulfide(2−), except that misinterpretation of the first and third names as denoting NaS^- is improbable. In Ref. 2, the names 'hydrogensulfide(1−)' and 'monohydrogensulfide' for HS^- were proposed to avoid ambiguity. (However, in some languages there is no space in compositional names so that very delicate distinctions are required anyway.)

The strict definition of *hydrogen names* proposed here is meant to eliminate such confusion by imposing the requirements:

(i) that 'hydrogen' be attached to the rest of the name,
(ii) that the number of hydrogens *must* be specified by a multiplicative prefix,
(iii) that the anionic part be placed in enclosing marks, and
(iv) that the charge of the *total* structure being named is specified.

Hydrogen names constructed in this way cannot be mistaken for other types of name.

The only acceptable exceptions to the above format for hydrogen names are the few particular abbreviated anion names listed in Section IR-8.5.

In a few cases, no confusion can arise, and the distinction between compositional name and hydrogen name is not as important, most notably for the hydrogen halides. Thus, HCl can equally unambiguously be named 'hydrogen chloride' (compositional name) and 'hydrogen(chloride)' (hydrogen name).

Examples 1, 3–6 and 14 above demonstrate that homo- and heteropolyoxoacids and their partially dehydronated forms can be given hydrogen names once the corresponding anions have been named. Examples 4–6 each feature three alternatives. The first two names are both fully additive for the anion part and correspond to different ways of dissecting the structure into ligands and central atoms. The last names, involving the prefixes 'silico' and 'phospho', are examples of a common semi-systematic nomenclature which is not recommended for general use because it requires complex conventions in order to be unambiguous.

Rules for naming very complicated homo- and heteropolyoxoanions are given in Chapter II-1 of Ref. 3.

Note that Examples 10–14 above show how one may easily name transition metal compounds that have been named as acids in the past. Names such as permanganic acid, dichromic acid, *etc.*, are not included in the present recommendations because they represent an area where it is difficult to systematize and decide what to include, and where the names are not needed for organic nomenclature, as opposed to the corresponding 'acid' names for acids of main group elements.

Finally, note that usage is different from the above in the names of salts and partial esters of organic polyvalent acids, where 'hydrogen' is always cited as a separate word just before the anion name, *e.g.* potassium hydrogen phthalate or ethyl hydrogen phthalate.

IR-8.5 ABBREVIATED HYDROGEN NAMES FOR CERTAIN ANIONS

A few common anionic species have names which can be regarded as short forms of hydrogen names formed according to the above method. These names, all in one word without explicit indication of the molecular charge, and without the enclosing marks, are accepted due to their brevity and long usage and because they are not ambiguous. It is strongly recommended that this list be viewed as limiting due to the ambiguities that may arise in many other cases. (See the discussion in Section IR-8.4.)

Anion	*Accepted simplified hydrogen name*	*Hydrogen name*
$H_2BO_3^-$	dihydrogenborate	dihydrogen(trioxidoborate)(1−)
HBO_3^{2-}	hydrogenborate	hydrogen(trioxidoborate)(2−)
HSO_4^-	hydrogensulfate	hydrogen(tetraoxidosulfate)(1−)
HCO_3^-	hydrogencarbonate	hydrogen(trioxidocarbonate)(1−)
$H_2PO_4^-$	dihydrogenphosphate	dihydrogen(tetraoxidophosphate)(1−)
HPO_4^{2-}	hydrogenphosphate	hydrogen(tetraoxidophosphate)(2−)
$HPHO_3^-$	hydrogenphosphonate	hydrogen(hydridotrioxidophosphate)(1−)
$H_2PO_3^-$	dihydrogenphosphite	dihydrogen(trioxidophosphate)(1−)
HPO_3^{2-}	hydrogenphosphite	hydrogen(trioxidophosphate)(2−)
HSO_4^-	hydrogensulfate	hydrogen(tetraoxidosulfate)(1−)
HSO_3^-	hydrogensulfite	hydrogen(trioxidosulfate)(1−)

IR-8.6 FUNCTIONAL REPLACEMENT NAMES FOR DERIVATIVES OF OXOACIDS

In functional replacement nomenclature, substitution of =O or −OH groups in parent oxoacids (such as O → S, O → OO, OH → Cl, *etc.*) is indicated by the use of infixes or prefixes as exemplified below (see Ref. 1, Section P–67.1).

Replacement operation	*Prefix*	*Infix*
$OH \rightarrow NH_2$	amid(o)	amid(o)
$O \rightarrow OO$	peroxy	peroxo
$O \rightarrow S$	thio	thio
$O \rightarrow Se$	seleno	seleno
$O \rightarrow Te$	telluro	telluro
$OH \rightarrow F$	fluoro	fluorid(o)
$OH \rightarrow Cl$	chloro	chlorid(o)
$OH \rightarrow Br$	bromo	bromid(o)
$OH \rightarrow I$	iodo	iodid(o)
$OH \rightarrow CN$	cyano	cyanid(o)

Example 5 in Section IR-8.1 demonstrates the use of the infixes for $OH \rightarrow Cl$ and $O \rightarrow S$ to arrive at the name 'arsonochloridothioic *O*-acid' for the derived parent HAsCl(OH)S = [AsClH(OH)S], required for naming the organic derivative:

EtAsCl(OH)S ethylarsonochloridothioic *O*-acid.

Functional replacement names may, of course, be used for the derived parent acids themselves. However, this amounts to introducing an additional system which is not needed in inorganic nomenclature. As mentioned above, additive and substitutive nomenclature can always be used.

Example:

1. HAsCl(OH)S = [AsClH(OH)S]
   chloridohydridohydroxidosulfidoarsenic (additive), or
   chloro(hydroxy)-λ^5-arsanethione (substitutive)

Nevertheless, in Table IR-8.2 several inorganic species are listed which can be regarded as derived from species in Table IR-8.1 by various replacement operations, and for which the common names are in fact derived by the above prefix method (*e.g.* 'thiosulfuric acid').

A problem that would arise with the general use of the prefix variant of functional replacement names is illustrated by the thio acids. The names trithiocarbonic acid, tetrathiophosphoric acid, *etc.*, would lead to anion names trithiocarbonate, tetrathiophosphate, *etc.*, which appear to be additive names but are incorrect as such because the ligand prefix is now 'sulfido' or 'sulfanediido' [thus giving trisulfidocarbonate(2−), tetrasulfidophosphate(3−), *etc.*]. Section P-65.2 of Ref. 1 prescribes the infix-based name carbonotrithioic acid, leading to the anion name carbonotrithioate, which will not be mistaken for an additive name.

A few examples of other functional nomenclature are also included in Table IR-8.2 (*e.g.* phosphoryl chloride, sulfuric diamide). These particular names are well entrenched and can still be used, but this type of nomenclature is not recommended for compounds other than those shown. Again, additive and substitutive names may always be constructed, as exemplified in the Table.

Table IR-8.2 *Acceptable common names, functional replacement names, and systematic (additive) names for some functional replacement derivatives of oxoacids*

This Table gives acceptable common names, functional replacement names (see Section IR-8.6) and systematic (additive) names for compounds related to oxoacids in Table IR-8.1 and certain isomers and corresponding anions. The examples given are derived by formal replacement of an O atom/O atoms, or of an OH group/OH groups, by (an)other atom(s) or group(s).

Formulae are in some cases given in the classical format with the 'acid' (oxygen- or chalcogen-bound) hydrogen atoms listed first (*e.g.* $H_2S_2O_3$). In most cases formulae are also (or only) written as coordination entities, assembled according to the principles of Chapter IR-7 (*e.g.* '$H_2S_2O_3 = [SO(OH)_2S]$').

Formula	*Acceptable common name*	*Functional replacement name*	*Systematic (additive) name*
$HNO_4 = [NO_2(OOH)]$	peroxynitric acid[a]	peroxynitric acid[a]	(dioxidanido)dioxidonitrogen
$NO_4^- = [NO_2(OO)]^-$	peroxynitrate[a]	peroxynitrate[a]	dioxidoperoxidonitrate(1−)
$[NO(OOH)]$	peroxynitrous acid[a]	peroxynitrous acid[a]	(dioxidanido)oxidonitrogen
$[NO(OO)]^-$	peroxynitrite[a]	peroxynitrite[a]	oxidoperoxidonitrate(1−)
$NO_2NH_2 = N(NH_2)O_2$	nitramide	nitric amide	amidodioxidonitrogen, or dihydrido-1κ^2*H*-dioxido-2κ^2*O*-dinitrogen(*N*—*N*)
$H_3PO_5 = [PO(OH)_2(OOH)]$	peroxyphosphoric acid[a]	phosphoroperoxoic acid	(dioxidanido)dihydroxidooxidophosphorus
$[PO_5]^{3-} = [PO_3(OO)]^{3-}$	peroxyphosphate[a]	phosphoroperoxoate	trioxidoperoxidophosphate(3−)
$[PCl_3O]$	phosphoryl trichloride, or phosphorus trichloride oxide	phosphoryl trichloride	trichloridooxidophosphorus
$H_4P_2O_8 = [(HO)_2P(O)OOP(O)(OH)_2]$	peroxydiphosphoric acid[a]	2-peroxydiphosphoric acid	μ-peroxido-1κ*O*,2κ*O*′-bis(dihydroxidooxidophosphorus)
$[P_2O_8]^{4-} = [O_3POOPO_3]^{4-}$	peroxydiphosphate[a]	2-peroxydiphosphate	μ-peroxido-1κ*O*,2κ*O*′-bis(trioxidophosphate)(4−)
$H_2SO_5 = [SO_2(OH)(OOH)]$	peroxysulfuric acid[a]	sulfuroperoxoic acid	(dioxidanido)hydroxidodioxidosulfur
$[SO_5]^{2-} = [SO_3(OO)]^{2-}$	peroxysulfate[a]	sulfuroperoxoate	trioxidoperoxidosulfate(2−)
$H_2S_2O_8 = [(HO)S(O)_2OOS(O)_2(OH)]$	peroxydisulfuric acid[a]	2-peroxydisulfuric acid	μ-peroxido-1κ*O*,2κ*O*′-bis(hydroxidodioxidosulfur)
$[S_2O_8]^{2-} = [O_3SOOSO_3]^{2-}$	peroxydisulfate[a]	2-peroxydisulfate	μ-peroxido-1κ*O*,2κ*O*′-bis(trioxidosulfate)(2−)
$H_2S_2O_3 = [SO(OH)_2S]$	thiosulfuric acid	sulfurothioic *O*-acid	dihydroxidooxidosulfidosulfur
$H_2S_2O_3 = [SO_2(OH)(SH)]$	thiosulfuric acid	sulfurothioic *S*-acid	hydroxidodioxidosulfanidosulfur
$S_2O_3^{2-} = [SO_3S]^{2-}$	thiosulfate	sulfurothioate	trioxidosulfidosulfate(2−)
$H_2S_2O_2 = [S(OH)_2S]$	thiosulfurous acid	sulfurothious *O*-acid	dihydroxidosulfidosulfur
$H_2S_2O_2 = [SO(OH)(SH)]$	thiosulfurous acid	sulfurothious *S*-acid	hydroxidooxidosulfanidosulfur

(*continued*)

Table IR-8.2 *Continued*

Formula	*Acceptable common name*	*Functional replacement name*	*Systematic (additive) name*
$[SO_2S]^{2-}$	thiosulfite	sulfurothioite	dioxidosulfidosulfate(2−)
$SO_2Cl_2 = [SCl_2O_2]$	sulfuryl dichloride, or sulfur dichloride dioxide	sulfuryl dichloride	dichloridodioxidosulfur
$SOCl_2 = [SCl_2O]$	thionyl dichloride, or sulfur dichloride oxide	sulfurous dichloride	dichloridooxidosulfur
$[S(NH_2)O_2(OH)]$	sulfamic acid	sulfuramidic acid	amidohydroxidodioxidosulfur
$[S(NH_2)_2O_2]$	sulfuric diamide	sulfuric diamide	diamidodioxidosulfur
$HSCN = [C(N)(SH)]$	thiocyanic acid		nitridosulfanidocarbon
$HNCS = [C(NH)S]$	isothiocyanic acid		imidosulfidocarbon
SCN^-	thiocyanate		nitridosulfidocarbonate(1−)

[a] These names were given with the prefix ‘peroxo’ rather than ‘peroxy’ in Ref. 4 (Rule 5.22). However, in Ref. 2 names with the prefix ‘peroxo’ were dismissed, with no reason given, and no other prefixes were provided instead. The names with the prefix ‘peroxy’ continue to be in frequent use. Furthermore, the general rule in functional replacement nomenclature (Ref. 1, Sec. P-15.5) is that the replacement *prefix* for the replacement −O− → −OO− is, indeed, ‘peroxy’ (as opposed to the *infix* for this replacement, which is ‘peroxo’). In view of this, the names with the prefix ‘peroxy’ are listed here. For most mononuclear oxoacids, the present rules in Ref. 1 (Section P-67.1) prescribe using the infix method for systematic names; in those cases the resulting names are given in the second column here. The prefix method is used for nitric and nitrous acids and dinuclear oxoacids, as also seen here.

IR-8.7 REFERENCES

1. *Nomenclature of Organic Chemistry, IUPAC Recommendations*, eds. W.H. Powell and H. Favre, Royal Society of Chemistry, in preparation.
2. *Nomenclature of Inorganic Chemistry, IUPAC Recommendations 1990*, ed. G.J. Leigh, Blackwell Scientific Publications, Oxford, 1990.
3. *Nomenclature of Inorganic Chemistry II, IUPAC Recommendations 2000*, eds. J.A. McCleverty and N.G. Connelly, Royal Society of Chemistry, 2001.
4. *IUPAC Nomenclature of Inorganic Chemistry, Second Edition, Definitive Rules 1970*, Butterworths, London, 1971.

IR-9 Coordination Compounds

CONTENTS

IR-9.1 INTRODUCTION

IR-9.1.1 General

This Chapter presents the definitions and rules necessary for formulating and naming coordination compounds. Key terms such as coordination entity, coordination polyhedron, coordination number, chelation and bridging ligands are first defined and the role of additive nomenclature explained (see also Chapter IR-7).

These definitions are then used to develop rules for writing the names and formulae of coordination compounds. The rules allow the composition of coordination compounds to be described in a way that is as unambiguous as possible. The names and formulae provide information about the nature of the central atom, the ligands that are attached to it, and the overall charge on the structure.

Stereochemical descriptors are then introduced as a means of identifying or distinguishing between the diastereoisomeric or enantiomeric structures that may exist for a compound of any particular composition.

The description of the configuration of a coordination compound requires first that the coordination geometry be specified using a polyhedral symbol (Section IR-9.3.2.1). Once this is done the relative positions of the ligands around the coordination polyhedron are specified using the configuration index (Section IR-9.3.3). The configuration index is a sequence of ligand priority numbers produced by following rules specific to each coordination geometry. If required, the chirality of a coordination compound can be described, again using ligand priority numbers (Section IR-9.3.4). The ligand priority numbers used in these descriptions are based on the chemical composition of the ligands. A detailed description of the rules by which they are obtained is provided in Section P-91 of Ref. 1, but an outline is given in Section IR-9.3.5.

IR-9.1.2 Definitions

IR-9.1.2.1 *Background*

The development of coordination theory and the identification of a class of compounds called coordination compounds began with the historically significant concepts of primary and secondary valence.

Primary valencies were obvious from the stoichiometries of simple compounds such as $NiCl_2$, $Fe_2(SO_4)_3$ and $PtCl_2$. However, new materials were frequently observed when other, independently stable substances, *e.g.* H_2O, NH_3 or KCl, were added to these simple compounds giving, for example, $NiCl_2{\cdot}4H_2O$, $Co_2(SO_4)_3{\cdot}12NH_3$ or $PtCl_2{\cdot}2KCl$. Such species were called complex compounds, in recognition of the stoichiometric complications they represented, and were considered characteristic of certain metallic elements. The number of species considered to be added to the simple compounds gave rise to the concept of secondary valence.

Recognition of the relationships between these complex compounds led to the formulation of coordination theory and the naming of coordination compounds using additive nomenclature. Each coordination compound either is, or contains, a coordination entity (or complex) that consists of a central atom to which other groups are bound.

While these concepts have usually been applied to metal compounds, a wide range of other species can be considered to consist of a central atom or central atoms to which a number of other groups are bound. The application of additive nomenclature to such species is briefly described and exemplified in Chapter IR-7, and abundantly exemplified for inorganic acids in Chapter IR-8.

IR-9.1.2.2 *Coordination compounds and the coordination entity*

A coordination compound is any compound that contains a coordination entity. A coordination entity is an ion or neutral molecule that is composed of a central atom, usually that of a metal, to which is attached a surrounding array of other atoms or groups of atoms, each of which is called a ligand. Classically, a ligand was said to satisfy either a secondary or a primary valence of the central atom and the sum of these valencies (often equal to the number of ligands) was called the *coordination number* (see Section IR-9.1.2.6). In formulae, the coordination entity is enclosed in square brackets whether it is charged or uncharged (see Section IR-9.2.3.2).

Examples:

1. $[Co(NH_3)_6]^{3+}$
2. $[PtCl_4]^{2-}$
3. $[Fe_3(CO)_{12}]$

IR-9.1.2.3 *Central atom*

The central atom is the atom in a coordination entity which binds other atoms or groups of atoms (ligands) to itself, thereby occupying a central position in the coordination entity. The central atoms in $[NiCl_2(H_2O)_4]$, $[Co(NH_3)_6]^{3+}$ and $[PtCl_4]^{2-}$ are nickel, cobalt and platinum, respectively. In general, a name for a (complicated) coordination entity will be more easily produced if more central atoms are chosen (see Section IR-9.2.5) and the connectivity of the structure is indicated using the kappa convention (see Section IR-9.2.4.2).

IR-9.1.2.4 *Ligands*

The ligands are the atoms or groups of atoms bound to the central atom. The root of the word is often converted into other forms, such as to ligate, meaning to coordinate as a ligand, and the derived participles, ligating and ligated. The terms 'ligating atom' and 'donor atom' are used interchangeably.

IR-9.1.2.5 *Coordination polyhedron*

It is standard practice to regard the ligand atoms directly attached to the central atom as defining a coordination polyhedron (or polygon) about the central atom. Thus $[Co(NH_3)_6]^{3+}$ is an octahedral ion and $[PtCl_4]^{2-}$ is a square planar ion. In such cases, the coordination number will be equal to the number of vertices in the coordination polyhedron. This may not hold true in cases where one or more ligands coordinate to the central atom through two or more contiguous atoms. It may hold if the contiguous atoms are treated as a single ligand occupying one vertex of the coordination polyhedron.

Examples:

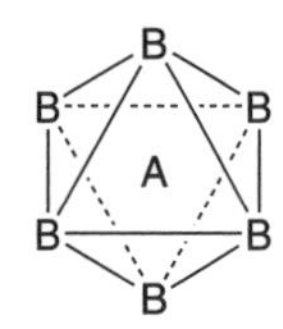

1. octahedral coordination polyhedron

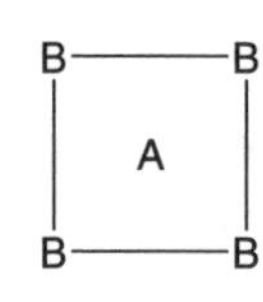

2. square planar coordination polygon

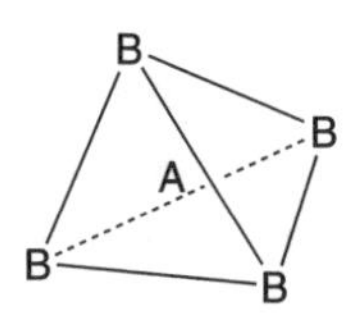

3. tetrahedral coordination polyhedron

IR-9.1.2.6 *Coordination number*

For coordination compounds, the coordination number equals the number of σ-bonds between ligands and the central atom. Note that where both σ- and π-bonding occurs between the ligating atom and the central atom, *e.g.* with ligands such as CN^-, CO, N_2 and PMe_3, the π-bonds are not considered in determining the coordination number.

IR-9.1.2.7 *Chelation*

Chelation involves coordination of more than one non-contiguous σ-electron pair donor atom from a given ligand to the same central atom. The number of such ligating atoms in a single chelating ligand is indicated by the adjectives bidentate[2], tridentate, tetradentate, pentadentate, *etc.* (see Table IV* for a list of multiplicative prefixes). The number of donor atoms from a given ligand attached to the same central atom is called the denticity.

Examples:

1. bidentate chelation

2. bidentate chelation

3. tridentate chelation

4. tetradentate chelation

The cyclic structures formed when more than one donor atom from the same ligand is bound to the central atom are called chelate rings, and the process of coordination of these donor atoms is called chelation.

* Tables numbered with a Roman numeral are collected together at the end of this book.

If a potentially bidentate ligand, such as ethane-1,2-diamine, coordinates to two metal ions, it does *not* chelate but coordinates in a monodentate fashion to each metal ion, forming a connecting link or bridge.

Example:

1. $[(H_3N)_5Co(\mu\text{-}NH_2CH_2CH_2NH_2)Co(NH_3)_5]^{6+}$

Alkenes, arenes and other unsaturated molecules attach to central atoms, using some or all of their multiply bonded atoms, to give organometallic complexes. While there are many similarities between the nomenclature of coordination and organometallic compounds, the latter differ from the former in clearly definable ways. Organometallic complexes are therefore treated separately in Chapter IR-10.

IR-9.1.2.8 *Oxidation state*

The oxidation state of a central atom in a coordination entity is defined as the charge it would bear if all the ligands were removed along with the electron pairs that were shared with the central atom. It is represented by a Roman numeral. It must be emphasized that oxidation state is an index derived from a simple and formal set of rules (see also Sections IR-4.6.1 and IR-5.4.2.2) and that it is not a direct indicator of electron distribution. In certain cases, the formalism does not give acceptable central atom oxidation states. Because of such ambiguous cases, the net charge on the coordination entity is preferred in most nomenclature practices. The following examples illustrate the relationship between the overall charge on a coordination entity, the number and charges of ligands, and the derived central atom oxidation state.

	Formula	*Ligands*	*Central atom oxidation state*
1.	$[Co(NH_3)_6]^{3+}$	6 NH_3	III
2.	$[CoCl_4]^{2-}$	4 Cl^-	II
3.	$[MnO_4]^-$	4 O^{2-}	VII
4.	$[MnFO_3]$	3 O^{2-}+1 F^-	VII
5.	$[Co(CN)_5H]^{3-}$	5 CN^-+1 H^-	III
6.	$[Fe(CO)_4]^{2-}$	4 CO	−II

IR-9.1.2.9 *Coordination nomenclature: an additive nomenclature*

When coordination theory was first developed, coordination compounds were considered to be formed by addition of independently stable compounds to a simple central compound. They were therefore named on the basis of an additive principle, where the names of the added compounds and the central simple compound were combined. This principle remains the basis for naming coordination compounds.

The name is built up around the central atom name, just as the coordination entity is built up around the central atom.

Example:

1. Addition of ligands to a central atom:

$$Ni^{2+} + 6H_2O \longrightarrow [Ni(OH_2)_6]^{2+}$$

Addition of ligand names to a central atom name:

hexaaquanickel(II)

This nomenclature then extends to more complicated structures where central atoms (and their ligands) are added together to form polynuclear species from mononuclear building blocks. Complicated structures are usually more easily named by treating them as polynuclear species (see Section IR-9.2.5).

IR-9.1.2.10 *Bridging ligands*

In polynuclear species a ligand can also act as a bridging group, by forming bonds to two or more central atoms simultaneously. Bridging is indicated in names and formulae by adding the symbol μ as a prefix to the ligand formula or name (see Section IR-9.2.5.2).

Bridging ligands link central atoms together to produce coordination entities having more than one central atom. The number of central atoms joined into a single coordination entity by bridging ligands or direct bonds between central atoms is indicated by using the terms dinuclear, trinuclear, tetranuclear, *etc.*

The bridging index is the number of central atoms linked by a particular bridging ligand (see Section IR-9.2.5.2). Bridging can be through one atom or through a longer array of atoms.

Example:

1.

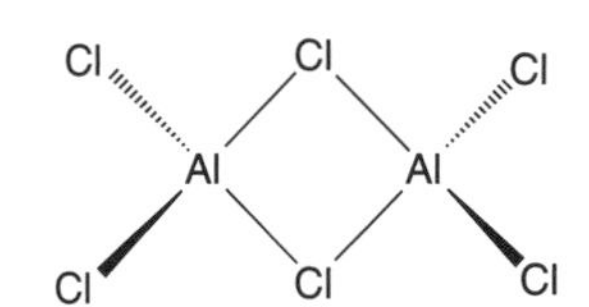

$[Al_2Cl_4(\mu\text{-}Cl)_2]$ or $[Cl_2Al(\mu\text{-}Cl)_2AlCl_2]$

di-μ-chlorido-tetrachlorido-1κ^2*Cl*,2κ^2*Cl*-dialuminium

IR-9.1.2.11 *Metal–metal bonds*

Simple structures that contain a metal–metal bond are readily described using additive nomenclature (see Section IR-9.2.5.3), but complications arise for structures that involve three or more central atoms. Species that contain such clusters of central atoms are treated in Sections IR-9.2.5.6 and IR-9.2.5.7.

Examples:

1. $[Br_4ReReBr_4]^{2+}$
   bis(tetrabromidorhenium)(*Re*—*Re*)(2+)

2. $[(OC)_5\overset{1}{Re}\overset{2}{Co}(CO)_4]$
   nonacarbonyl-1κ^5*C*,2κ^4*C*-rheniumcobalt(*Re*—*Co*)

IR-9.2 DESCRIBING THE CONSTITUTION OF COORDINATION COMPOUNDS

IR-9.2.1 General

Three main methods are available for describing the constitution of compounds: one can draw structures, write names or write formulae. A drawn structure contains information about the structural components of the molecule as well as their stereochemical relationships. Unfortunately, such structures are not usually suitable for inclusion in text. Names and formulae are therefore used to describe the constitution of a compound.

The name of a coordination compound provides detailed information about the structural components present. However, it is important that the name can be easily interpreted unambiguously. For that reason, there should be rules that define how the name is constructed. The following sections detail these rules and provide examples of their use.

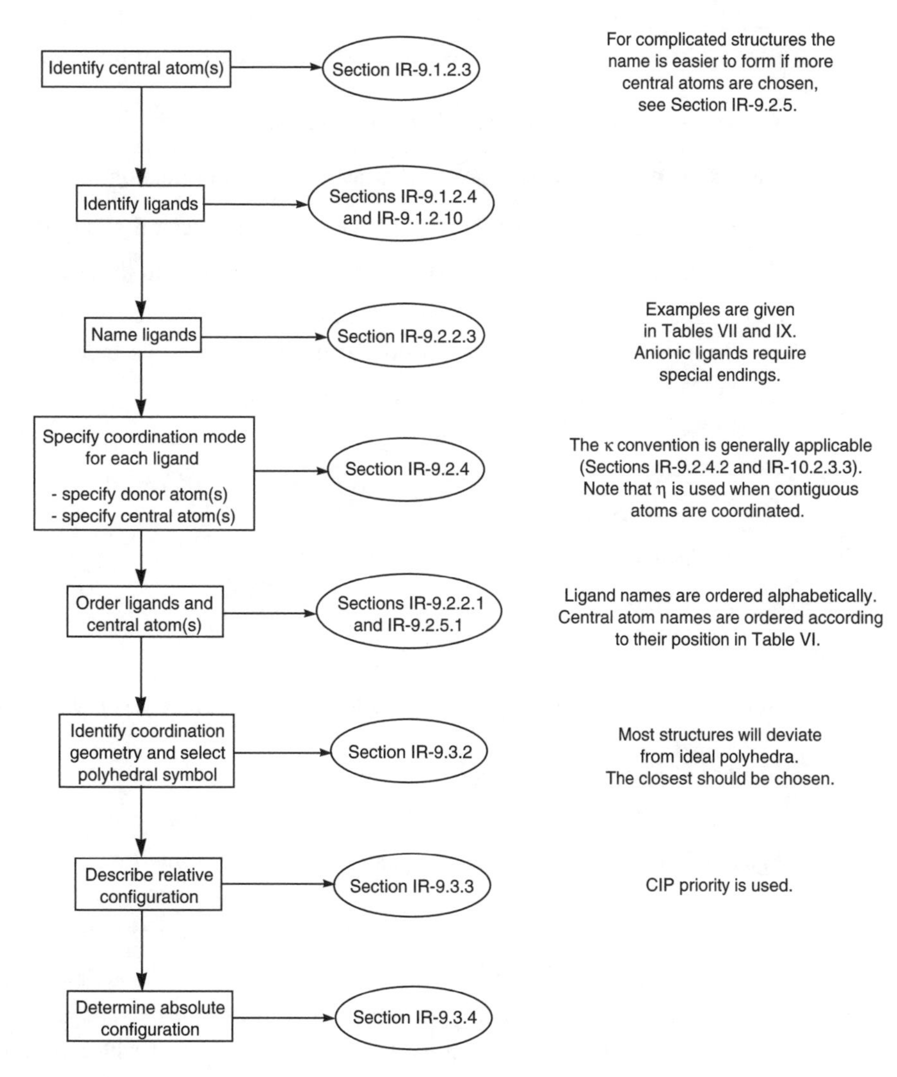

Figure IR–9.1 *Stepwise procedure for naming coordination compounds.*

The flowchart shown in Figure IR-9.1 illustrates a general procedure for producing a name for a coordination compound. Sections containing the detailed rules, guidelines and examples relevant to each stage of the procedure are indicated.

The name of a compound can, however, be rather long and its use may be inconvenient. In such circumstances a formula provides a shorthand method of representing the compound. Rules are provided in order to make the use of formulae more straightforward. It should be noted that, because of their abbreviated form, it is often not possible to provide as much information about the structure of a compound in its formula as can be provided by its name.

IR-9.2.2 Names of coordination compounds

The systematic names of coordination entities are derived by following the principles of additive nomenclature, as outlined in Chapter IR-7. Thus, the groups that surround the central atom or structure must be identified in the name. They are listed as prefixes to the name of the central atom (see Section IR-9.2.2.1) along with any appropriate multipliers (see Section IR-9.2.2.2). These prefixes are usually derived in a simple way from the ligand names (see Section IR-9.2.2.3). Names of anionic coordination entities are furthermore given the ending 'ate'.

IR-9.2.2.1 *Sequences of ligands and central atoms within names*

The following general rules are used when naming coordination compounds:

(i) ligand names are listed before the name(s) of the central atom(s),

(ii) no spaces are left between parts of the name that refer to the same coordination entity,

(iii) ligand names are listed in alphabetical order (multiplicative prefixes indicating the number of ligands are not considered in determining that order),

(iv) the use of abbreviations in *names* is discouraged.

Examples:

1. $[CoCl(NH_3)_5]Cl_2$
   penta**a**mmine**c**hloridocobalt(2+) chloride
2. $[AuXe_4]^{2+}$
   tetraxenonidogold(2+)

Additional rules which apply to polynuclear compounds are dealt with in Section IR-9.2.5.

IR-9.2.2.2 *Number of ligands in a coordination entity*

Two kinds of multiplicative prefix are available for indicating the number of each type of ligand within the name of the coordination entity (see Table IV).

(i) Prefixes di, tri, *etc.* are generally used with the names of simple ligands. Enclosing marks are not required.

(ii) Prefixes bis, tris, tetrakis, *etc.* are used with complex ligand names and in order to avoid ambiguity. Enclosing marks (the nesting order of which is described in Section IR-2.2) must be placed around the multiplicand.

For example, one would use diammine for $(NH_3)_2$, but bis(methylamine) for $(NH_2Me)_2$, to make a distinction from dimethylamine. There is no elision of vowels or use of a hyphen, *e.g.* in tetraammine and similar names.

IR-9.2.2.3 *Representing ligands in names*

Systematic and alternative names for some common ligands are given in Tables VII and IX. Table VII contains the names of common organic ligands whereas Table IX contains the names of other simple molecules and ions that may act as ligands. The general features are as follows:

(i) Names of anionic ligands, whether inorganic or organic, are modified to end in 'o'. In general, if the anion name ends in 'ide', 'ite' or 'ate', the final 'e' is replaced by 'o', giving 'ido', 'ito' and 'ato', respectively. In particular, alcoholates, thiolates, phenolates, carboxylates, partially dehydronated amines, phosphanes, *etc.* are in this category. Also, it follows that halide ligands are named fluorido, chlorido, bromido and iodido, and coordinated cyanide is named cyanido.

In its complexes, except for those of molecular hydrogen, hydrogen is always treated as anionic. 'Hydrido' is used for hydrogen coordinating to all elements including boron.[3]

(ii) Names of neutral and cationic ligands, including organic ligands,[4] are used without modification (even if they carry the endings 'ide', 'ite' or 'ate'; see Examples 8 and 14 below).

(iii) Enclosing marks are required for neutral and cationic ligand names, for names of inorganic anionic ligands containing multiplicative prefixes (such as triphosphato), for compositional names (such as carbon disulfide), for names of substituted organic ligands (even if there is no ambiguity in their use), and wherever necessary to avoid ambiguity. However, common ligand names such as aqua, ammine, carbonyl, nitrosyl, methyl, ethyl, *etc.*, do not require enclosing marks, unless there is ambiguity when they are absent.

(iv) Ligands binding to metals through carbon atoms are treated in Chapter IR-10 on organometallic compounds.

Examples:

	Formula	*Ligand name*
1.	Cl^-	chlorido
2.	CN^-	cyanido
3.	H^-	hydrido[3]
4.	D^- or $^2H^-$	deuterido[3] or [^{2}H]hydrido[3]
5.	$PhCH_2CH_2Se^-$	2-phenylethane-1-selenolato
6.	$MeCOO^-$	acetato or ethanoato
7.	Me_2As^-	dimethylarsanido
8.	$MeCONH_2$	acetamide (*not* acetamido)
9.	$MeCONH^-$	acetylazanido or acetylamido (*not* acetamido)
10.	$MeNH_2$	methanamine

11. $MeNH^-$	methylazanido, or methylamido, or methanaminido (*cf.* Example 3 of Section IR-6.4.6)
12. $MePH_2$	methylphosphane
13. $MePH^-$	methylphosphanido
14. MeOS(O)OH	methyl hydrogen sulfite
15. $MeOS(O)O^-$	methyl sulfito, or methanolatodioxidosulfato(1−)

IR-9.2.2.4 *Charge numbers, oxidation numbers and ionic proportions*

The following methods can be used to assist in describing the composition of a compound:

(i) The oxidation number of the central atom in a coordination entity may be indicated by a Roman numeral appended in parentheses to the central atom name (including the ending 'ate', if applicable), but only if the oxidation state can be defined without ambiguity. When necessary a negative sign is placed before the number. Arabic zero indicates the oxidation number zero.

(ii) Alternatively, the charge on a coordination entity may be indicated. The net charge is written in arabic numbers, with the number preceding the charge sign, and enclosed in parentheses. It follows the name of the central atom (including the ending 'ate', if applicable) without the intervention of a space.

(iii) The proportions of ionic entities in a coordination compound may be given by using multiplicative prefixes. (See Section IR-5.4.2.1.)

Examples:

1. $K_4[Fe(CN)_6]$
   potassium hexacyanidoferrate(II), or
   potassium hexacyanidoferrate(4−), or
   tetrapotassium hexacyanidoferrate
2. $[Co(NH_3)_6]Cl_3$
   hexaamminecobalt(III) chloride
3. $[CoCl(NH_3)_5]Cl_2$
   pentaamminechloridocobalt(2+) chloride
4. $[CoCl(NH_3)_4(NO_2)]Cl$
   tetraamminechloridonitrito-κ*N*-cobalt(III) chloride
5. $[PtCl(NH_2Me)(NH_3)_2]Cl$
   diamminechlorido(methanamine)platinum(II) chloride
6. $[CuCl_2\{O{=}C(NH_2)_2\}_2]$
   dichloridobis(urea)copper(II)
7. $K_2[PdCl_4]$
   potassium tetrachloridopalladate(II)
8. $K_2[OsCl_5N]$
   potassium pentachloridonitridoosmate(2−)

9. $Na[PtBrCl(NH_3)(NO_2)]$
sodium amminebromidochloridonitrito-κ*N*-platinate(1−)
10. $[Fe(CNMe)_6]Br_2$
hexakis(methyl isocyanide)iron(II) bromide
11. $[Co(en)_3]Cl_3$
tris(ethane-1,2-diamine)cobalt(III) trichloride

IR-9.2.3 Formulae of coordination compounds

A (line) formula of a compound is used to provide basic information about the constitution of the compound in a concise and convenient manner. Different applications may require flexibility in the writing of formulae. Thus, on occasion it may be desirable to violate the following guidelines in order to provide more information about the structure of the compound that the formula represents. In particular, this is the case for dinuclear compounds where a great deal of structural information can be provided by relaxing the ordering principles outlined in Section IR-9.2.3.1. (See also Section IR-9.2.5, particularly Section IR-9.2.5.5.)

IR-9.2.3.1 *Sequence of symbols within the coordination formula*

(i) The central atom symbol(s) is (are) listed first.

(ii) The ligand symbols (line formulae, abbreviations or acronyms) are then listed in alphabetical order (see Section IR-4.4.2.2).[5] Thus, CH_3CN, MeCN and NCMe would be ordered under C, M and N respectively, and CO precedes Cl because single letter symbols precede two letter symbols. The placement of the ligand in the list does not depend on the charge of the ligand.

(iii) More information is conveyed by formulae that show ligands with the donor atom nearest the central atom; this procedure is recommended wherever possible, even for coordinated water.

IR-9.2.3.2 *Use of enclosing marks*

The formula for the entire coordination entity, whether charged or not, is enclosed in square brackets. When ligands are polyatomic, their formulae are enclosed in parentheses. Ligand abbreviations are also usually enclosed in parentheses. The nesting order of enclosing marks is as given in Sections IR-2.2 and IR-4.2.3. Square brackets are used only to enclose coordination entities, and parentheses and braces are nested alternately.

Examples 1–11 in Section IR-9.2.2.4 illustrate the use of enclosing marks in formulae. Note also that in those examples there is no space between representations of ionic species within a formula.

IR-9.2.3.3 *Ionic charges and oxidation numbers*

If the formula of a charged coordination entity is to be written without that of any counter-ion, the charge is indicated outside the square bracket as a right superscript, with the number before the sign. The oxidation number of a central atom may be represented by a Roman numeral, which should be placed as a right superscript on the element symbol.

Examples:

1. $[PtCl_6]^{2-}$
2. $[Cr(OH_2)_6]^{3+}$
3. $[Cr^{III}(NCS)_4(NH_3)_2]^-$
4. $[Cr^{III}Cl_3(OH_2)_3]$
5. $[Fe^{-II}(CO)_4]^{2-}$

IR-9.2.3.4 *Use of abbreviations*

Abbreviations can be used to represent complicated organic ligands in formulae (although they should not normally be used in names). When used in formulae they are usually enclosed in parentheses.

Guidelines for the formulation of ligand abbreviations are given in Section IR-4.4.4; examples of such abbreviations are listed alphabetically in Table VII with diagrams of most shown in Table VIII.

In cases where coordination occurs through one of several possible donor atoms of a ligand, an indication of that donor atom may be desirable. This may be achieved in names through use of the kappa convention (see Section IR-9.2.4.2) in which the Greek lower case kappa (κ) is used to indicate the donor atom. To some extent, this device may also be used in formulae. For example, if the glycinate anion (gly) coordinates only through the nitrogen atom, the abbreviation of the ligand would be shown as gly-κ*N*, as in the complex $[M(gly\text{-}\kappa N)_3X_3]$.

IR-9.2.4 **Specifying donor atoms**

IR-9.2.4.1 *General*

There is no need to specify the donor atom of a ligand that has only one atom able to form a bond with a central atom. However, ambiguity may arise when there is more than one possible donor atom in a ligand. It is then necessary to specify which donor atom(s) of the ligand is (are) bound to the central atom. This includes cases where a ligand can be thought of as being formed by removal of H^+ from a particular site in a molecule or ion. For example, acetylacetonate, $MeCOCHCOMe^-$, has the systematic ligand name 2,4-dioxopentan-3-ido, which does not, however, imply bonding to the central atom from the central carbon atom in the ligand. The donor atom can be specified as shown in IR-9.2.4.2.

The only cases where specification of the donor atom is not required for a ligand that can bind to a central atom in more than one way are:

monodentate O-bound carboxylate groups
monodentate C-bound cyanide (ligand name 'cyanido')
monodentate C-bound carbon monoxide (ligand name 'carbonyl')
monodentate N-bound nitrogen monoxide (ligand name 'nitrosyl').

By convention, in these cases the ligand names imply the binding mode shown.

The following sections detail the means by which donor atoms are specified. The kappa (κ) convention, introduced in Section IR-9.2.4.2, is general and can be used for systems of great complexity. In some cases it may be simplified to the use of just the donor atom symbol (see Section IR-9.2.4.4).

These systems may be used in names, but they are not always suitable for use in formulae. The use of donor atom symbols is possible in the formulae of simple systems (see Section IR-9.2.3.4), but care must be taken to avoid ambiguity. The kappa convention is not generally compatible with the use of ligand abbreviations.

These methods are normally used only for specifying bonding between the central atom and isolated donor atoms. The eta (η) convention is used for any cases where the central atom is bonded to contiguous donor atoms within one ligand (see IR-10.2.5.1). Most examples of this latter kind are organometallic compounds (Chapter IR-10) but the example below shows its use for a coordination compound.

Example:

1.

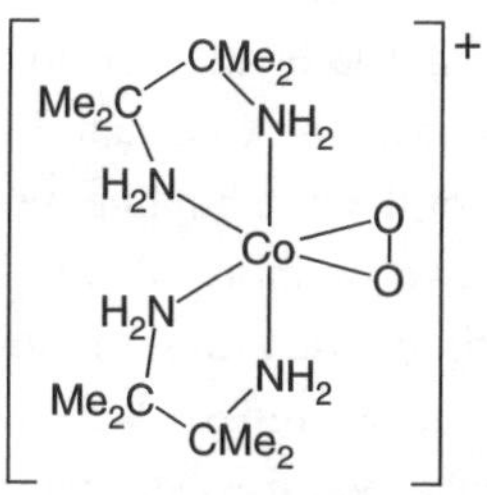

bis(2,3-dimethylbutane-2,3-diamine)(η^2-peroxido)cobalt(1+)

IR-9.2.4.2 *The kappa convention*

Single ligating atoms are indicated by the italicized element symbol preceded by a Greek kappa, κ. These symbols are placed after the portion of the ligand name that represents the ring, chain or substituent group in which the ligating atom is found.

Example:

1. $[NiBr_2(Me_2PCH_2CH_2PMe_2)]$
   dibromido[ethane-1,2-diylbis(dimethylphosphane-κ*P*)]nickel(II)

Multiplicative prefixes which apply to a ligand or portions of a ligand also apply to the donor atom symbols. In some cases this may require the use of an alternative ligand name, *e.g.* where multiplicative prefixes can no longer be used because the ligation of otherwise equivalent portions of the ligand is different. Several examples of this are given below.

Simple examples are thiocyanato-κ*N* for nitrogen-bonded NCS and thiocyanato-κ*S* for sulfur-bonded NCS. Nitrogen-bonded nitrite is named nitrito-κ*N* and oxygen-bonded nitrite is named nitrito-κ*O*, as in pentaamminenitrito-κ*O*-cobalt(III).

For ligands with several ligating atoms linearly arranged along a chain, the order of κ symbols should be successive, starting at one end. The choice of end is based upon alphabetical order if the ligating atoms are different, *e.g.* cysteinato-κ*N*,κ*S*; cysteinato-κ*N*,κ*O*.

Donor atoms of a particular element may be distinguished by adding a right superscript numerical locant to the italicized element symbol or, in simple cases (such as Example 3 below), a prime or primes.

Superscript numerals, on the other hand, are based on an appropriate numbering of some or all of the atoms of the ligand, such as numbering of the skeletal atoms in parent hydrides, and allow the position of the bond(s) to the central atom to be specified even in quite complex cases. In the simple case of acetylacetonate, $MeCOCHCOMe^{-}$, mentioned above, the ligand name 2,4-dioxopentan-3-ido-κC^3 would imply ligation by the central carbon atom in the pentane skeleton (see also Example 4 below).

In some cases, standard nomenclature procedures do not provide locants for the donor atoms in question. In such cases simple *ad hoc* procedures may be applicable. For example, for the ligand $(CF_3COCHCOMe)^{-}$, the name 1,1,1-trifluoro-2,4-dioxopentan-3-ido-κ*O* could be used to refer to coordination, through oxygen, of the CF_3CO portion of the molecule, while coordination by MeCO would be identified by 1,1,1-trifluoro-2,4-dioxopentan-3-ido-κ*O*′. The prime indicates that the MeCO oxygen atom is associated with a higher locant in the molecule than the CF_3CO oxygen atom. The oxygen atom of the CF_3CO portion of the ligand is attached to C2, while that of MeCO is attached to C4. Alternatively, the name could be modified to 1,1,1-trifluoro-2-(oxo-κ*O*)-4-oxopentan-3-ido and 1,1,1-trifluoro-2-oxo-4-(oxo-κ*O*)pentan-3-ido, respectively, for the two binding modes above.

In cases where two or more identical ligands (or parts of a polydentate ligand) are involved, a superscript is used on κ to indicate the number of such ligations. As mentioned above, any multiplicative prefixes for complex entities are presumed to operate on the κ symbol as well. Thus, one uses the partial name '...bis(2-amino-κ*N*-ethyl)... ' and not '...bis(2-amino-κ^2*N*-ethyl)...' in Example 2 below. Examples 2 and 3 use tridentate chelation by the linear tetraamine ligand *N,N*′-bis(2-aminoethyl)ethane-1,2-diamine to illustrate these rules.

Examples:

2.

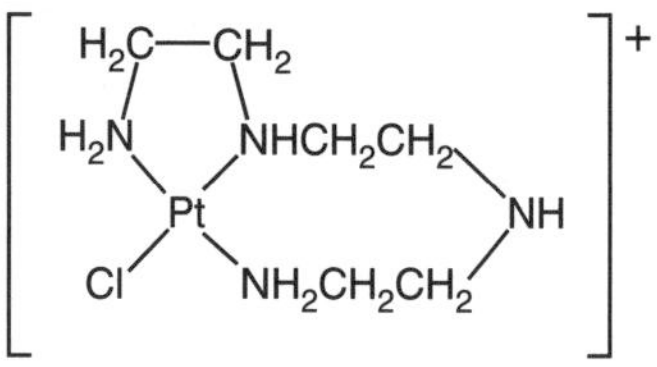

[*N,N*′-bis(2-amino-κ*N*-ethyl)ethane-1,2-diamine-κ*N*]chloridoplatinum(II)

3.

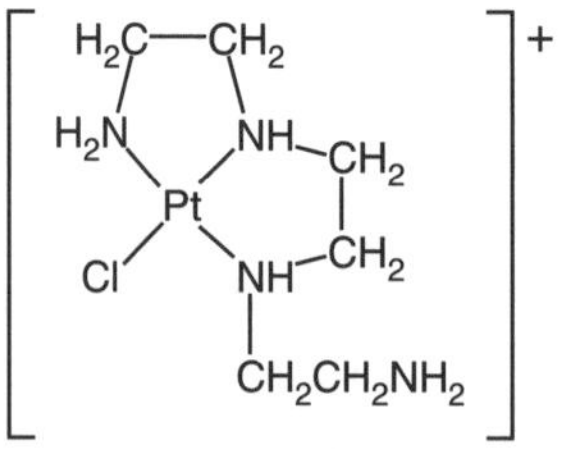

[*N*-(2-amino-κ*N*-ethyl)-*N*′-(2-aminoethyl)ethane-1,2-diamine-κ^2*N,N*′] chloridoplatinum(II)

Example 2 illustrates how coordination by the two terminal primary amino groups of the ligand is indicated by placing the kappa index after the substituent group name and within the effect of the 'bis' doubling prefix. The appearance of the simple index κ*N* after the 'ethane-1,2-diamine' indicates the binding by only one of the two equivalent secondary amino nitrogen atoms.

Only one of the primary amines is coordinated in Example 3. This is indicated by not using the doubling prefix 'bis', repeating (2-aminoethyl), and inserting the κ index only in the first such unit, *i.e.* (2-amino-κ*N*-ethyl). The involvement of both of the secondary ethane-1,2-diamine nitrogen atoms in chelation is indicated by the index $\kappa^2 N,N'$.

Tridentate chelation by the tetrafunctional macrocycle in Example 4 is shown by the kappa index following the ligand name. The ligand locants are required in order to distinguish this complex from those where the central atom is bound to other combinations of the four potential donor atoms.

Example:

4.

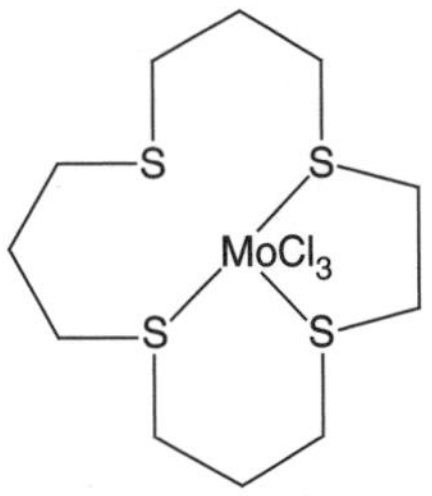

trichlorido(1,4,8,12-tetrathiacyclopentadecane-$\kappa^3 S^{1,4,8}$)molybdenum, or
trichlorido(1,4,8,12-tetrathiacyclopentadecane-$\kappa^3 S^1,S^4,S^8$)molybdenum

Well-established modes of chelation of the (ethane-1,2-diyldinitrilo)tetraacetato ligand (edta), namely bidentate, tetradentate and pentadentate, are illustrated in Examples 5–8. The multiplicative prefix 'tetra' used in Example 5 cannot be used in Examples 6 and 7 because of the need to avoid ambiguity about which acetate arms are coordinated to the central atom. In such cases the coordinated fragments are cited before the uncoordinated fragments in the ligand name. Alternatively, a modified name may be used, as in Example 7, where the use of the preferred IUPAC name *N,N′*-ethane-1,2-diylbis[*N*-(carboxymethyl)glycine] (see Section P-44.4 of Ref. 1) is demonstrated.

Examples:

5.

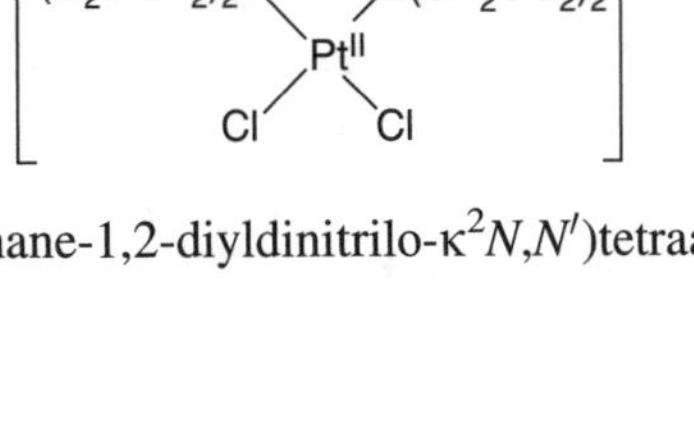

dichlorido[(ethane-1,2-diyldinitrilo-$\kappa^2 N,N'$)tetraacetato]platinate(4−)

6.

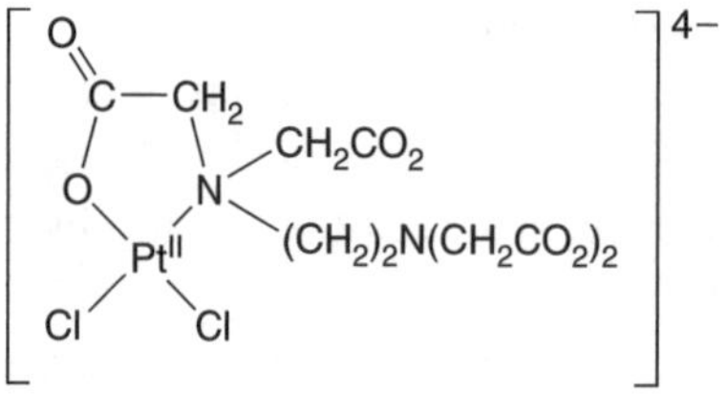

dichlorido[(ethane-1,2-diyldinitrilo-κ*N*)(acetato-κ*O*)triacetato]platinate(II)

7.

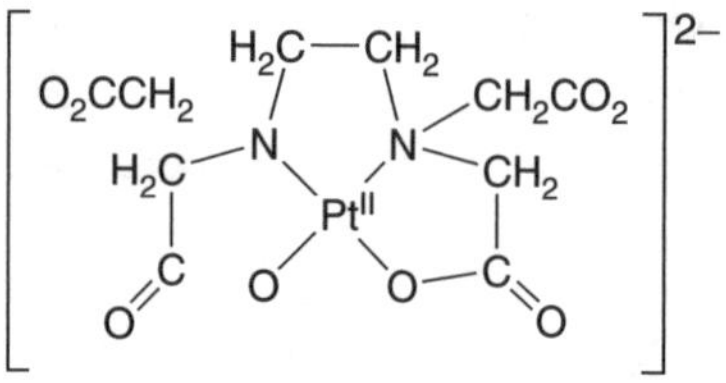

[(ethane-1,2-diyldinitrilo-κ^2*N*,*N*′)(*N*,*N*′-diacetato-κ^2*O*,*O*′)(*N*,*N*′-diacetato)]platinate(2−), or
{*N*,*N*′-ethane-1,2-diylbis[*N*-(carboxylatomethyl)glycinato-κ*O*,κ*N*]}platinate(2−)

8.

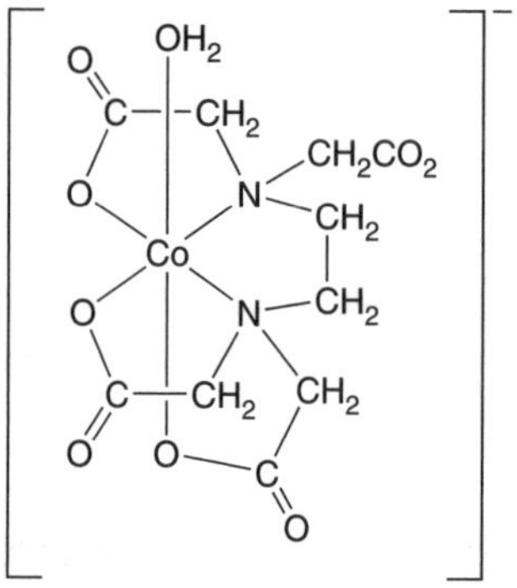

aqua[(ethane-1,2-diyldinitrilo-κ^2*N*,*N*′)tris(acetato-κ*O*)acetato]cobaltate(1−),
or aqua[*N*-{2-[bis(carboxylato-κ*O*-methyl)amino-κ–]ethyl}-
N-(carboxylato-κ*O*-methyl)glycinato-κ–]cobaltate(1−)

A compound of edta in which one amino group is not coordinated while all four carboxylato groups are bound to a single metal ion would bear the ligand name (ethane-1,2-diyldinitrilo-κ*N*)tetrakis(acetato-κ*O*) within the name of the complex.

The mixed sulfur–oxygen cyclic polyether 1,7,13-trioxa-4,10,16-trithiacyclooctadecane might chelate to alkali metals only through its oxygen atoms and to second-row transition elements only through its sulfur atoms. The corresponding kappa indexes for such chelate complexes would be $\kappa^3 O^1,O^7,O^{13}$ and $\kappa^3 S^4,S^{10},S^{16}$, respectively.

Examples 9–11 illustrate three modes of chelation of the ligand *N*-[*N*-(2-aminoethyl)-*N*′,*S*-diphenylsulfonodiimidoyl]benzenimidamide. The use of kappa indexes allows these binding modes (and others) to be distinguished and identified, in spite of the abundance of heteroatoms that could coordinate.

Examples:

9.

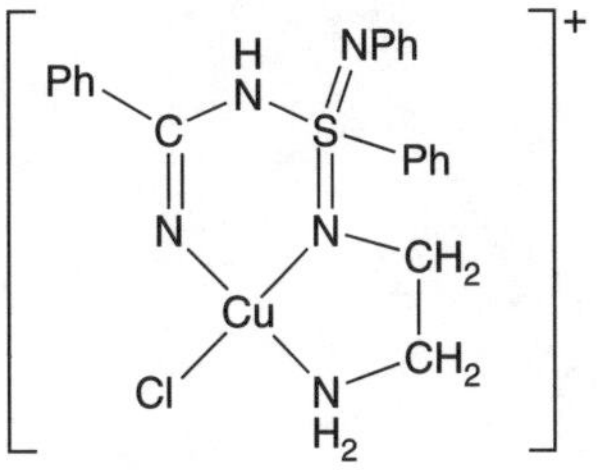

{*N*-[*N*-(2-amino-κ*N*-ethyl)-*N*′,*S*-diphenylsulfonodiimidoyl-κ*N*]benzenimidamide-κ*N*′}chloridocopper(II)

10.

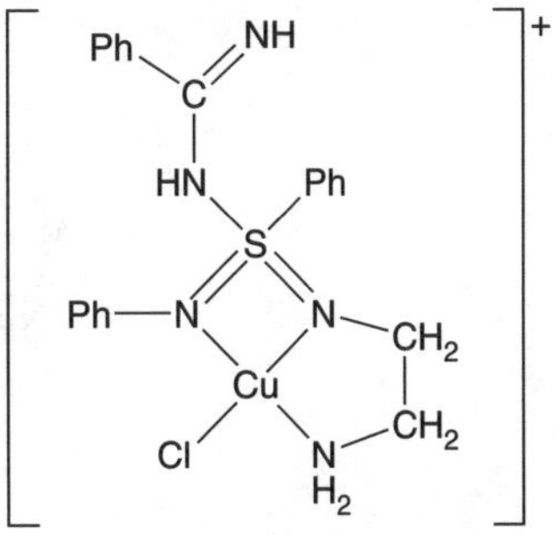

{*N*-[*N*-(2-amino-κ*N*-ethyl)-*N*′,*S*-diphenylsulfonodiimidoyl-κ^2N,N']benzenimidamide}chloridocopper(II)

11.

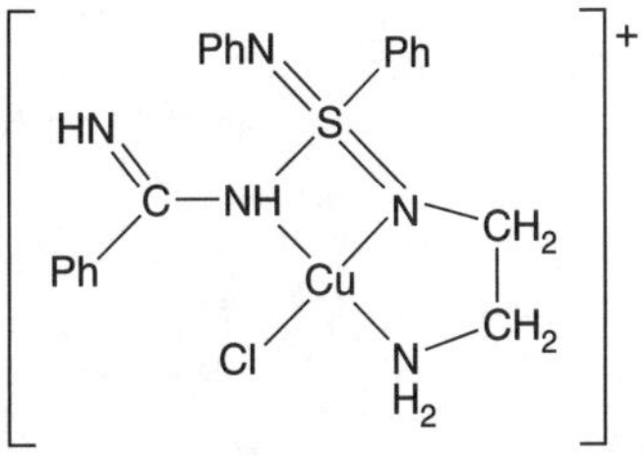

{*N*-[*N*-(2-amino-κ*N*-ethyl)-*N*′,*S*-diphenylsulfonodiimidoyl-κ*N*]benzenimidamide-κ*N*}chloridocopper(II)

The distinction between the names in Examples 9 and 11 rests on the conventional priming of the imino nitrogen atom in the benzenimidamide functional group. The prime differentiates the imino benzenimidamide nitrogen atom from that which is substituted (and unprimed at the beginning of the name).

The use of donor atom locants on the atomic symbols to indicate point of ligation is again illustrated by the two isomeric bidentate modes of binding of the macrocycle 1,4,7-triazecane (or 1,4,7-triazacyclodecane) (Examples 12 and 13). Conveying the formation of the five-membered chelate ring requires the index κ^2N^1,N^4, while the six-membered chelate ring requires the index κ^2N^1,N^7. Example 14 shows that due to the local nature of the locants used with κ, the same locant and atomic symbol may appear several times, referring to different parts of the ligand.

Examples:

12.

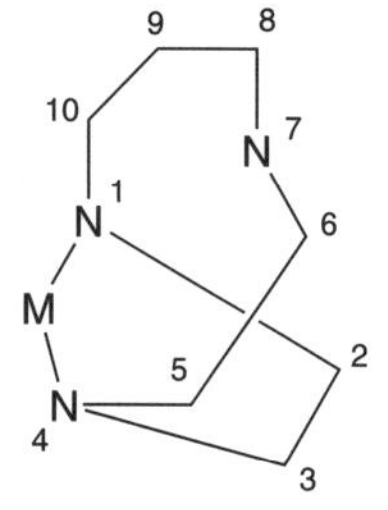

9 8
10
N 7
N 1
6
M
5 2
N
4
3

$\kappa^2 N^1, N^4$

13.

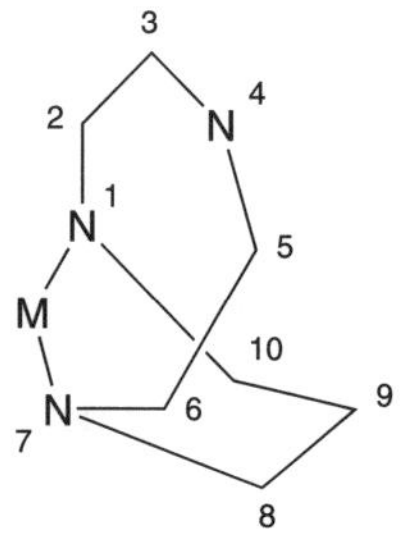

3
2 N 4
N 1
5
M
10
9
N 6
7
8

$\kappa^2 N^1, N^7$

14.

NH2
N 2
O 6 1
NH3 5 4 N 3
H3N—Pt—N 7
9 N
O 5′ CH2OH
N 7 8
8 1′ 4′
6 5
HN 1 9 2′ 3′
2 4 N
3 O O−
O P
H2N N 1′ 4′
O
2′ 3′ 5′ O 5′
CH2 O CH2
OH O—P—O 3′
O− 4′
2′ O
1′
N O
N
NH2

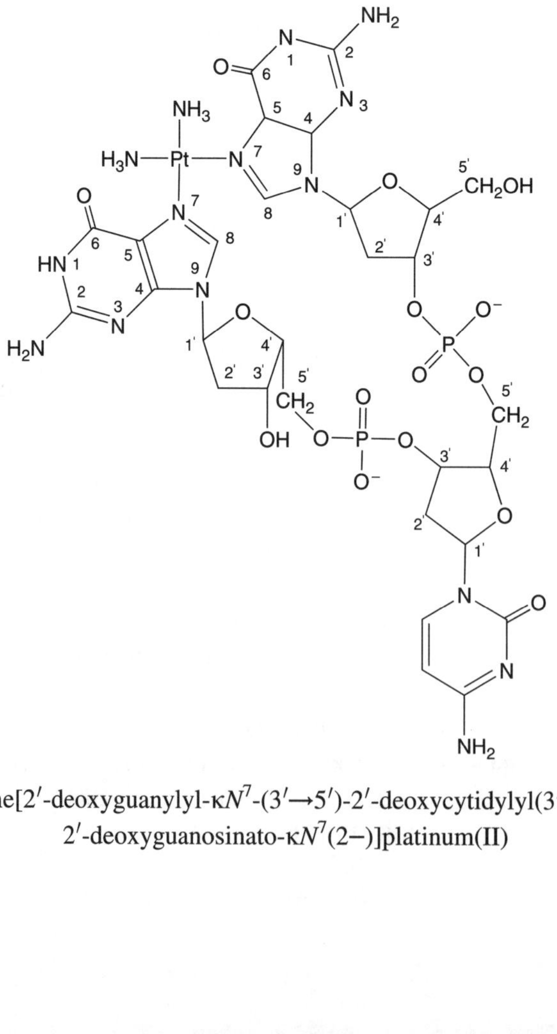

diammine[2′-deoxyguanylyl-κN^7-(3′→5′)-2′-deoxycytidylyl(3′→5′)-2′-deoxyguanosinato-κN^7(2−)]platinum(II)

IR-9.2.4.3 *Comparison of the eta and kappa conventions*

The eta convention (Section IR-10.2.5.1) is applied in cases where contiguous donor atoms within a given ligand are involved in bonding to a central atom. Thus, it is used only when there is more than one ligating atom, and the term η^1 is not used. The contiguous atoms are often the same element, but need not be.

The kappa convention is used to specify bonding from isolated donor atoms to one or more central atoms.

In cases where two or more identical ligands (or parts of a polydentate ligand) are bound to a central atom, a superscript is used on κ to indicate the number of donor atom-to-central atom bonds.

IR-9.2.4.4 *Use of donor atom symbol alone in names*

In certain cases the kappa convention may be simplified. Donor atoms of a ligand may be denoted by adding only the italicized symbol(s) for the donor atom (or atoms) to the end of the name of the ligand. Thus, for the 1,2-dithiooxalate anion, ligand names such as 1,2-dithiooxalato-κ*S*,κ*S′* and 1,2-dithiooxalato-κ*O*,κ*S* may, with no possibility of confusion, be shortened to 1,2-dithiooxalato-*S*,*S′* and 1,2-dithiooxalato-*O*,*S*, respectively. Other examples are thiocyanato-*N* and thiocyanato-*S*, and nitrito-*N* and nitrito-*O*.

IR-9.2.5 **Polynuclear complexes**

IR-9.2.5.1 *General*

Polynuclear inorganic complexes exist in a bewildering array of structural types, such as ionic solids, molecular polymers, extended assemblies of oxoanions, chains and rings, bridged metal complexes, and homonuclear and heteronuclear clusters. This section primarily treats the nomenclature of bridged metal complexes and homonuclear and heteronuclear clusters. Coordination polymers are treated extensively elsewhere.[6]

As a general principle, as much structural information as possible should be presented when writing the formula or name of a polynuclear complex. However, polynuclear complexes may have structures so large and extended as to make a rational structure-based nomenclature impractical. Furthermore, their structures may be undefined or not suitably elucidated. In such cases, the principal function of the name or formula is to convey the stoichiometric proportions of the various moieties present.

In the present and following sections, particular complexes are often used as examples several times to show how they may be named differently according to whether only stoichiometry is to be specified or partial or complete structural information is to be included.

Ligands in polynuclear complexes are cited in alphabetical order both in formulae and names. The number of each ligand is specified by subscript numerical multipliers in formulae (Sections IR-9.2.3.1 to IR-9.2.3.4) and by appropriate multiplicative prefixes in names (Sections IR-9.2.2.1 to IR-9.2.2.3). The number of central atoms of a given kind, if greater than one, is indicated similarly.

Note, however, that the rules for formula writing may be relaxed in various ways in order better to display particular features of the structures in question. Use is made of this flexibility in many examples below.

Example:

1. $[Rh_3H_3\{P(OMe)_3\}_6]$
   trihydridohexakis(trimethyl phosphite)trirhodium

If there is more than one element designated as a central atom, these elements are ranked according to the order in which they appear in Table VI. The later an element appears in the sequence of Table VI, the earlier it comes in the list of central atom symbols in the formula as well as in the list of central atom names in the name of the complex.

Example:

2. $[ReCo(CO)_9]$ nonacarbonylrheniumcobalt

For anionic species, the ending 'ate' and the charge number (see Section IR-5.4.2.2) are added after the central atom list which is enclosed in parentheses if more than one element is involved.

Examples:

3. $[Cr_2O_7]^{2-}$ heptaoxidodichromate(2−)
4. $[Re_2Br_8]^{2-}$ octabromidodirhenate(2−)
5. 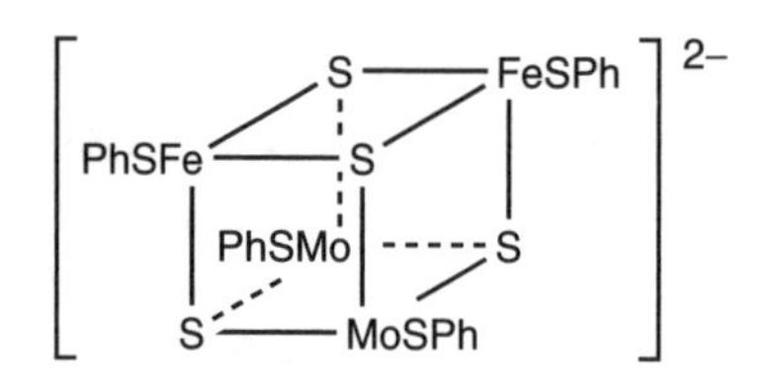

$[Mo_2Fe_2S_4(SPh)_4]^{2-}$
tetrakis(benzenethiolato)tetrakis(sulfido)(dimolybdenumdiiron)ate(2−)

Although not extensively exemplified here, it is worth noting that the formalism developed below for polynuclear complexes is applicable also to (formal) complexes in which the central atoms are not metals.

Example:

6. $[PSO_7]^{2-}$ heptaoxido(phosphorussulfur)ate(2−)

A number of oxoacids and related species are given such names in Chapter IR-8 and Table IX.

The symbol kappa, κ, was introduced, in Section IR-9.2.4.2, in order to specify the ligating atoms in polyatomic ligands. This use also applies to such ligands when they appear in polynuclear complexes. However, the symbol κ then assumes a new function, namely that of specifying which ligating atoms bind to which central atom. In order to do this, the central atoms must be identified, *i.e.* by assigning numbers to these atoms according to the order in which they appear in the central atom list. (The later the central atom elements appear in Table VI, the lower the numbers they are assigned.)

Additional rules are needed when there is more than one central atom of the same element (see Sections IR-9.2.5.5 and IR-9.2.5.6) except if the presence of symmetry in the structure makes two or more of the central atoms equivalent (see, for example, Section IR-9.2.5.4) and the name eventually generated is independent of the numbering.

The central atom numbers are then used as locants for the ligating atoms and are placed to the left of each kappa symbol. Individual kappa designators, *i.e.* kappa symbols with a numerical superscript (as applicable), central atom locant and ligator atom symbol, are separated by commas.

Examples:

7. $[(OC)_5\overset{1}{Re}\overset{2}{Co}(CO)_4]$
   nonacarbonyl-1κ^5*C*,2κ^4*C*-rheniumcobalt

8. $[Cl_4\overset{1}{Re}\overset{2}{Re}Cl_4]^{2-}$
   octachlorido-1κ^4*Cl*,2κ^4*Cl*-dirhenate(2−)

In these two examples, structural information indicated by the formulae is not communicated by the names. In fact, any polynuclear complex must either contain at least one ligand binding to more than one central atom (a bridging ligand) or contain a bond between two central atoms. In order to specify these aspects of the structure in names, further devices are needed. These are introduced in the following two sections.

IR-9.2.5.2 *Bridging ligands*

Bridging ligands, as far as they can be specified, are indicated by the Greek letter μ appearing before the ligand symbol or name and separated from it by a hyphen; the conventions applied were briefly introduced in IR-9.1.2.10. In names, the whole term, *e.g.* μ-chlorido, is separated from the rest of the name by hyphens, as in ammine-μ-chlorido-chlorido, *etc.*, unless the bridging ligand name is contained within its own set of enclosing marks. If the bridging ligand occurs more than once, multiplicative prefixes are employed, as in tri-μ-chlorido-chlorido, or as in bis(μ-diphenylphosphanido), if more complex ligand names are involved.

Bridging ligands are listed in alphabetical order together with the other ligands, but in names a bridging ligand is cited before a corresponding non-bridging ligand, as in di-μ-chlorido-tetrachlorido. In formulae, bridging ligands are placed after terminal ligands of the same kind. Thus, in both names and formulae bridging ligands are placed further away from the central atoms than are terminal ligands of the same kind.

Example:

1. $[Cr_2O_6(\mu\text{-}O)]^{2-}$ μ-oxido-hexaoxidodichromate(2−)

The bridging index *n*, the number of coordination centres connected by a bridging ligand, is placed as a right subscript. The bridging index 2 is not normally indicated. Multiple bridging is listed in descending order of complexity, *e.g.* μ_3-oxido-di-μ-oxido-trioxido. For ligand names requiring enclosing marks, μ is contained within those marks.

The kappa convention is used together with μ when it is necessary to specify which central atoms are bridged, and through which donor atoms. The kappa descriptor counts all donor atom-to-central atom bonds so that in Example 2 below the descriptor 1:2:3κ^3*S* specifies all three bonds from the sulfur atom bridging central atoms 1, 2 and 3.

Example:

2.

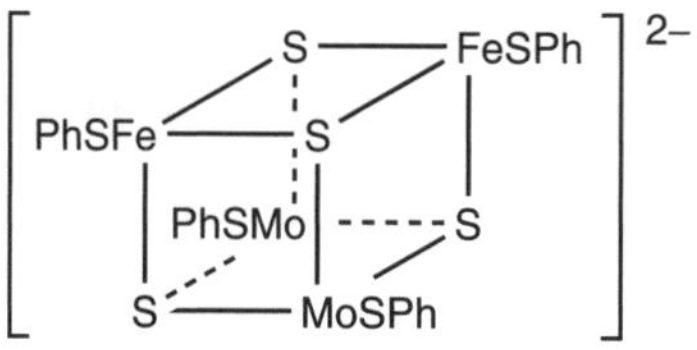

$[Mo_2Fe_2S_4(SPh)_4]^{2-}$

tetrakis(benzenethiolato)-1κ*S*,2κ*S*,3κ*S*,4κ*S*-tetra-μ_3-sulfido-1:2:3κ^3*S*;1:2:4κ^3*S*;1:3:4κ^3*S*;2:3:4κ^3*S*-(dimolybdenumdiiron)ate(2−)

Here, the two molybdenum atoms are numbered 1 and 2 and the two iron atoms 3 and 4 according to the rule in Section IR-9.2.5.1. Due to the symmetry of the compound, it is not necessary to distinguish between 1 and 2 or between 3 and 4.

Example:

3. $[O_3S(\mu\text{-}O_2)SO_3]^{2-}$ μ-peroxido-1κ*O*,2κ*O'*-hexaoxidodisulfate(2−)

When single ligating atoms bind to two or more central atoms, the central atom locants are separated by a colon. For example, tri-μ-chlorido-1:2κ^2*Cl*;1:3κ^2*Cl*;2:3κ^2*Cl*- indicates that there are three bridging chloride ligands and they bridge between central atoms 1 and 2, 1 and 3, and 2 and 3. Note that because of the use of the colon, sets of bridge locants are separated here by semicolons rather than commas.

Example:

4.

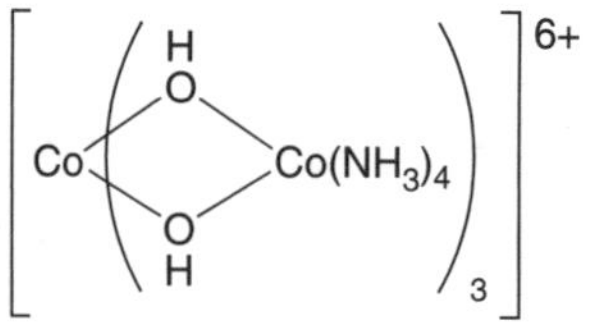

$[Co\{(\mu\text{-}OH)_2Co(NH_3)_4\}_3]^{6+}$

dodecaammine-1κ^4*N*,2κ^4*N*,3κ^4*N*-hexa-μ-hydroxido-1:4κ^4*O*;2:4κ^4*O*;3:4κ^4*O*-tetracobalt(6+)

The central atom locants given in this example are assigned by following the rules in Sections IR-9.2.5.5 and IR-9.2.5.6. In this case, the central cobalt atom is assigned the locant 4.

Example:

5.

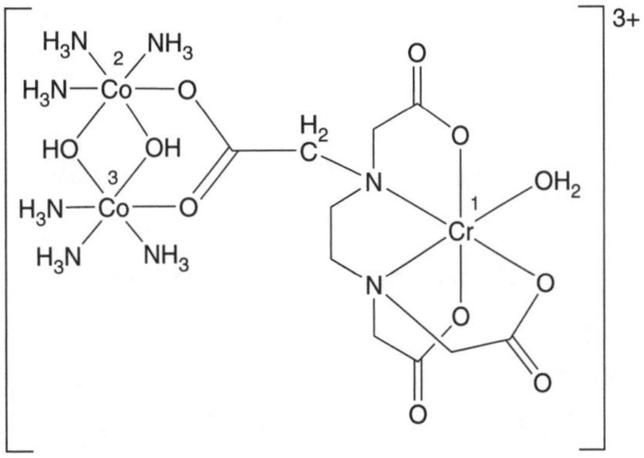

hexaammine-$2\kappa^{3}N,3\kappa^{3}N$-aqua-$1\kappa O$-{$\mu_3$-(ethane-1,2-diyldinitrilo-$1\kappa^{2}N,N'$)-tetraacetato-$1\kappa^{3}O^{1},O^{2},O^{3}{:}2\kappa O^{4}{:}3\kappa O^{4'}$}-di-$\mu$-hydroxido-$2{:}3\kappa^{4}O$-chromiumdicobalt(3+)

In this name, the obvious numbering ($1,1',2,2',3,3',4,4'$) of the oxygen ligating atoms of the four carboxylate groups is tacitly assumed.

IR-9.2.5.3 *Metal–metal bonding*

Metal–metal bonding or, more generally, bonding between central atoms in complexes, may be indicated in names by placing italicized atomic symbols of the appropriate central atoms, separated by an 'em' dash and enclosed in parentheses, after the list of central atom names and before the ionic charge. The central atom element symbols are placed in the same order as the central atoms appear in the name (*i.e.* according to Table VI, with the first element reached when following the arrow being placed last). The number of such bonds is indicated by an arabic numeral placed before the first element symbol and separated from it by a space. For the purpose of nomenclature, no distinction is made between different bond orders. If there is more than one central atom of an element present in the structure, and it is necessary to indicate which of them is involved in the bond in question (because they are inequivalent), the central atom locant (see Section IR 9.2.5.6) can be placed as a superscript immediately after the element symbol, as shown in Example 4.

Examples:

1.

$[Cl_4\overset{1}{Re}\overset{2}{Re}Cl_4]^{2-}$

octachlorido-$1\kappa^{4}Cl,2\kappa^{4}Cl$-dirhenate(*Re*—*Re*)(2−)

2.

$[(OC)_5\overset{1}{Re}\overset{2}{Co}(CO)_4]$

nonacarbonyl-$1\kappa^{5}C,2\kappa^{4}C$-rheniumcobalt(*Re*—*Co*)

3.

$Cs_3[Re_3Cl_{12}]$

caesium dodecachlorido-*triangulo*-trirhenate(3 *Re*—*Re*)(3−)

4. 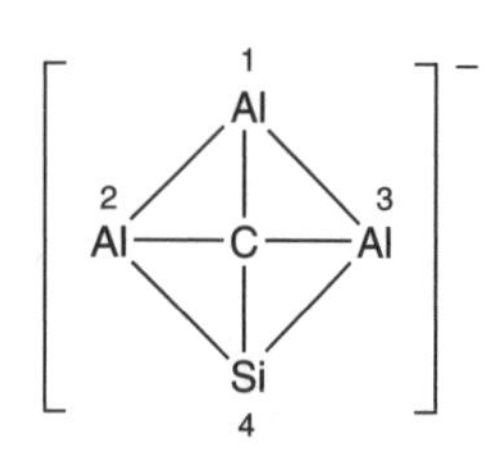

μ_4-carbido-*quadro*-(trialuminiumsilicon)ate $(Al^1—Al^2)$ $(Al^1—Al^3)(Al^2—Si)(Al^3—Si)$(1−)

(Examples 3 and 4 include the structural descriptors *triangulo* and *quadro* which are introduced below in Section IR-9.2.5.7.) Note that the name in Example 3 does not specify which chloride ligands bind to which central atoms.

IR-9.2.5.4 *Symmetrical dinuclear entities*

For symmetrical dinuclear entities, the name may be simplified by employing multiplicative prefixes.

Examples:

1. $[Re_2Br_8]^{2-}$
   bis(tetrabromidorhenate)(*Re*—*Re*)(2−)
2. $[Mn_2(CO)_{10}]$
   bis(pentacarbonylmanganese)(*Mn*—*Mn*)
3. $[\{Cr(NH_3)_5\}_2(\mu\text{-}OH)]^{5+}$
   μ-hydroxido-bis(pentaamminechromium)(5+)
4. $[\{PtCl(PPh_3)\}_2(\mu\text{-}Cl)_2]$
   di-μ-chlorido-bis[chlorido(triphenylphosphane)platinum]
5. $[\{Fe(NO)_2\}_2(\mu\text{-}PPh_2)_2]$
   bis(μ-diphenylphosphanido)bis(dinitrosyliron)
6. $[\{Cu(py)\}_2(\mu\text{-}O_2CMe)_4]$
   tetrakis(μ-acetato-κ*O*:κ*O*′)bis[(pyridine)copper(II)]

In some cases multiplicative prefixes may also be used to simplify names of unsymmetrical complexes (see Example 5 in Section IR-9.2.5.5).

IR-9.2.5.5 *Unsymmetrical dinuclear entities*

The name of an unsymmetrical dinuclear species will result from following the general rules described in Sections IR-9.2.5.1 to IR-9.2.5.3.

Example:

1. $[ClHgIr(CO)Cl_2(PPh_3)_2]$
   carbonyl-1κ*C*-trichlorido-1κ^2*Cl*,2κ*Cl*-bis(triphenylphosphane-1κ*P*)iridiummercury(*Ir*—*Hg*)

In this example, iridium is reached last on following the arrow shown in Table VI. It is therefore listed before mercury in the name and is given the central atom locant 1.

The only remaining problem is to number the central atoms in cases where they are the same but have different coordination environments. In this case, the central atom with the larger coordination number is given the lower number (locant), if applicable. If the coordination numbers are equal, the central atom with the greater number of ligands or ligating atoms represented earlier in the name is given the lower number (locant). Thus, in Example 2 the chromium atom with five of the nine ammine ligands attached is given priority number 1.

Examples:

2. $[(H_3N)_5\overset{1}{Cr}(\mu\text{-}OH)\overset{2}{Cr}(NH_2Me)(NH_3)_4]^{5+}$
   nonaammine-1κ^5*N*,2κ^4*N*-μ-hydroxido-(methanamine-2κ*N*)dichromium(5+)
3. $[(H_3N)_3\overset{1}{Co}(\mu\text{-}NO_2)(\mu\text{-}OH)_2\overset{2}{Co}(NH_3)_2(py)]^{3+}$
   pentaammine-1κ^3*N*,2κ^2*N*-di-μ-hydroxido-μ-nitrito-1κ*N*:2κ*O*-(pyridine-2κ*N*)dicobalt(3+)
4. $[(bpy)(H_2O)\overset{1}{Cu}(\mu\text{-}OH)_2\overset{2}{Cu}(bpy)(SO_4)]$
   aqua-1κ*O*-(2,2′-bipyridine-1κ^2*N*,*N*′)(2,2′-bipyridine-2κ^2*N*,*N*′)-di-μ-hydroxido-(sulfato-2κ*O*)dicopper(II)

In some cases, it is not necessary to number explicitly the two differently coordinated central atoms to arrive at a name, as shown in Example 5. Note the use of a multiplicative prefix to simplify the name, as also demonstrated in Section IR-9.2.5.4 for fully symmetrical structures.

Example:

5. $[\{Co(NH_3)_3\}_2(\mu\text{-}NO_2)(\mu\text{-}OH)_2]^{3+}$
   di-μ-hydroxido-μ-nitrito-κ*N*:κ*O*-bis(triamminecobalt)(3+)

IR-9.2.5.6 *Trinuclear and larger structures*

The methods described in the preceding sections for naming ligands and designating ligating atoms are general, and applicable irrespective of the nuclearity (the number of central atoms involved). However, in most cases numbering of the central atoms is needed in order to construct a systematic additive name for a coordination entity. Obtaining such a numbering is the part of the naming process which becomes increasingly complex in the general case as the nuclearity increases. This section suggests general procedures for assigning locant numbers to central atoms.

If no two central atoms are the same element, locant numbers for the central atoms and the order they appear in the name can be determined using Table VI. The first central atom reached on following the arrow in the Table receives the highest locant number, while the last reached is given the locant 1. This method can also be applied to systems where there is more than one of a given type of central atom, provided there is symmetry present in the structure that makes all of the central atoms of a given element equivalent. Indeed, in the extreme case, it may not be necessary to assign locants at all, provided all the central atoms are equivalent.

Examples:

1. $[Be_4(\mu_4\text{-}O)(\mu\text{-}O_2CMe)_6]$
   hexakis(μ-acetato-κ*O*:κ*O′*)-μ_4-oxido-*tetrahedro*-tetraberyllium
2. $[Os_3(CO)_{12}]$
   dodecacarbonyl-1κ^4*C*,2κ^4*C*,3κ^4*C*-*triangulo*-triosmium(3 *Os*—*Os*)

(The descriptors *tetrahedro* and *triangulo* are introduced in Section IR-9.2.5.7.)

Another such case is Example 5 in Section IR-9.2.5.2 where it is immaterial which of the two cobalt atoms is given number 2 and which one number 3. The systematic name will be the same.

The proposed *general procedure* for constructing a coordination-type additive name for a polynuclear entity is as follows:

(i) Identify the central atoms and ligands.
(ii) Name the ligands, including κ, η and μ designators (except for the central atom locants). Note that ligand names may have to be modified if κ, η or μ symbols apply only to some portions of the ligand that are otherwise equivalent (and described by a multiplicative prefix such as 'tri' or 'tris').
(iii) Place ligand names in alphabetical order.
(iv) Assign central atom locants by applying the following rules:

   (a) Apply the element sequence of Table VI. The later an element is met when following the arrows, the lower its locant number. This criterion will determine the numbering if all central atoms are different elements. Locants may be assigned to atoms of the same element by applying the next rules.
   (b) Within each class of identical central atoms, assign lower locant numbers to central atoms with higher coordination numbers.
   (c) Proceed through the alphabetical list of ligand names. Examine the names or name parts specifying ligating atoms explicitly (as in a κ or η designator) or implicitly (as in the ligand name 'carbonyl'). As soon as a subset of ligating atoms is met which is not evenly distributed among the central atoms still awaiting the assignment of distinct locant numbers, the central atoms with the most ligating atoms of this kind are given the lowest numbers available. This process of sequential examination of the ligands is continued until all central atoms have been assigned locants or all ligands have been considered.
   (d) Any central atoms that are inequivalent and have not yet been assigned distinct locant numbers will differ only in the other central atoms to which they are directly bonded. The locant numbers of these directly bonded neighbouring central atoms are compared and the central atom with the lowest-locant neighbouring atoms is given the lowest of the remaining possible locants (see Example 9 below).

Note that the central atom locants assigned using these rules need not coincide with those assigned when using other types of nomenclature such as substitutive nomenclature (*cf.* Chapter IR-6), if that is applicable, or the nomenclature systems described in Chapters II-1 or II-5 of Ref. 7.

Example:

3.

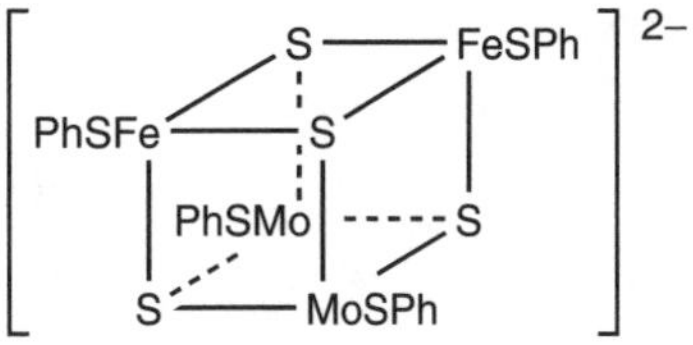

$[Mo_2Fe_2S_4(SPh)_4]^{2-}$

tetrakis(benzenethiolato)-1κ*S*,2κ*S*,3κ*S*,4κ*S*-tetra-μ_3-sulfido-
1:2:3κ^3*S*;1:2:4κ^3*S*;1:3:4κ^3*S*;2:3:4κ^3*S*-(dimolybdenumdiiron)ate(2−)

Using the rules above, no distinction is obtained between the two molybdenum atoms or between the two iron atoms. However, no distinction is needed.

Example:

4.

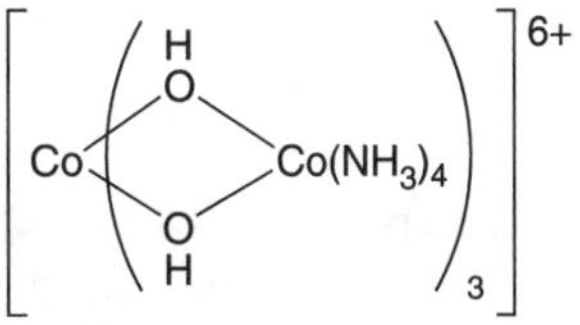

$[Co\{(\mu\text{-}OH)_2Co(NH_3)_4\}_3]^{6+}$
dodecaammine-1κ^4*N*,2κ^4*N*,3κ^4*N*-hexa-μ-hydroxido-
1:4κ^4*O*;2:4κ^4*O*;3:4κ^4*O*-tetracobalt(6+)

Rules (a) and (b) do not result in a distinction between the four cobalt atoms. By rule (c), however, the three peripheral cobalt atoms are assigned numbers 1, 2 and 3 because they carry the ammine ligands appearing first in the name, and the central cobalt atom is thus number 4. This is all that is required to construct the name, because of the symmetry of the complex.

Examples:

5.

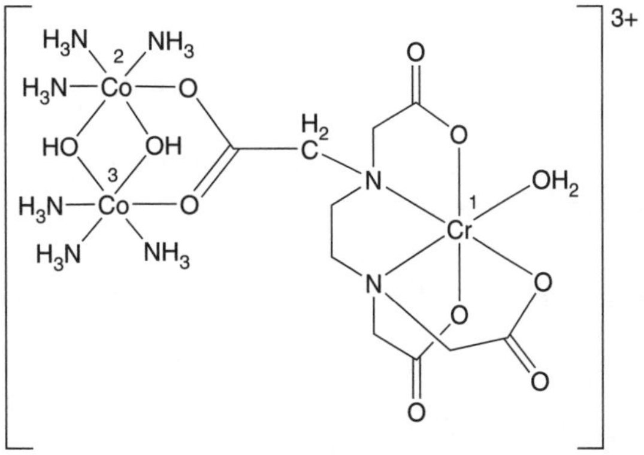

hexaammine-2κ^3*N*,3κ^3*N*-aqua-1κ*O*-{μ_3-(ethane-1,2-diyldinitrilo-1κ^2*N*,*N*′)-
tetraacetato-1$\kappa^3$$O^1$,$O^2$,$O^3$:2κ$O^4$:3κ$O^{4'}$}-di-μ-hydroxido-2:3$\kappa^4$*O*-
chromiumdicobalt(3+)

6.

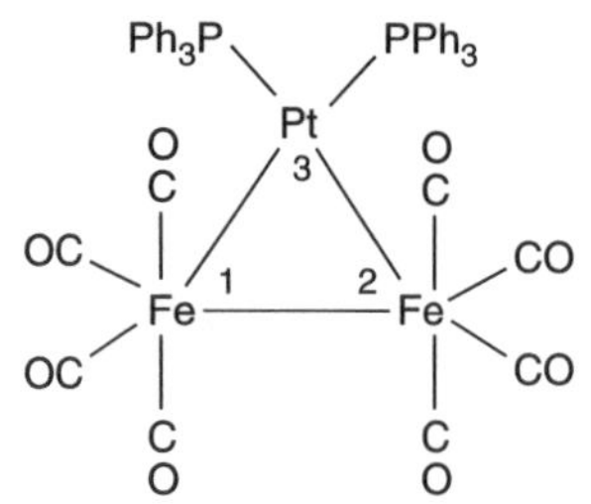

octacarbonyl-1κ^4C,2κ^4C-bis(triphenylphosphane-3κ*P*)-*triangulo*-diironplatinum(*Fe*—*Fe*)(2 *Fe*—*Pt*)

7. $[Os_3(CO)_{12}(SiCl_3)_2]$

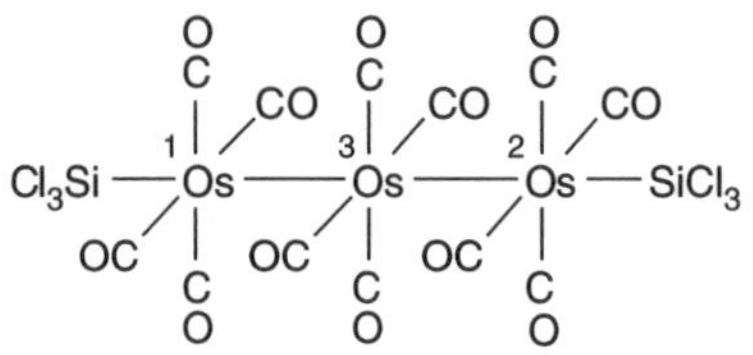

dodecacarbonyl-1κ^4C,2κ^4C,3κ^4C-bis(trichlorosilyl)-1κ*Si*,2κ*Si*-triosmium(Os^1—Os^3)(Os^2—Os^3)

All three osmium atoms have four carbonyl ligands. The two osmium atoms with trichlorosilyl ligands are assigned central atom locants 1 and 2, as these ligands are the first that are not evenly distributed. Symmetry in the structure means that the locants 1 and 2 can be assigned either way around.

Example:

8.

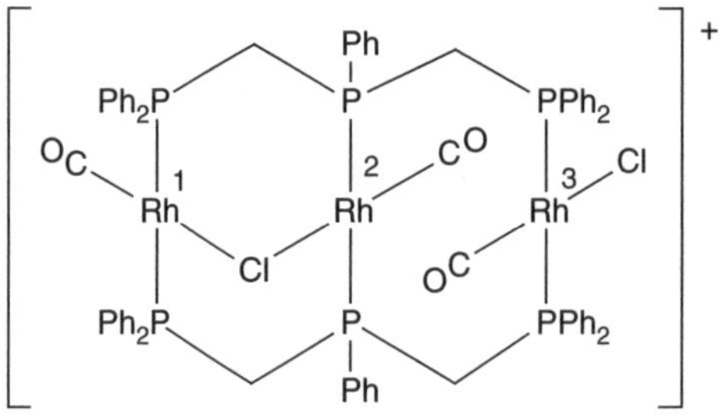

tricarbonyl-1κ*C*,2κ*C*,3κ*C*-μ-chlorido-1:2κ^2*Cl*-chlorido-3κ*Cl*-bis{μ_3-bis[(diphenylphosphanyl)methyl]-1κ*P*:3κ*P′*-phenylphosphane-2κ*P*}trirhodium(1+)

or, using the preferred IUPAC name[1] for the phosphane ligand:

tricarbonyl-1κ*C*,2κ*C*,3κ*C*-μ-chlorido-1:2κ^2*Cl*-chlorido-3κ*Cl*-bis{μ_3-[phenylphosphanediyl-1κ*P*-bis(methylene)]bis(diphenylphosphane)-2κ*P′*:3κ*P″*}trirhodium(1+)

Example 8 illustrates how using different (equally systematic) names for ligands may result in different additive names and different locant numberings.

In the first name, the first place where the rhodium atoms can be identified as being inequivalent is at the kappa term associated with the μ-chlorido ligand. Thus, the chloride-bridged rhodium atoms must be assigned the central atom locants 1 and 2 (although which is which is not known at this stage), and the other rhodium atom must be assigned the locant 3. The next difference in the name that relates to central atom 1 or 2 is the diphenylphosphanyl κ term. Those portions of the ligand are bound to the end rhodium atoms and not to the middle rhodium atom. Since one of the end rhodium atoms is already given the locant 3, from the earlier difference, the other rhodium atom must be assigned locant 1, and the middle atom is left with locant 2.

For the second name, the locant 3 is assigned in the same way, but the middle Rh atom should be assigned locant 1 as it now appears earlier in the ligand name (in the κ term for phosphanediyl).

Example:

9.

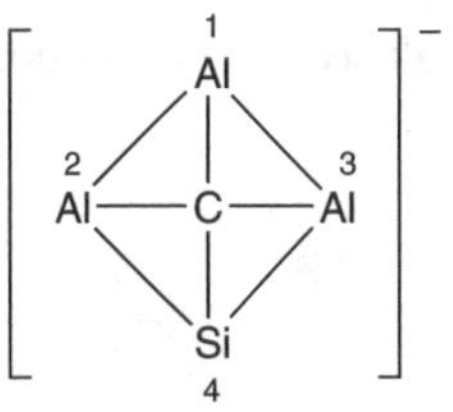

μ_4-carbido-*quadro*-(trialuminiumsilicon)ate(*Al*^{1}—*Al*^{2})(*Al*^{1}—*Al*^{3})(*Al*^{2}—*Si*)(*Al*^{3}—*Si*)(1−)

In this example the central atom locants are assigned as follows. Rule (a), above, results in the silicon atom being assigned locant 4. The coordination numbers and ligand distribution are the same for the three aluminium atoms, which only differ in which other central atoms they are bonded to. The numbering of the aluminium atoms follows from rule (d) above.

The prefix '*cyclo*', italicized and cited before all ligands, may be used for monocyclic compounds.

Example:

10.

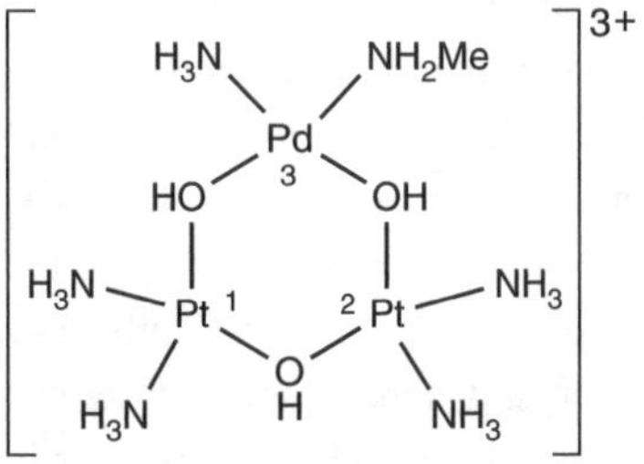

cyclo-pentaammine-1κ^2N,2κ^2N,3κN-tri-μ-hydroxido-1:2κ^2O;1:3κ^2O;2:3κ^2O-(methanamine-3κN)diplatinumpalladium(3+)

The two platinum atoms are equivalent and receive lower central atom locants than palladium by rule (a).

Examples:

11.

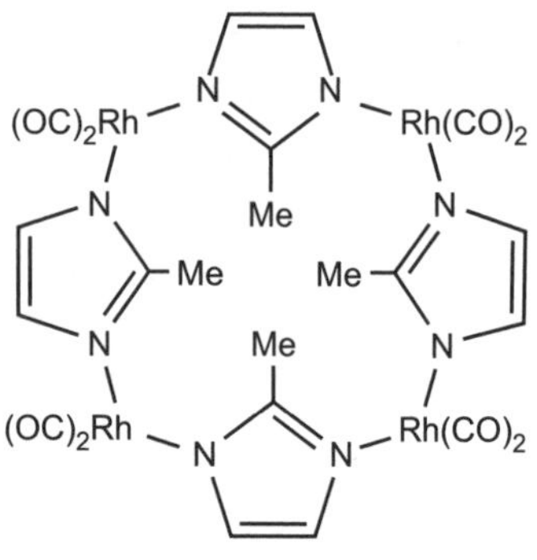

cyclo-tetrakis(μ-2-methylimidazolido-κN^1:κN^3)tetrakis(dicarbonylrhodium)

12.

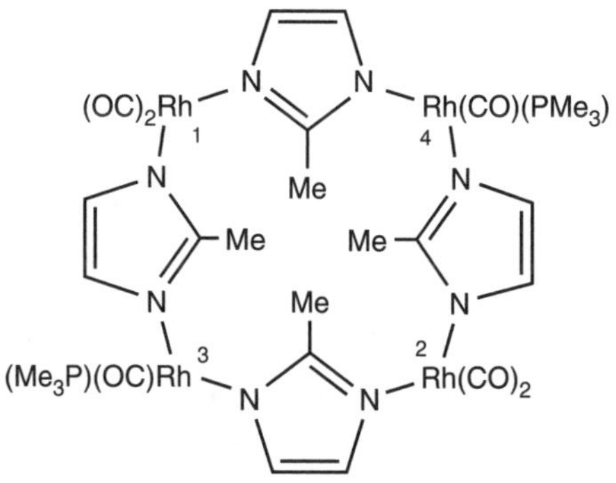

cyclo-hexacarbonyl-1κ^2C,2κ^2C,3κC,4κC-tetrakis(μ-2-methyl-1*H*-imidazol-l-ido)-1:3κ$^2N^1$:N^3;1:4κ$^2N^3$:N^1;2:3κ$^2N^3$:N^1;2:4κ$^2N^1$:N^3-bis(trimethylphosphane)-3κP,4κP-tetrarhodium

IR-9.2.5.7 *Polynuclear clusters: symmetrical central structural units*

The structural features of complex polynuclear entities may be communicated using the concept of a *central structural unit* (CSU). Only the metal atoms are considered for this purpose. For nonlinear clusters, descriptors such as *triangulo*, *tetrahedro* and *dodecahedro* are used to describe central structural units in simple cases, as has already been exemplified above. However, synthetic chemistry has advanced far beyond the limited range of central structural units associated with this usage. A more comprehensive CSU descriptor and a numbering system, the CEP (Casey, Evans, Powell) system, has been developed specifically for fully triangulated polyboron polyhedra (deltahedra).[8] These CEP descriptors may be used in general as systematic alternatives to the traditional descriptors for fully triangulated polyhedra (deltahedra). Examples are listed in Table IR-9.1.

Table IR-9.1 Structural descriptors

Number of atoms in CSU	*Descriptor*	*Point group*	*CEP descriptor*
3	*triangulo*	D_{3h}	
4	*quadro*	D_{4h}	
4	*tetrahedro*	T_d	[T_d-(13)-Δ^4-*closo*]
5		D_{3h}	[D_{3h}-(131)-Δ^6-*closo*]
6	*octahedro*	O_h	[O_h-(141)-Δ^8-*closo*]
6	*triprismo*	D_{3h}	
8	*antiprismo*	S_6	
8	*dodecahedro*	D_{2d}	[D_{2d}-(2222)-Δ^6-*closo*]
8	*hexahedro (cube)*	O_h	
12	*icosahedro*	I_h	[I_h-(1551)-Δ^{20}-*closo*]

In brief, the numbering of the CSU is based on locating a reference axis and planes of atoms perpendicular to the reference axis. The reference axis is the axis of highest rotational symmetry. Select that end of the reference axis with a single atom (or smallest number of atoms) in the first plane to be numbered. Orient the CSU so that the first position to receive a locant in the first plane with more than one atom is in the twelve o'clock position. Assign locant numbers to the axial position or to each position in the first plane, beginning at the twelve o'clock position and moving in either the clockwise or anticlockwise direction. From the first plane move to the next position and continue numbering in the same direction (clockwise or anticlockwise), always returning to the twelve o'clock position or the position nearest to it in the forward direction before assigning locants in that plane. Continue numbering in this manner until all positions are numbered.

A full discussion of numbering deltahedra may be found elsewhere.[8] The complete descriptor for the CSU should appear just before the central atom list. Where structurally significant, metal–metal bonds may be indicated (see Section IR-9.2.5.3 and examples below).

The chain or ring structure numbering in a CSU must be consecutive and only thereafter obey rules (a)–(d) given in Section IR 9.2.5.6. In Example 3 below, the CSU numbering in fact coincides with the numbering that would be reached using those rules alone.

Examples:

1. $[\{Co(CO)_3\}_3(\mu_3\text{-}CBr)]$
   (μ_3-bromomethanetriido)nonacarbonyl-*triangulo*-tricobalt(3 *Co—Co*), or
   (μ_3-bromomethanetriido)-*triangulo*-tris(tricarbonylcobalt)(3 *Co—Co*)

2. $[Cu_4(\mu_3\text{-}I)_4(PEt_3)_4]$
   tetra-μ_3-iodido-tetrakis(triethylphosphane)-*tetrahedro*-tetracopper, or
   tetra-μ_3-iodido-tetrakis(triethylphosphane)-[T_d-(13)-Δ^4-*closo*]-tetracopper

3. $[Co_4(CO)_{12}]$

tri-μ-carbonyl-1:2κ^2*C*;1:3κ^2*C*;2:3κ^2*C*-nonacarbonyl-1κ^2*C*,2κ^2*C*,3κ^2*C*,4κ^3*C*-[T_d-(13)-Δ^4-*closo*]-tetracobalt(6 *Co—Co*)

This compound has also been named, in Section II-5.3.3.3.6 of Ref. 7, using the chain and ring nomenclature (see Section IR-7.4). However, that name is based on a completely different numbering scheme.

Examples:

4. $[Mo_6S_8]^{2-}$
octa-μ$_3$-sulfido-*octahedro*-hexamolybdate(2−), or
octa-μ$_3$-sulfido-[O_h-(141)-Δ^8-*closo*]-hexamolybdate(2−)

5.

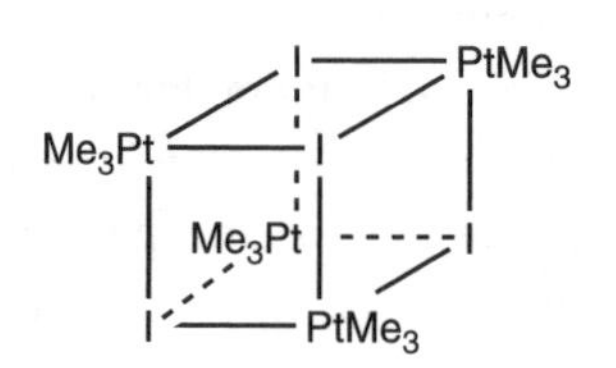

tetra-μ$_3$-iodido-1:2:3κ^3*I*;1:2:4κ^3*I*;1:3:4κ^3*I*;2:3:4κ^3*I*-dodecamethyl-1κ^3*C*,2κ^3*C*,3κ^3*C*,4κ^3*C*-*tetrahedro*-tetraplatinum(IV), or
tetra-μ$_3$-iodido-1:2:3κ^3*I*;1:2:4κ^3*I*;1:3:4κ^3*I*;2:3:4κ^3*I*-dodecamethyl-1κ^3*C*,2κ^3*C*,3κ^3*C*,4κ^3*C*-[T_d-(13)-Δ^4-*closo*]-tetraplatinum(IV)

6. $[(HgMe)_4(\mu_4\text{-}S)]^{2+}$
μ$_4$-sulfido-tetrakis(methylmercury)(2+), or
tetramethyl-1κ*C*,2κ*C*,3κ*C*,4κ*C*-μ$_4$-sulfido-*tetrahedro*-tetramercury(2+), or
tetramethyl-1κ*C*,2κ*C*,3κ*C*,4κ*C*-μ$_4$-sulfido-[T_d-(13)-Δ^4-*closo*]-tetramercury(2+)

IR-9.3 DESCRIBING THE CONFIGURATION OF COORDINATION ENTITIES

IR-9.3.1 **Introduction**

Once the constitution of a coordination entity has been defined, it remains to describe the spatial relationships between the structural components of the molecule or ion. Molecules that differ only in the spatial distribution of the components are known as stereoisomers. Stereoisomers that are mirror images of one another are called enantiomers (sometimes these have been called optical isomers), while those that are not are called diastereoisomers (or geometrical isomers). This is an important distinction in chemistry as, in general, diastereoisomers exhibit different physical, chemical and spectroscopic properties from one another, while enantiomers exhibit identical properties (except in the presence of other chiral entities). It is instructive to consider an everyday analogy in order to establish how the configuration of a molecule (and the embedded spatial relationships) can be described.

Using the terminology introduced above, left and right hands may be regarded as enantiomers of one another, since they are different (non-superimposable), but they are mirror images of each other. In both cases the thumbs are adjacent to the index finger, and the components of each hand are similarly disposed relative to all the other parts of that hand. If the thumb and index finger of a right hand were to be exchanged, the resulting hand could be considered to be a diastereoisomer of the normal right hand (and it too would have an enantiomer, resulting from a similar exchange on a left hand). The key point is that the relative positions of the components of diastereoisomers (the normal right hand and the modified one) are different.

In order to describe the hand fully the components (four fingers, one thumb and the central part of the hand) must be identified, the points of attachment available on the hand, and the relative positions of the fingers and thumb around the hand, must be described and whether the hand is 'left' or 'right' must be specified. The last three steps deal with the configuration of the hand.

In the case of a coordination compound, the name and formula describe the ligands and central atom(s). Describing the configuration of such a coordination compound requires consideration of three factors:

(i) coordination geometry – identification of the overall shape of the molecule;
(ii) relative configuration – description of the relative positions of the components of the molecule, *i.e.* where the ligands are placed around the central atom(s) in the identified geometry;
(iii) absolute configuration – identification of which enantiomer is being specified (if the mirror images are non-superimposable).

The next three sections deal with these steps in turn. A more detailed discussion of the configuration of coordination compounds can be found elsewhere.[9]

IR-9.3.2 Describing the coordination geometry

IR-9.3.2.1 *Polyhedral symbol*

Different geometrical arrangements of the atoms attached to the central atom are possible for all coordination numbers greater than one. Thus, two-coordinate species may involve a linear or a bent disposition of the ligands and central atom. Similarly, three-coordinate species may be trigonal planar or trigonal pyramidal, and four-coordinate species may be square planar, square pyramidal or tetrahedral. The coordination polyhedron (or polygon in planar molecules) may be denoted in the name by an affix called the *polyhedral symbol*. This descriptor distinguishes isomers differing in the geometries of their coordination polyhedra.

The polyhedral symbol must be assigned before any other spatial features can be considered. It consists of one or more capital italic letters derived from common geometric terms which denote the idealized geometry of the ligands around the coordination centre, and an arabic numeral that is the coordination number of the central atom.

Distortions from idealized geometries commonly occur. However, it is normal practice to relate molecular structures to idealized models. The polyhedral symbol is used as an affix, enclosed in parentheses and separated from the name by a hyphen. The polyhedral symbols for the most common geometries for coordination numbers 2 to 9 are given in Table IR-9.2 and the corresponding structures and/or polyhedra are shown in Table IR-9.3.

Table IR-9.2 *Polyhedral symbols*[a]

Coordination polyhedron	*Coordination number*	*Polyhedral symbol*
linear	2	*L*-2
angular	2	*A*-2
trigonal plane	3	*TP*-3
trigonal pyramid	3	*TPY*-3
T-shape	3	*TS*-3
tetrahedron	4	*T*-4
square plane	4	*SP*-4
square pyramid	4	*SPY*-4
see-saw	4	*SS*-4
trigonal bipyramid	5	*TBPY*-5
square pyramid	5	*SPY*-5
octahedron	6	*OC*-6
trigonal prism	6	*TPR*-6
pentagonal bipyramid	7	*PBPY*-7
octahedron, face monocapped	7	*OCF*-7
trigonal prism, square-face monocapped	7	*TPRS*-7
cube	8	*CU*-8
square antiprism	8	*SAPR*-8
dodecahedron	8	*DD*-8
hexagonal bipyramid	8	*HBPY*-8
octahedron, *trans*-bicapped	8	*OCT*-8
trigonal prism, triangular-face bicapped	8	*TPRT*-8
trigonal prism, square-face bicapped	8	*TPRS*-8
trigonal prism, square-face tricapped	9	*TPRS*-9
heptagonal bipyramid	9	*HBPY*-9

[a] Strictly, not all geometries can be represented by polyhedra.

Table IR-9.3 *Polyhedral symbols, geometrical structures and/or polyhedra*

Three-coordination

trigonal plane	trigonal pyramid	T-shape
TP-3	*TPY*-3	*TS*-3

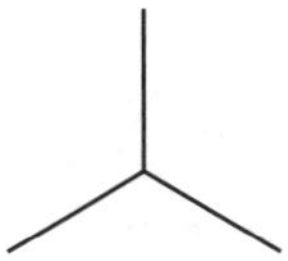

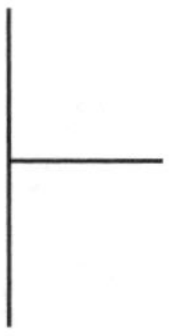

Four-coordination

tetrahedron	square plane
T-4	*SP*-4

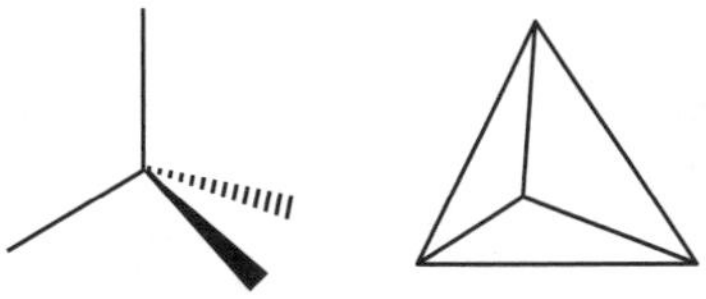

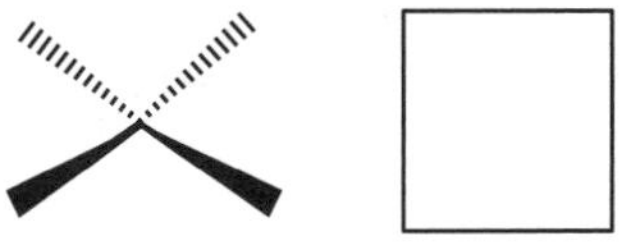

square pyramid	see-saw
SPY-4	*SS*-4

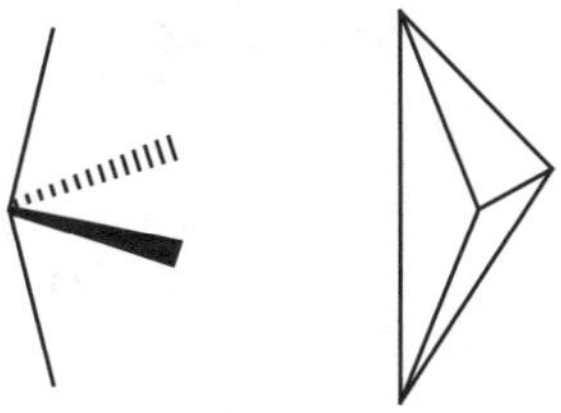

Five-coordination

trigonal bipyramid	square pyramid
TBPY-5	*SPY*-5

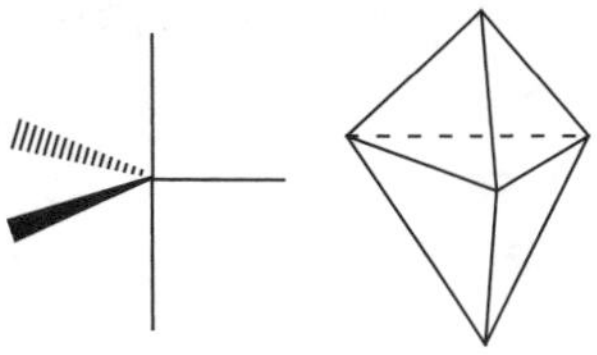

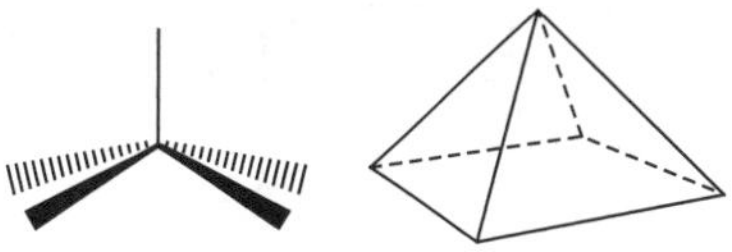

Six-coordination

octahedron	trigonal prism
OC-6	*TPR*-6

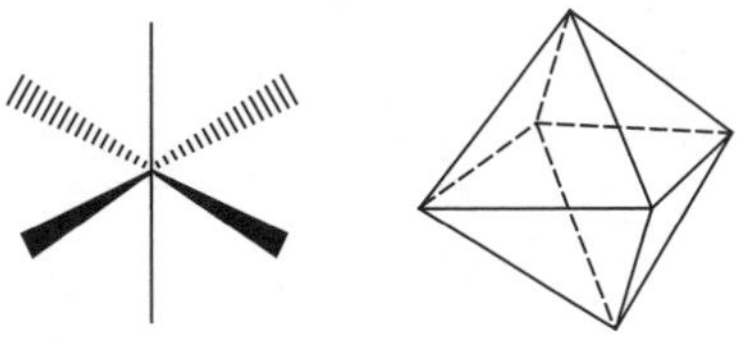

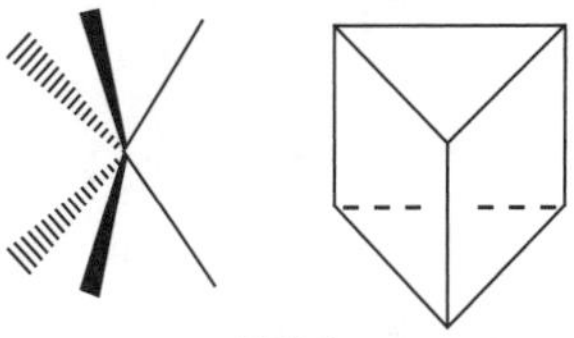

Table IR-9.3 *Continued*

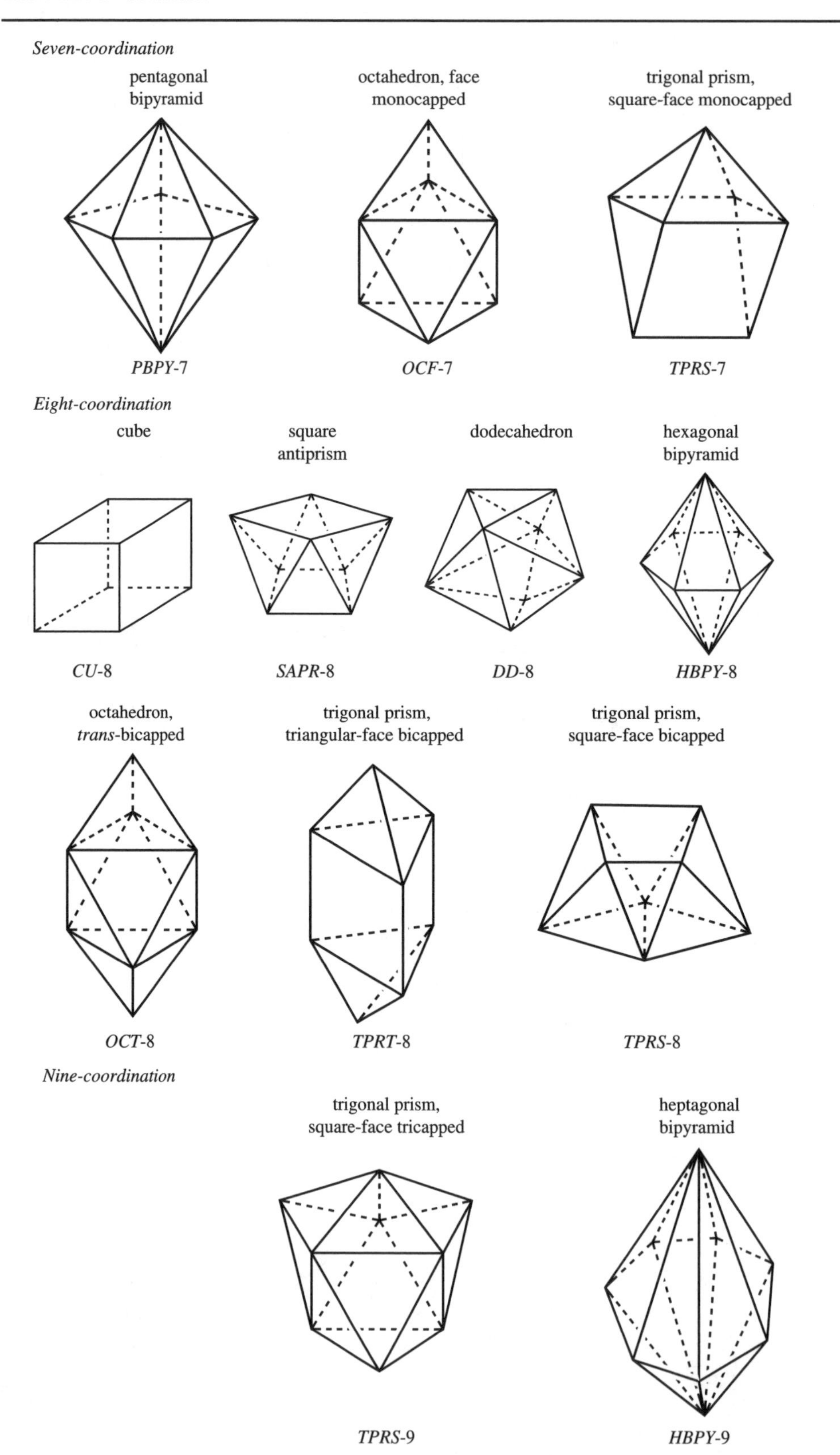
Seven-coordination
pentagonal bipyramid
octahedron, face monocapped
trigonal prism, square-face monocapped
PBPY-7
OCF-7
TPRS-7
Eight-coordination
cube
square antiprism
dodecahedron
hexagonal bipyramid
CU-8
SAPR-8
DD-8
HBPY-8
octahedron, *trans*-bicapped
trigonal prism, triangular-face bicapped
trigonal prism, square-face bicapped
OCT-8
TPRT-8
TPRS-8
Nine-coordination
trigonal prism, square-face tricapped
heptagonal bipyramid
TPRS-9
HBPY-9

IR-9.3.2.2 *Choosing between closely related geometries*

For real molecules or ions, the stereochemical descriptor should be based on the nearest idealized geometry. However, some idealized geometries are closely related [*e.g.* square planar (*SP*-4), four-coordinate square pyramidal (*SPY*-4), see-saw (*SS*-4), and tetrahedral (*T*-4); T-shaped (*TS*-3), trigonal planar (*TP*-3), and trigonal pyramidal (*TPY*-3)] and care may therefore be required in making the choice.

The following approach is useful in determining the polyhedral symbol for four-coordinate structures. The key is to consider the locations of the central atom and the coordinating atoms in relation to each other. If all five atoms are in (or are close to being in) the same plane, then the molecule should be treated as square planar. If the four coordinating atoms are in a plane, but the central atom is significantly displaced from the plane, then the square pyramidal geometry is appropriate. If the four coordinating atoms do not lie in (or close to) a plane, then a polyhedron can be defined by joining all four coordinating atoms together with lines. If the central atom lies inside this polyhedron the molecule should be regarded as tetrahedral, otherwise, it should be regarded as having a see-saw structure.

T-shaped and trigonal planar molecules both have a central atom that lies in (or close to) the plane defined by the coordinating atoms. They differ in that the angles between the three coordinating atoms are approximately the same in the trigonal planar structure, while one angle is much larger than the other two in a T-shaped molecule. The central atom lies significantly out of the plane in a trigonal pyramidal structure.

IR-9.3.3 Describing configuration – distinguishing between diastereoisomers

IR-9.3.3.1 *General*

The placement of ligands around the central atom must be described in order to identify a particular diastereoisomer. There are a number of common terms (*e.g. cis, trans, mer* and *fac*) used to describe the relative locations of ligands in simple systems. However, they can be used only when a particular geometry is present (*e.g.* octahedral or square planar), and when there are only two kinds of donor atom present (*e.g.* Ma_2b_2 in a square planar complex, where M is a central atom and 'a' and 'b' are types of donor atom).

Several methods have been used to distinguish between diastereoisomers in more complex systems. Thus, stereoisomers resulting from the coordination of linear tetradentate ligands have often been identified as *trans*, *cis*-α, or *cis*-β,[10] and those resulting from coordination of macrocyclic tetradentate ligands have their own system.[11] The scope of most of these nomenclatures is generally quite limited, but a proposal with wider application in the description of complexes of polydentate ligands has been made more recently.[12]

Clearly a general method is required in order to distinguish between diastereoisomers of compounds in which either other geometries or more than two kinds of donor atoms are present. The *configuration index* has been developed for this purpose. The next section outlines the method by which a configuration index is obtained for a compound, and the following sections give details for particular geometries. Commonly used terms are included for each geometry discussed.

IR-9.3.3.2 *Configuration index*

Once the coordination geometry has been specified by the polyhedral symbol, it becomes necessary to identify which ligands (or donor atoms) occupy particular coordination positions. This is achieved through the use of the configuration index which is a series of digits identifying the positions of the ligating atoms on the vertices of the coordination polyhedron. The configuration index has the property that it distinguishes between diastereoisomers. It appears within the parentheses enclosing the polyhedral symbol (see Section IR-9.3.2.1), following that symbol and separated from it by a hyphen.

Each donor atom must be assigned a priority number based on the rules developed by Cahn, Ingold and Prelog (the CIP rules).[13] These priority numbers are then used to form the configuration index for the compound. The application of the CIP rules to coordination compounds is discussed in detail in Section IR-9.3.5 but, in general, donor atoms that have a higher atomic number have higher priority than those that have a lower atomic number.

The presence of polydentate ligands may require the use of primes on some of the numbers in the configuration index. The primes are used to indicate either that donor atoms are not part of the same polydentate ligand as those that have unprimed priority numbers, or that the donor atoms belong to different parts of a polydentate ligand that are related by symmetry. A primed priority number means that that donor atom has lower priority than the same kind of donor atom without a prime on the priority number. More detail on the 'priming convention' can be found in Section IR-9.3.5.3.

IR-9.3.3.3 *Square planar coordination systems (SP-4)*

The terms *cis* and *trans* are used commonly as prefixes to distinguish between stereoisomers in square planar systems of the form [Ma_2b_2], where M is the central atom, and 'a' and 'b' are different types of donor atom. Similar donor atoms occupy coordination sites adjacent to one another in the *cis* isomer, and opposite to one another in the *trans* isomer. The *cis-trans* terminology is not adequate to distinguish between the three isomers of a square planar coordination entity [Mabcd], but could be used, in principle, for an [Ma_2bc] system (where the terms *cis* and *trans* would refer to the relative locations of the similar donor atoms). This latter use is not recommended.

The configuration index for a square planar system is placed after the polyhedral symbol (*SP*-4). It is the single digit which is the priority number for the ligating atom *trans* to the ligating atom of priority number 1, *i.e.* the priority number of the ligating atom *trans* to the most preferred ligating atom.

Examples:

1. Priority sequence: a > b > c > d
   Priority number sequence: 1 < 2 < 3 < 4

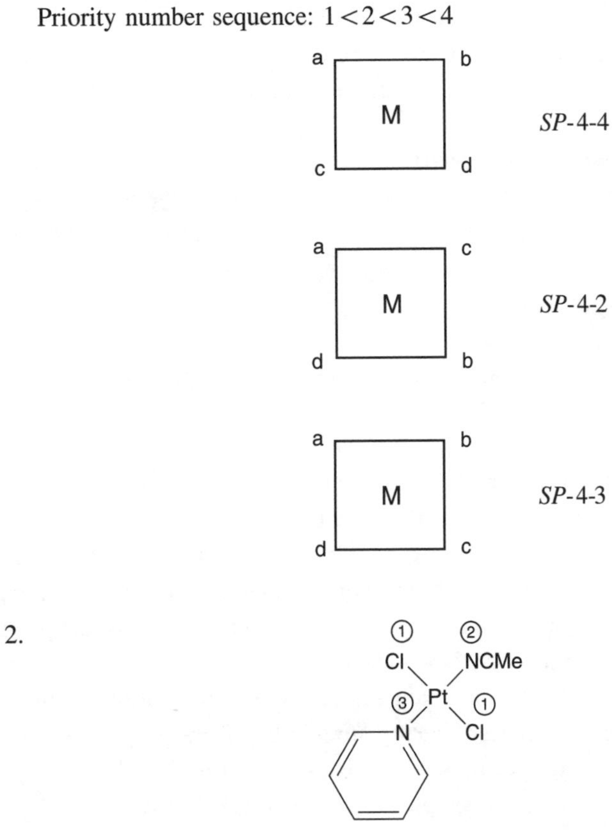

2.

(*SP*-4-1)-(acetonitrile)dichlorido(pyridine)platinum(II)

If there are two possibilities, as in Example 3, the configuration index is the priority number with the higher numerical value. Both the priority 2 ligand (acetonitrile) and the priority 3 ligand (pyridine) are *trans* to a priority 1 ligand (chloride). The higher numerical value (3) is chosen for the configuration index. This choice is sometimes referred to as having been made according to the principle of *trans* maximum difference, *i.e.* that the difference between the numerical values of the priority numbers of the ligands should be as large as possible.

Example:

3.

(*SP*-4-3)-(acetonitrile)dichlorido(pyridine)platinum(II)

IR-9.3.3.4 *Octahedral coordination systems (OC-6)*

The terms *cis* and *trans* are used commonly as prefixes to distinguish between stereoisomers in octahedral systems of the form $[Ma_2b_4]$, where M is the central atom, and 'a' and 'b' are different types of donor atom, and in certain similar systems. The 'a' donors occupy adjacent coordination sites in the *cis* isomer, and opposite coordination sites in the *trans* isomer (Example 1).

The terms *mer* (meridional) and *fac* (facial) are used commonly to distinguish between stereoisomers of complexes of the form $[Ma_3b_3]$. In the *mer* isomer (Example 2) the two groups of three similar donors each lie on a meridian of the coordination octahedron, in planes that also contain the central atom. In the *fac* isomer (Example 3) the two groups of three similar donors each occupy coordination sites on the corners of a face of the coordination octahedron.

The configuration index of an octahedral system follows the polyhedral symbol (*OC*-6) and consists of two digits.

The first digit is the priority number of the ligating atom *trans* to the ligating atom of priority number 1, *i.e.* the priority number of the ligating atom *trans* to the most preferred ligating atom. If there is more than one ligating atom of priority 1, then the first digit is the priority number of the *trans* ligand with the highest numerical value (remembering that a primed number will be of higher numerical value than the corresponding unprimed number).

These two ligating atoms, the priority 1 atom and the (lowest priority) atom *trans* to it, define the *reference axis* of the octahedron.

The second digit of the configuration index is the priority number of the ligating atom *trans* to the most preferred ligating atom in the plane that is perpendicular to the reference axis. If there is more than one such ligating atom in that plane, the priority number of the *trans* atom having the largest numerical value is selected.

Examples:

1.

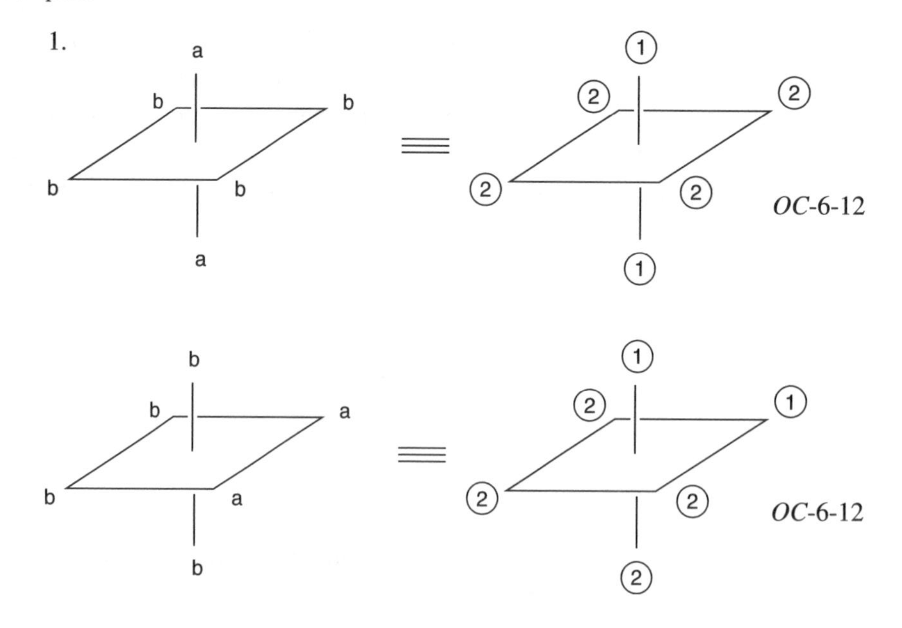

2.

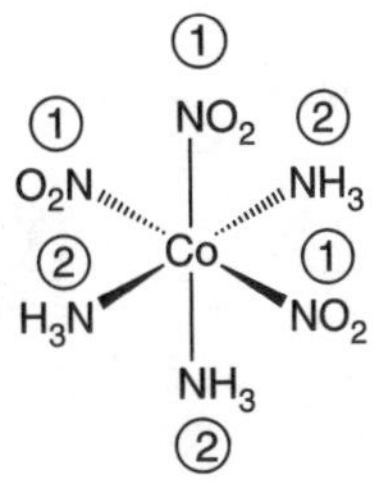

mer-[Co(NH$_3$)$_3$(NO$_2$)$_3$]
(*OC*-6-21)-triamminetrinitrito-κ^3*N*-cobalt(III)

3.

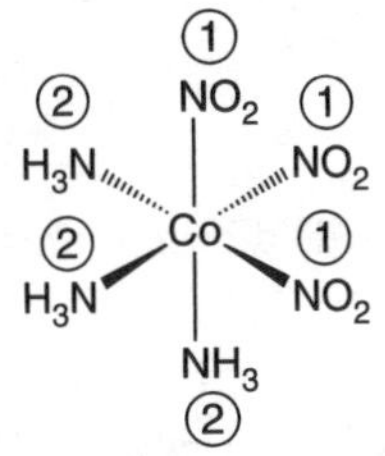

fac-[Co(NH$_3$)$_3$(NO$_2$)$_3$]
(*OC*-6-22)-triamminetrinitrito-κ^3*N*-cobalt(III)

4.

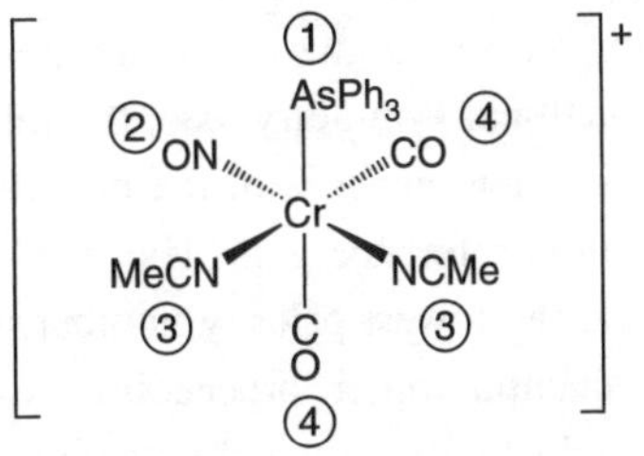

(*OC*-6-43)-bis(acetonitrile)dicarbonylnitrosyl(triphenylarsane)chromium(1+)

IR-9.3.3.5 *Square pyramidal coordination systems (SPY-4, SPY-5)*

The configuration index of an *SPY*-5 system consists of two digits. The first digit is the priority number of the ligating atom on the C_4 symmetry axis (the reference axis) of the idealized pyramid. The second digit is the priority number of the ligating atom *trans* to the ligating atom with the lowest priority number in the plane perpendicular to the C_4 symmetry axis. If there is more than one such atom in the perpendicular plane, then the second digit is chosen to have the highest numerical value.

The configuration index of an *SPY*-4 system is a single digit that is chosen in the same way as the second digit of *SPY*-5 systems. The configuration index of a four-coordinate square pyramidal system will therefore be the same as that for the square planar structure that would result from the ligands and the central atom being coplanar. The difference between the structures is described by the polyhedral symbol rather than by the configuration index.

Examples:

1.

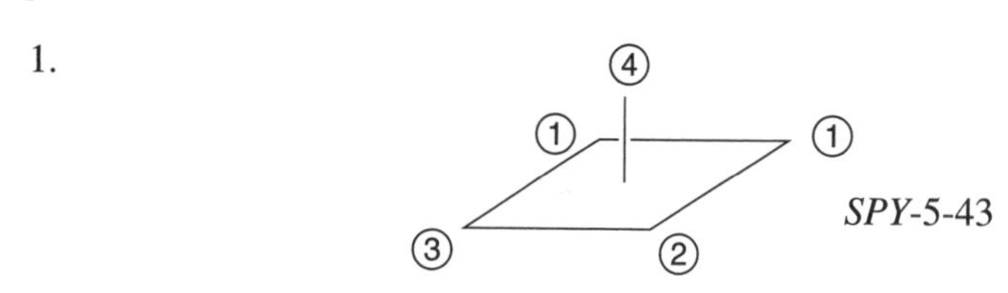

SPY-5-43

2.

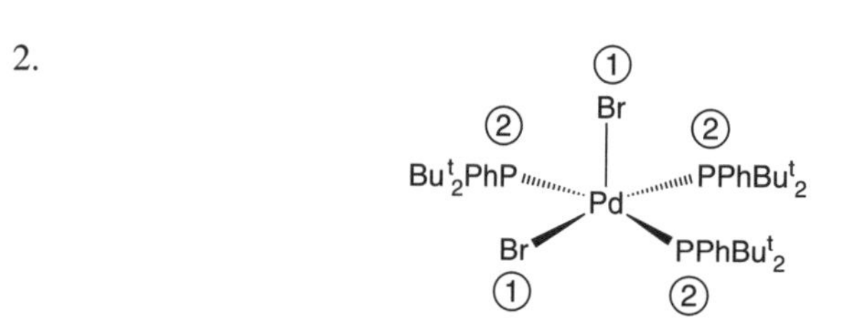

(*SPY*-5-12)-dibromidotris[di-*tert*-butyl(phenyl)phosphane]palladium

IR-9.3.3.6 *Bipyramidal coordination systems (TBPY-5, PBPY-7, HBPY-8 and HBPY-9)*

The configuration index for bipyramidal coordination systems follows the appropriate polyhedral symbol, and consists of two segments separated by a hyphen, except for the trigonal bipyramid where the second segment is not required and is therefore omitted. The first segment has two digits which are the priority numbers of the ligating atoms on the highest order rotational symmetry axis, the reference axis. The lower number is cited first.

The second segment consists of the priority numbers of the ligating atoms in the plane perpendicular to the reference axis. The first digit is the priority number for the preferred ligating atom, *i.e.* the lowest priority number in the plane. The remaining priority numbers are cited in sequential order proceeding around the projection of the structure either clockwise or anticlockwise, in whichever direction gives the lower numerical sequence. The lowest numerical sequence is that having the lower number at the first point of difference when the numbers are compared digit by digit from one end to the other.

Examples:

1. Trigonal bipyramid (*TBPY*-5)

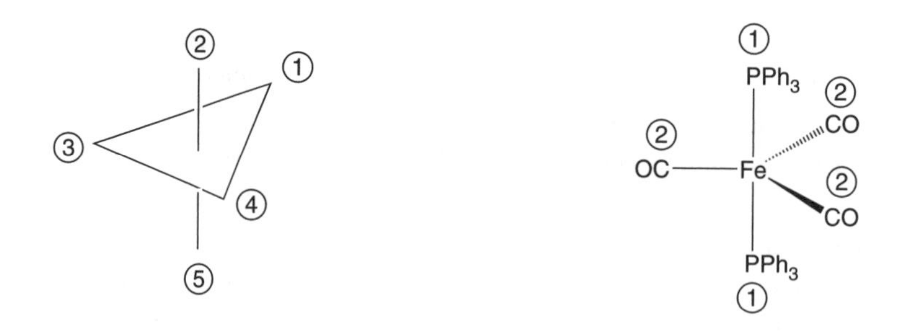

TBPY-5-25 (*TBPY*-5-11)-tricarbonylbis(triphenylphosphane)iron

2. Pentagonal bipyramid (*PBPY*-7)

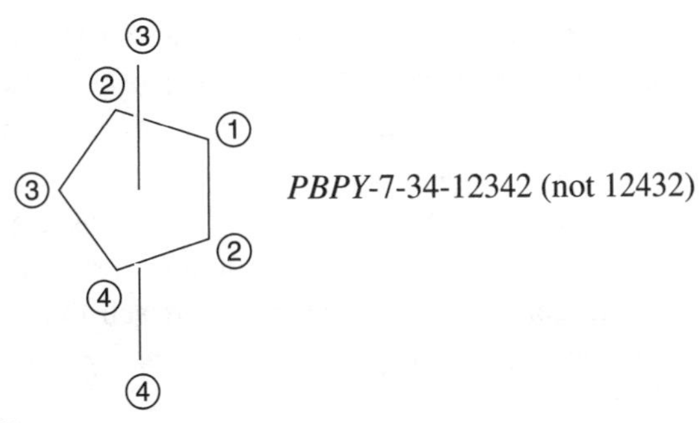

PBPY-7-34-12342 (not 12432)

IR-9.3.3.7 *T-shaped systems (TS-3)*

The configuration index for T-shaped systems follows the polyhedral symbol and consists of a single digit, the priority number of the ligating atom on the stem of the T (as opposed to the crosspiece of the T).

IR-9.3.3.8 *See-saw systems (SS-4)*

The configuration index for see-saw systems consists of two digits, the priority numbers of the two ligating atoms separated by the largest angle. The number of lower numerical value is cited first.

Examples:

1. 2.

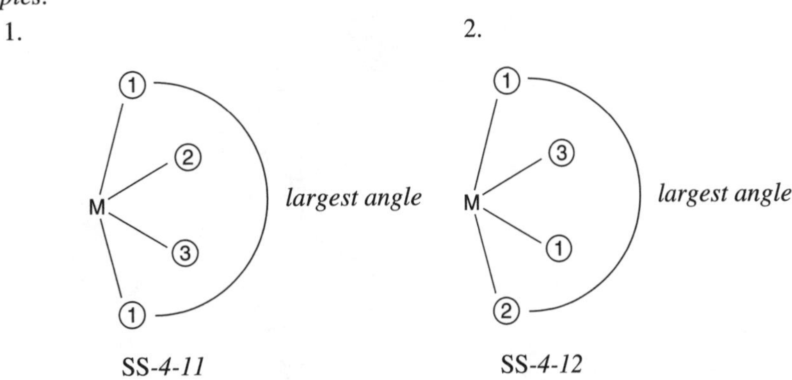

SS-4-*11* SS-4-*12*

IR-9.3.4 Describing absolute configuration – distinguishing between enantiomers

IR-9.3.4.1 *General*

There are two well-established, but fundamentally different, systems for distinguishing between two enantiomers (stereoisomers that are mirror images of one another). The first, based on the chemical constitution of the compound, involves the *R/S* convention used for describing tetrahedral centres and the closely related *C/A* convention used for other polyhedra. The *R/S* and *C/A* conventions use the priority sequence referred to in Section IR-9.3.3.2, and detailed in Section IR-9.3.5, where the ligating atoms are assigned a priority number based (usually) on their atomic number and their substituents.

The second is based on the geometry of the molecule and makes use of the skew-lines convention; it is usually applied only to octahedral complexes. The two enantiomers are identified by the symbols Δ and Λ in this system. The *C/A* nomenclature is not required for those chelate complexes where the skew-lines convention is completely unambiguous (see Sections IR-9.3.4.11 to 9.3.4.14).

IR-9.3.4.2 *The R/S convention for tetrahedral centres*

The convention used to describe the absolute configurations of tetrahedral centres was originally developed for carbon atom centres (see Ref. 13 and Section P-91 of Ref. 1) but can be used for any tetrahedral centre. There is no need to alter the rules in treating tetrahedral metal complexes.

The symbol *R* is assigned if the cyclic sequence of priority numbers, proceeding from highest priority, is clockwise when the viewer is looking down the vector from the tetrahedral centre to the least preferred substituent (the substituent having the priority number with the highest numerical value, *i.e.* 4). An anticlockwise cyclic sequence is assigned the symbol *S*.

This system is most often used in conjunction with configuration internally in ligands but can be applied equally well to tetrahedral metal centres. It has also been useful for pseudotetrahedral organometallic complexes when, for example, cyclopentadienyl ligands are treated as if they were monodentate ligands of high priority.

Example:

1.

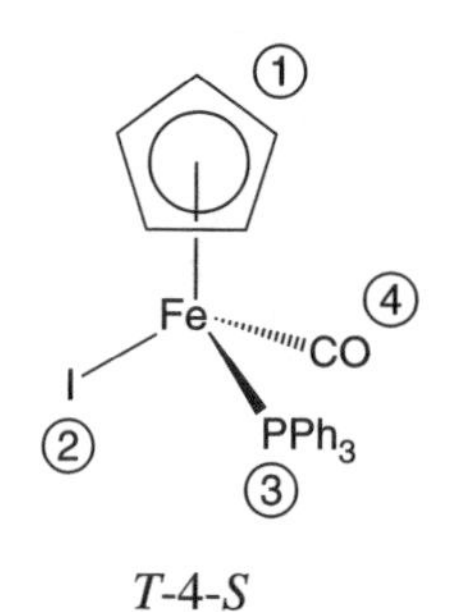

T-4-*S*

IR-9.3.4.3 *The R/S convention for trigonal pyramidal centres*

Molecules containing a trigonal pyramidal centre (*TPY*-3) may exist as a pair of stereoisomers. The configuration of this centre can be described in a similar way to that of a tetrahedral centre. This is achieved through notional placement of a 'phantom atom' of low priority in the coordination site that would create a tetrahedral centre from a trigonal pyramidal centre. The centre can then be identified as *R* or *S* by the methods described above.

The use of some bonding theories leads to the placement of a lone pair on a trigonal pyramidal centre. If this is done, the absolute configuration of the centre is also described by the *R/S* convention, in this case by placing the 'phantom atom' in the site that is occupied by the lone pair. Examples of this practice may be found in the description of absolute configurations for sulfoxides in which the alkyl substituents are different.

IR-9.3.4.4 *The C/A convention for other polyhedral centres*

The *R/S* convention makes use of priority numbers for the determination of chirality at tetrahedral centres, as detailed above. The same principles are readily extendable to geometries other than tetrahedral.[14] However, in order to avoid confusion, and to emphasize the unique aspects of the priority sequence systems as applied to coordination polyhedra, the symbols *R* and *S* are replaced by the symbols *C* and *A* when applied to other polyhedra.

The procedure for arriving at ligating atom priorities is detailed in Section IR-9.3.5. Once these priorities have been assigned, the reference axis (and direction) appropriate to the geometry is identified. The priority numbers of the ligating atoms coordinated in the plane perpendicular to the reference axis are then considered, viewing from the axial ligating atom of higher priority.

Beginning with the highest priority atom in the plane perpendicular to the reference axis, the clockwise and anticlockwise sequences of priority numbers are compared, and that with the lower number at the first point of difference is chosen. If the chosen sequence results from a clockwise reading of the priority numbers, then the structure is given the chirality symbol *C*, otherwise it is given the symbol *A*.

IR-9.3.4.5 *The C/A convention for trigonal bipyramidal centres*

The procedure is similar to that used for tetrahedral systems in the *R/S* convention, but it is modified because of the presence of a unique reference axis (running through the two axial donor atoms and the central atom).

The structure is oriented so that the viewer looks down the reference axis, with the more preferred donor atom (having a priority number with lower numerical value) closer to the viewer. Accordingly, the axial donor atom with the lower priority lies beyond the central atom. Using this orientation, the priority sequence of the three ligating atoms in the trigonal plane is examined. If the sequence proceeds from the highest priority to the lowest priority in a clockwise fashion, the chirality symbol *C* is assigned. Conversely, if the sequence from highest to lowest priority (from lowest numerical index to highest numerical index) is anticlockwise, the symbol *A* is assigned.

Examples:

1.

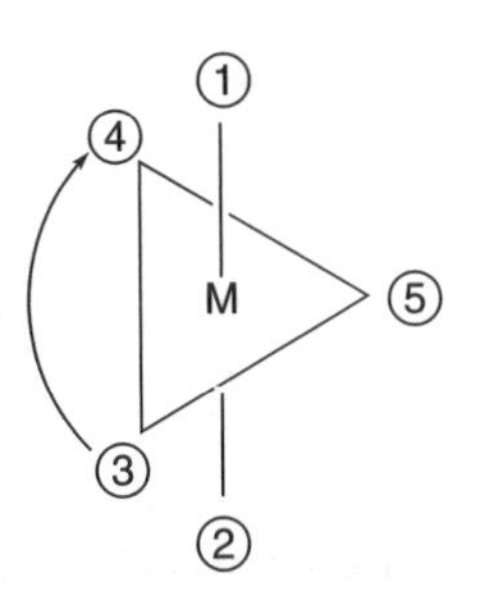

Chirality symbol = *C*

2.

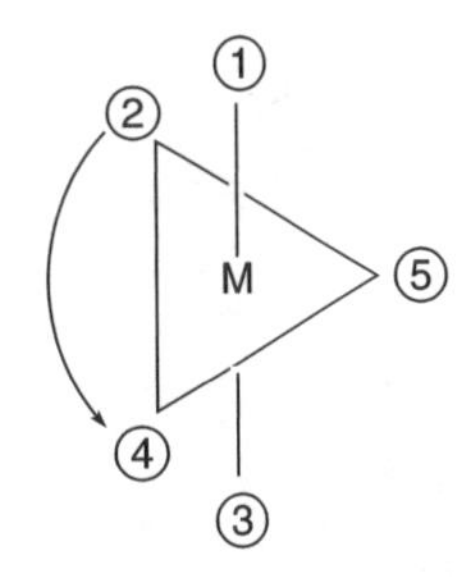

Chirality symbol = *A*

IR-9.3.4.6 *The C/A convention for square pyramidal centres*

A procedure similar to that described in Section IR-9.3.4.4 is used for square pyramidal structures. In the case of *SPY*-5 systems, the polyhedron is oriented so that the viewer looks along the formal C_4 axis, from the axial ligand toward the central atom. The priority numbers of the ligating atoms in the perpendicular plane are then considered, beginning with the highest priority atom (the one having the priority number of lowest numerical value). The clockwise and anticlockwise sequences of priority numbers are compared, and the structure is assigned the symbol *C* or *A* according to whether the clockwise (*C*) or anticlockwise (*A*) sequence is lower at the first point of difference.

The chirality of an *SPY*-4 system is defined in a similar way. In this case, the viewer looks along the formal C_4 axis in such a way that the ligands are further away than the central atom. The priority numbers are then used to assign the symbol *C* or *A*, as for the *SPY*-5 system.

Examples:

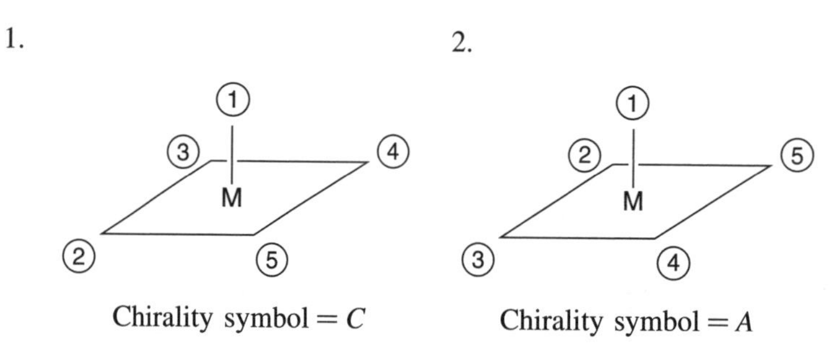

Chirality symbol = *C*　　Chirality symbol = *A*

IR-9.3.4.7 *The C/A convention for see-saw centres*

The absolute configurations of see-saw complexes can be described using the *C/A* system. The configuration index for see-saw systems consists of two digits, the priority numbers of the two ligands separated by the largest angle. The higher priority ligand of these two is identified and used as a point from which to view the two ligands not involved in the configuration index. If moving from the higher priority ligand to the lower (through the smaller angle) entails making a clockwise motion, the absolute configuration is assigned *C*. An anticlockwise direction results in the absolute configuration *A*.

Example:

1.

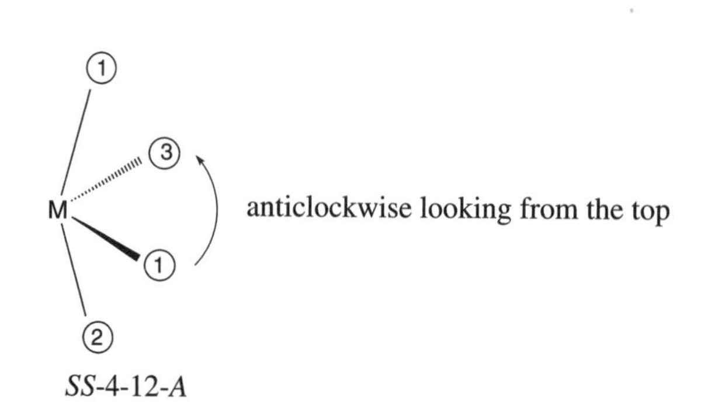

SS-4-12-*A*

IR-9.3.4.8 *The C/A convention for octahedral centres*

The absolute configurations of some octahedral complexes can be described using either the skew-line reference system (Section IR-9.3.4.11) or the *C/A* system. The first is used more commonly, but the *C/A* system is more general and may be used for most complexes. The skew-line reference system is only applicable to tris(bidentate), bis(bidentate) and closely related systems.

The reference axis for an octahedral centre is that axis containing the ligating atom of CIP priority 1 and the *trans* ligating atom of lowest possible priority (highest numerical value) (see Section IR-9.3.3.4). The atoms in the coordination plane perpendicular to the reference axis are viewed from the ligating atom having that highest priority (CIP priority 1) and the clockwise and anticlockwise sequences of priority numbers are compared. The structure is assigned the symbol *C* or *A*, according to whether the clockwise (*C*) or anticlockwise (*A*) sequence is lower at the first point of difference.

Examples:

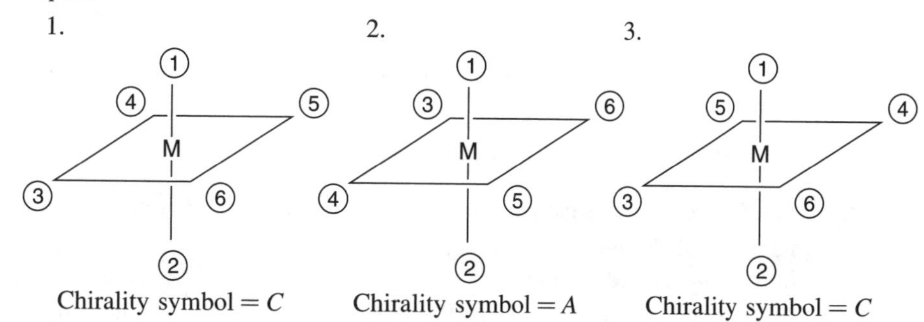

Chirality symbol = *C* Chirality symbol = *A* Chirality symbol = *C*

Example 4 shows the compound $[CoBr_2(en)(NH_3)_2]^+$ which has the polyhedral symbol *OC*-6 and the configuration index 32. The chirality symbol is *C*.

Example:

4.

Example 5 shows the complex $[Ru(CO)ClH(PMe_2Ph)_3]$ which has the descriptor *OC*-6-24-*A*. The chloride ligand has priority 1.

Example:

5.

The *C/A* assignment for polydentate ligands is illustrated by Example 6 which uses the priming convention developed in Section IR-9.3.5. Note that priority number 2 has higher priority than 2′.

Example:

6.

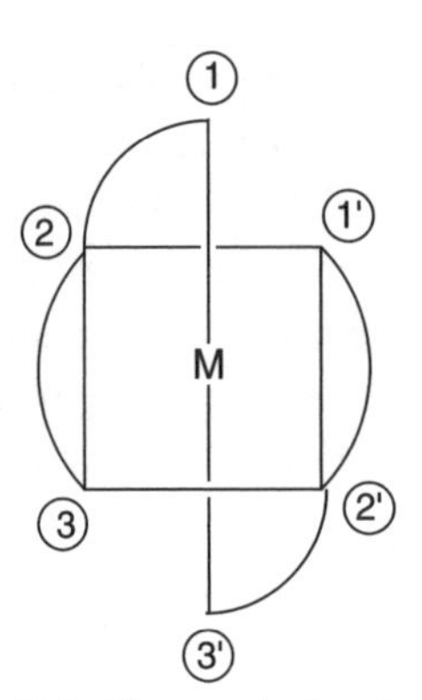

Chirality symbol = *A*

IR-9.3.4.9 *The C/A convention for trigonal prismatic centres*

For the trigonal prismatic system, the configuration index is derived from the CIP priority numbers of the ligating atoms opposite the triangular face containing the greater number of ligating atoms of highest CIP priority. The chirality symbol is assigned by viewing the trigonal prism from above the preferred triangular face and noting the direction of progression of the priority sequence for the less preferred triangular face.

Examples:

1.

Chirality symbol = *C*

2.

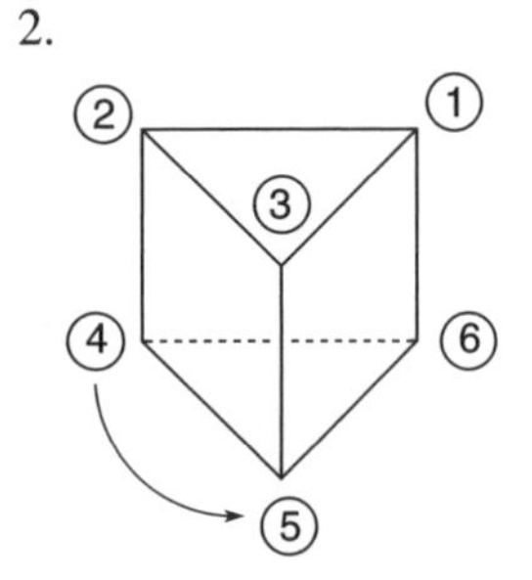

Chirality symbol = *A*

IR-9.3.4.10 *The C/A convention for other bipyramidal centres*

The procedure used for the trigonal bipyramid is appropriate for other bipyramidal structures. The structure is assigned the symbol *C* or *A*, according to whether the clockwise (*C*) or anticlockwise (*A*) sequence is lower at the first point of difference when the numbers are compared digit by digit from one end to the other (see Sections IR-9.3.4.5 and IR-9.3.4.6) and the molecule is viewed from the higher priority ligating atom on the reference axis.

Example:

1.

PBPY-7-12-11′1′33-*A*

IR-9.3.4.11 *The skew-lines convention*

Tris(bidentate) complexes constitute a general family for which a useful, unambiguous convention has been developed based on the orientation of skew lines which define a helix.

Examples 1 and 2 represent the *delta* (Δ) and *lambda* (Λ) forms of a complex such as $[Co(NH_2CH_2CH_2NH_2)_3]^{3+}$. The rules define the chiralities of two additional families of structures. These are the *cis*-bis(bidentate) octahedral structures and the conformations of certain chelate rings. It is possible to use the system described below for complexes of higher polydentate ligands, but additional rules are required.[15]

Examples:

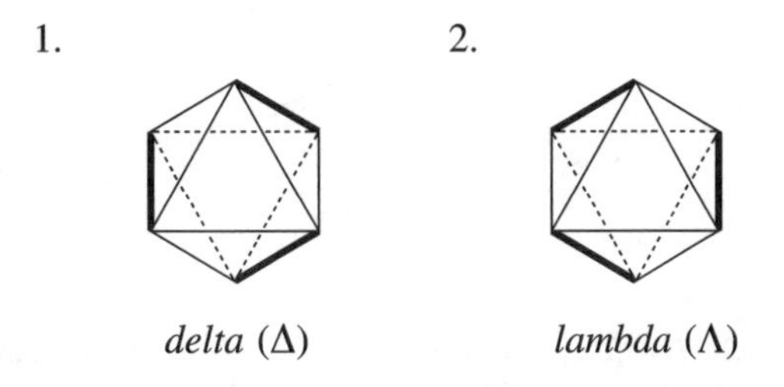

delta (Δ) *lambda* (Λ)

Two skew-lines which are not orthogonal possess the property of having one, and only one, normal in common. They define a helical system, as illustrated in Figures IR-9.1 and IR-9.2 (below). In Figure IR-9.1, one of the skew-lines, AA, determines the axis of a helix upon a cylinder whose radius is equal to the length of the common normal, NN, to the two skew-lines, AA and BB. The other of the skew-lines, BB, is a tangent to the helix at N and determines the pitch of the helix. In Figure IR-9.2, the two skew-lines AA and BB are seen in projection onto a plane orthogonal to their common normal.

Parts (a) of Figures IR-9.1 and IR-9.2 illustrate a right-handed helix to be associated with the Greek letter delta (Δ referring to configuration, δ to conformation). Parts (b) of Figures IR-9.1 and IR-9.2 illustrate a left-handed helix to be associated with the Greek letter lambda (Λ for configuration, λ for conformation). In view of the symmetry of the representation constituted by two skew-lines, the helix which the first line, say BB, determines around the second, AA, has the same chirality as that which AA determines around BB. As one of the lines is rotated about NN with respect to the other, inversion occurs when the lines are parallel or perpendicular (Figure IR-9.1).

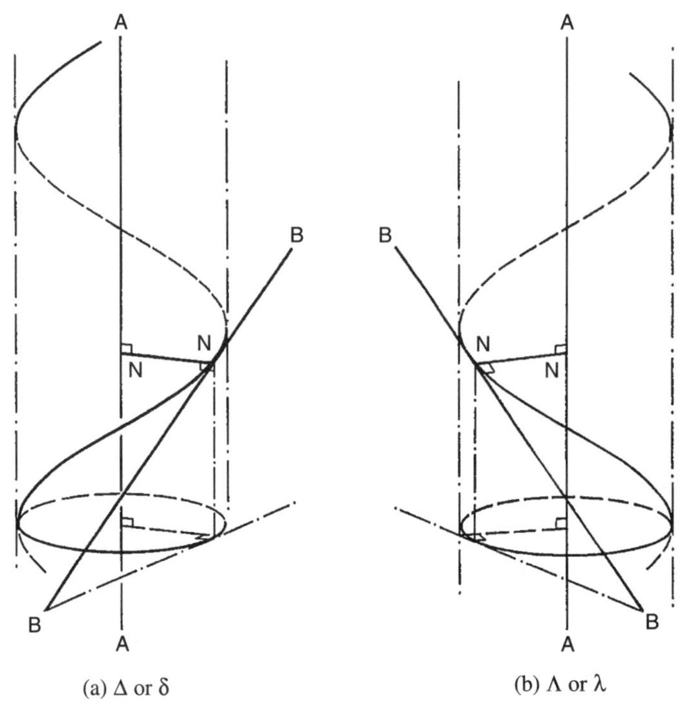

(a) Δ or δ (b) Λ or λ

Figure IR-9.1. *Two skew lines AA and BB which are not orthogonal define a helical system. In the Figure, AA is taken as the axis of a cylinder whose radius is determined by the common normal NN of the two skew-lines. The line BB is a tangent to the above cylinder at its crossing point with NN and defines a helix upon this cylinder. Cases (a) and (b) illustrate a right- and left-handed helix, respectively.*

(a) Δ or δ (b) Λ or λ

Figure IR-9.2. *The figure shows pairs of non-orthogonal skew-lines in projection upon a plane parallel to both lines. The full line BB is above the plane of the paper, the dotted line AA is below this plane. Case (a) corresponds to (a) of Figure IR-9.1 and defines a right-handed helix. Case (b) corresponds to (b) of Figure IR-9.1 and defines a left-handed helix.*

IR-9.3.4.12 *Application of the skew-lines convention to tris(bidentate) octahedral complexes*

Any two of the three chelate rings may be chosen to designate the configuration of tris(bidentate) coordination compounds. The donor atoms of each chelate ring define a line. Two such lines for a pair of chelate rings in the same complex define a helix, one line being the axis of the helix and the other a tangent to the helix at the normal common to the skew-lines. The tangent describes a right-handed (Δ) or a left-handed (Λ) helix with respect to the axis and thereby defines the chirality of that configuration.

IR-9.3.4.13 *Application of the skew-lines convention to bis(bidentate) octahedral complexes*

Figure IR-9.3(a) shows a common orientation of an octahedral tris(bidentate) structure projected onto a plane orthogonal to the three-fold axis of the structure. Figure IR-9.3(b) shows the same structure oriented to emphasize the skew-line relationship between a pair of chelate rings that can be used to define chirality. Figure IR-9.3(c) shows that the same convention can be used for the *cis*-bis(bidentate) complex. The two chelate rings define the two skew-lines that, in turn, define the helix and the chirality of the substance. The procedure is precisely the same as that described for the tris(bidentate) case, but only a single pair of chelate rings is available.

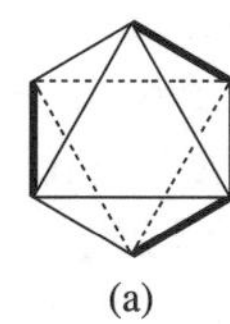

(a) (b) (c)

Figure IR-9.3. *Two orientations of a tris(bidentate) structure, (a) and (b), to show the chiral relationship between these two species and the bis(bidentate) structure (c).*

IR-9.3.4.14 *Application of the skew-lines convention to conformations of chelate rings*

In order to assign the chirality of a ring conformation, the line AA in Figure IR-9.2 is defined as that line joining the two ligating atoms of the chelate ring. The other line BB is that joining the two ring atoms which are neighbours to each of the ligating atoms. These two skew-lines define a helix in the usual way. The tangent describes a right-handed (δ) or a left-handed (λ) helix with respect to the axis and thereby defines the conformation in terms of the convention given in Figure IR-9.1. The relationship between the convention of Figure IR-9.2 and the usual representation of chelate ring conformation may be seen by comparing Figures IR-9.2 and IR-9.4.

Figure IR-9.4. δ-*Conformation of chelate rings: (a) five-membered; (b) six-membered.*

IR-9.3.5 Determining ligand priority

IR-9.3.5.1 *General*

The methods for differentiating between stereoisomers outlined earlier in this Chapter require the assignment of priorities for the ligand atoms attached to the central atom (*i.e.* the donor atoms). These priority numbers are then used in the configuration index, which describes the relative positions of the ligands, and in the assignment of the absolute configuration of the compound.

The following sections outline the method used to arrive at the priority numbers for a given set of donor atoms, and the ways that the basic rules have to be modified in order to describe adequately systems that include polydentate ligands. These modifications, which are collectively referred to as the priming convention, make use of primes on the priority numbers to indicate which donor atoms are grouped together within a particular polydentate ligand.

IR-9.3.5.2 *Priority numbers*

The procedure for assigning priorities in mononuclear coordination systems is based on the standard sequence rules developed for chiral carbon compounds by Cahn, Ingold and Prelog.[13] (See also Section P-91 of Ref. 1.) These CIP rules can be used quite generally for assigning priorities to groups attached to a central atom.

The essence of these rules, when applied to coordination compounds, is that the ligands attached to the central atom are compared to one another, beginning with the donor atom and then moving outwards in the structure. The comparison is made on the basis of atomic number and then, if required (*e.g.* when isotopes are being specified), atomic mass. Other properties may be used for subsequent comparisons, but the need for them is sufficiently rare that they need not be detailed here.

Once the ligands have been compared, the priority numbers are assigned as follows:

(i) identical ligands are assigned the same priority,
(ii) the ligand(s) with highest priority is (are) assigned the priority number 1; those with the next highest priority, 2; and so on.

Examples:

1.

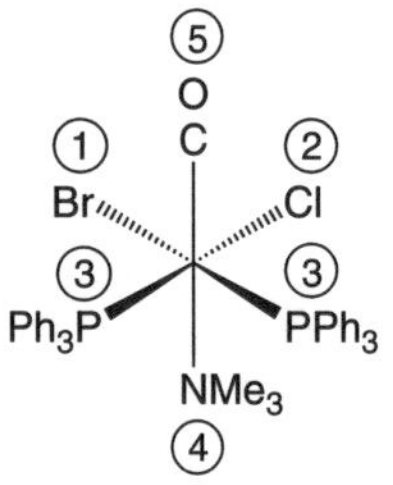

Priority sequence: Br > Cl > PPh_3, PPh_3 > NMe_3 > CO
Priority numbers sequence: 1 > 2 > 3, 3 > 4 > 5

2.

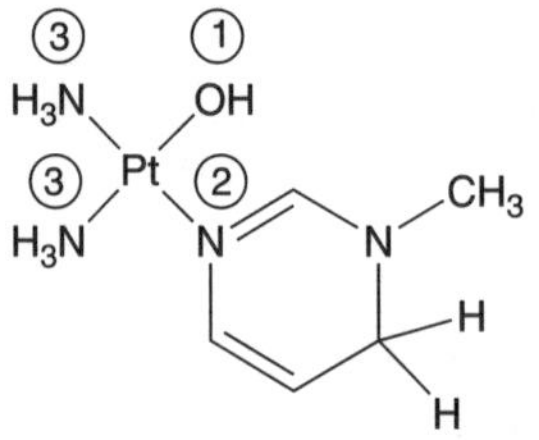

In Example 2, the heterocyclic ligand is given priority 2 since it has a lower atomic number donor atom than OH, and the substitution of the nitrogen donor ranks it above the ammine ligands.

3.

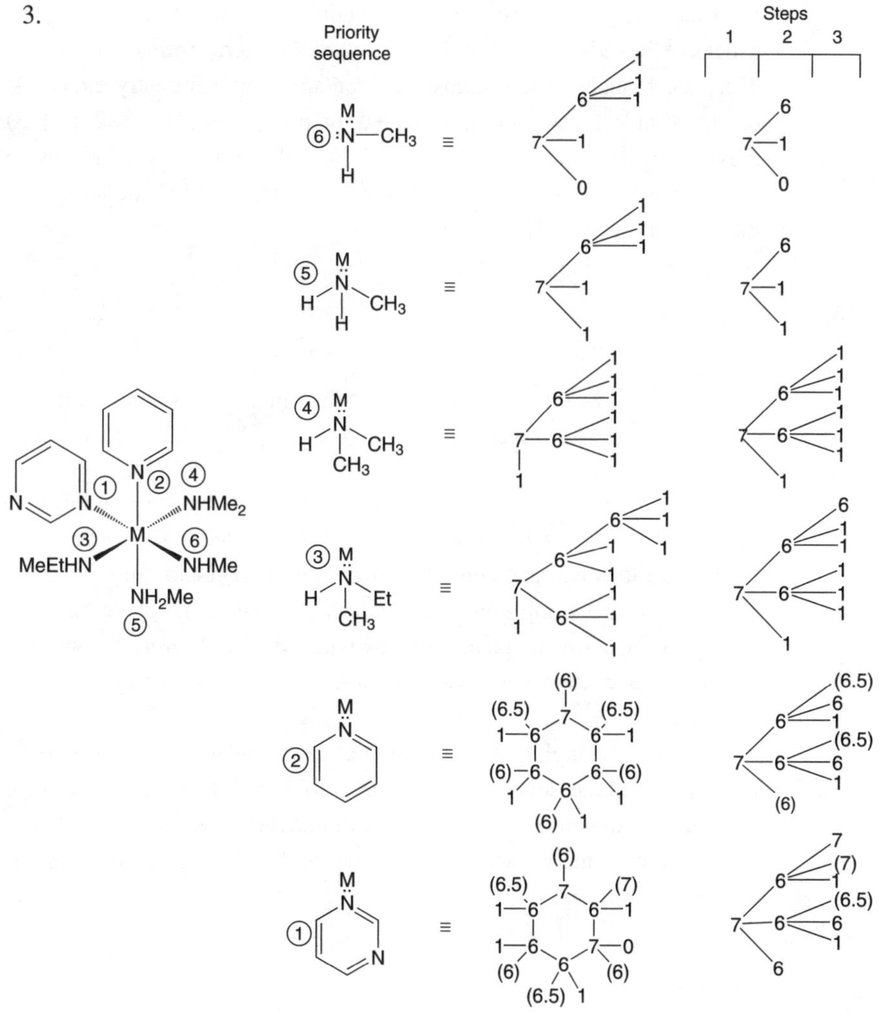

In Example 3, all the ligating atoms are nitrogen atoms. The key illustrates how proceeding along the branches of the ligand constituents allows priorities to be assigned. The numbers in columns 1, 2 and 3 on the right are the atomic numbers of the atoms in the structures, with those in brackets being used to take account of the presence of multiple bonds. The averaging techniques used in the case of resonance structures (last two ligands in the list) are given in the original paper.[13]

IR-9.3.5.3 *Priming convention*

The priming convention is required in order to avoid ambiguity when using the configuration index to describe the stereochemistry of systems that contain either more than one polydentate ligand of a particular kind, or a polydentate ligand that contains more than one coordinating fragment of a particular kind. This situation is found commonly with bis(tridentate) complexes, but also arises in more complicated cases. The need for this convention is best illustrated by example.

Bis(tridentate) complexes (*i.e.* octahedral complexes containing two identical linear tridentate ligands) may exist in three stereoisomeric forms, and there will be more if the tridentate ligands do not themselves contain some symmetry elements. The three isomers of the simplest case are represented below (Examples 1, 2 and 3), along with their polyhedral symbols (Section IR-9.3.2.1) and configuration indexes (Section IR-9.3.3.4). Complexes of *N*-(2-aminoethyl)ethane-1,2-diamine and iminodiacetate can be described by these diagrams.

N-(2-aminoethyl)ethane-1,2-diamine, or 2,2′-azanediylbis(ethan-1-amine)

iminodiacetate, or 2,2′-azanediyldiacetate

The need for the priming convention can be seen by considering what the configuration indexes of Examples 1 and 3 would be in the absence of the priming convention. The two ligands are identical and consist of two similar fragments fused together. If the primes are ignored, the two complexes have the same distributions of ligating atoms (four donors of priority 1 in a square plane, and two of priority 2 *trans* to one another). They would therefore have the same configuration index, even though they are clearly different complexes.

One way to highlight the difference between these two examples is to note that, in Example 1, all the donor atoms are *trans* to donors that are part of the other ligand. This is not true in Example 3. Using primes to indicate the groupings of donor atoms in particular ligands allows these two stereoisomers to be distinguished from one another by their configuration indexes.

Examples:

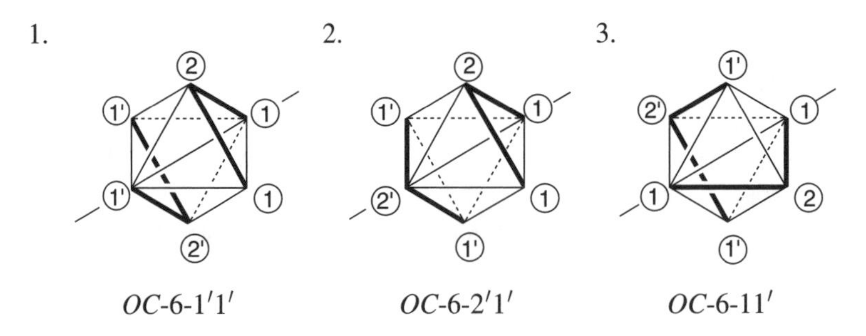

The priority numbers on one of the ligands are arbitrarily primed. The primed number is assigned lower priority than the corresponding unprimed number, but a higher priority than the next higher unprimed number. Thus 1′ has lower priority than 1, but higher than 2.

The technique also distinguishes between stereoisomers for complexes of higher polydentate ligands as indicated in Examples 4, 5 and 6 for linear tetradentate ligands such as *N*,*N*′-bis(2-aminoethyl)ethane-1,2-diamine. In this case, the donor atom priority numbers in half of the tetradentate ligand have been primed.

N,N'-bis(2-aminoethyl)ethane-1,2-diamine

Examples:

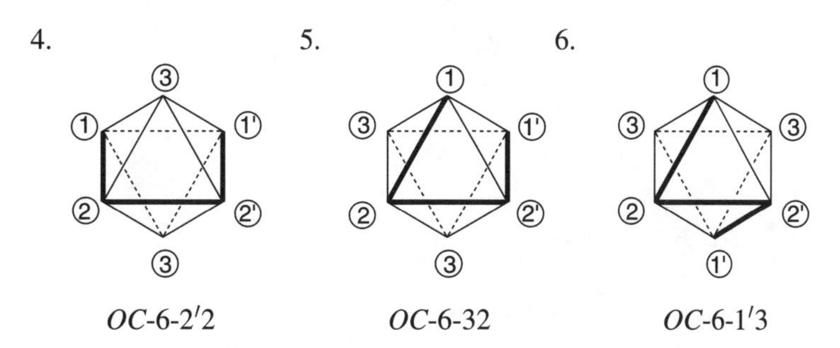

OC-6-2′2 *OC*-6-32 *OC*-6-1′3

Pentadentate and hexadentate ligands can be treated similarly. Examples 7 and 8 apply to stereoisomers of classic linear hexadentate ligands, whereas Examples 9 and 10 apply to ligands containing a branched structure.

Examples:

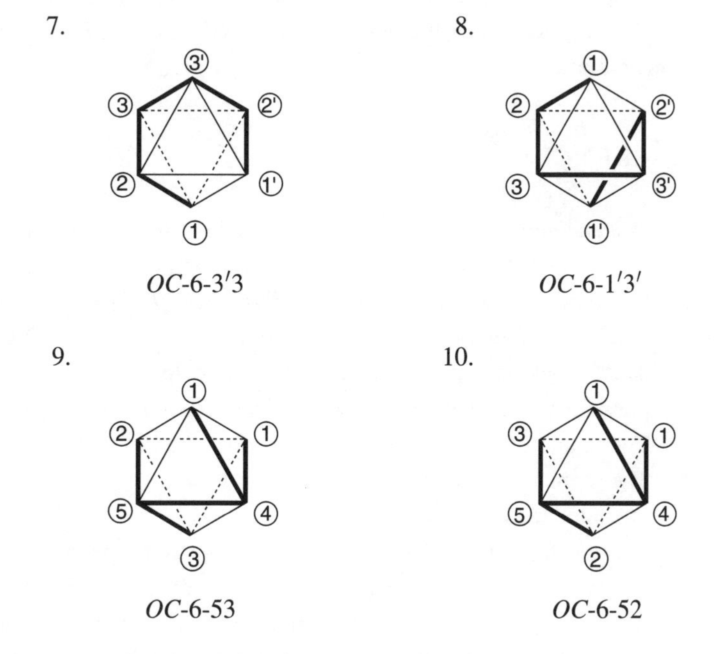

OC-6-3′3 *OC*-6-1′3′

OC-6-53 *OC*-6-52

Example 11 illustrates the use of priming for assigning absolute configuration in a non-octahedral structure. The chirality designation is determined by the system of assigning primes to ligands with the extra requirement that the symbol 1 on the top face is placed above the symbol 1″ on the bottom face. This produces the sequence shown and the chirality symbol *C* when viewing the prism from above the top face. The stereochemical descriptor is *TPR*-6-1″11′-*C*. The skew-lines method (Section I-9.3.4.11) could also be applied and would give in this case the descriptor Δ.

Example:

11.

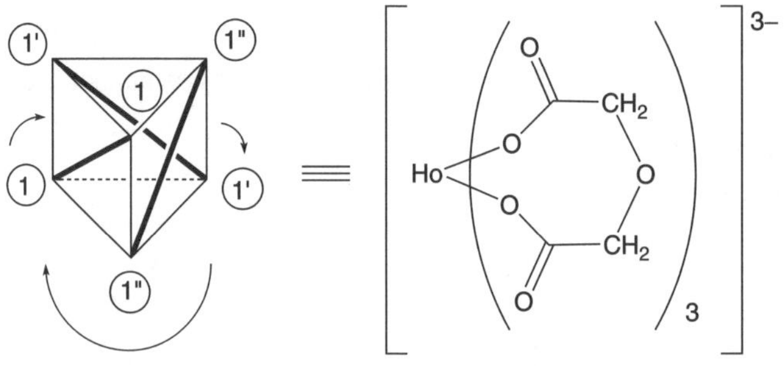

IR-9.4 FINAL REMARKS

This chapter has described the means by which coordination compounds can be named and formulated. These processes involve firstly identifying the central atom(s) and ligands (by name, formula or abbreviation, depending on the context), and secondly defining the nature of the attachment between the central atom(s) and the ligands. This latter step requires that the coordinating atoms in the ligand be identified (if there is any ambiguity), and that the spatial relationships between the ligands be described. The spatial relationships between the ligands are defined in terms of the coordination polyhedron (giving the polyhedral symbol) and the CIP priorities of the donor atoms (giving the configuration index and the absolute configuration).

IR-9.5 REFERENCES

1. *Nomenclature of Organic Chemistry, IUPAC Recommendations*, eds. W.H. Powell and H. Favre, Royal Society of Chemistry, in preparation.
2. In *Nomenclature of Inorganic Chemistry, IUPAC Recommendations 1990*, ed. G.J. Leigh, Blackwell Scientific Publications, Oxford, 1990, the term didentate was used rather than bidentate, for reasons of linguistic consistency. Reversion to the previously accepted term, bidentate, reflects common usage.
3. The names of the other hydrogen isotopes are discussed in Section IR-3.3.2.
4. The names of organic ligands should be assigned in accordance with IUPAC recommendations, see Ref. 1.
5. In order to simplify the rules and to resolve ambiguities that may arise when it is not clear whether a ligand is charged or not, the charge on a ligand is no longer considered in determining ligand order in the formulae of coordination compounds. (In *Nomenclature of Inorganic Chemistry, IUPAC Recommendations 1990*, ed. G.J. Leigh, Blackwell Scientific Publications, Oxford, 1990, anionic ligands were listed before neutral ligands.)
6. Chapter II-7 of *Nomenclature of Inorganic Chemistry II, IUPAC Recommendations 2000*, eds. J.A. McCleverty and N.G. Connelly, Royal Society of Chemistry, 2001.
7. *Nomenclature of Inorganic Chemistry II, IUPAC Recommendations 2000*, eds. J.A. McCleverty and N.G. Connelly, Royal Society of Chemistry, 2001.

8. J.B. Casey, W.J. Evans and W.H. Powell, *Inorg. Chem.*, **20,** 1333–1341 (1981).
9. A. von Zelewski, *Stereochemistry of Coordination Compounds*, John Wiley & Sons, Chichester, 1996.
10. A.M. Sargeson and G.H. Searle, *Inorg. Chem.*, **4,** 45–52 (1965); P.J. Garnett, D.W. Watts and J.I. Legg, *Inorg. Chem.*, **8,** 2534 (1969); P.F. Coleman, J.I. Legg and J. Steele, *Inorg. Chem.*, **9,** 937–944 (1970).
11. B. Bosnich, C.K. Poon and M.L. Tobe, *Inorg. Chem.*, **4,** 1102–1108 (1965); P.O. Whimp, M.F. Bailey and N.F. Curtis, *J. Chem. Soc.*, 1956–1963 (1970).
12. R.M. Hartshorn and D.A. House, *J. Chem. Soc., Dalton Trans.*, 2577–2588 (1998).
13. R.S. Cahn, C. Ingold and V. Prelog, *Angew. Chem., Int. Ed. Engl.*, **5**, 385–415 (1966); V. Prelog and G. Helmchen, *Angew. Chem., Int. Ed. Engl.*, **21,** 567–583 (1982).
14. M.F. Brown, B.R. Cook and T.E. Sloan, *Inorg. Chem.*, **7,** 1563–1568 (1978).
15. M. Brorson, T. Damhus and C.E. Schaeffer, *Inorg. Chem.*, **22,** 1569–1573 (1983).

IR-10 Organometallic Compounds

CONTENTS

IR-10.1 INTRODUCTION

The enormous growth in organometallic chemistry over the last fifty years and the discovery of new classes of compounds with unprecedented bonding modes has resulted in the need for additional nomenclature rules for organometallic compounds. This Chapter is therefore considerably expanded over Section I-10.9 of Ref. 1 and is largely based on the IUPAC recommendations published in 1999 for organometallic compounds of the transition elements.[2]

An organometallic compound is defined as any compound containing at least one bond between a metal atom and a carbon atom. The names of organometallic compounds should therefore accord with the rules of both organic and coordination chemistry nomenclature (even though these have tended to evolve separately).

The major part of this Chapter presents a system of nomenclature for transition element organometallic compounds, based on the additive nomenclature system introduced in Chapter IR-7 and applied to coordination compounds in Chapter IR-9 but incorporating, as far as possible, the rules for naming organic ligands.[3] Most importantly, further rules are formulated which unambiguously designate the special modes of bonding often found in organometallic compounds.

The later part of this Chapter briefly describes aspects of the naming of main group organometallic compounds, where the substitutive system of nomenclature (introduced in Chapter IR-6) is applied by substituting the appropriate parent hydrides of the elements of groups 13–16. The names of organometallic compounds of group 1 and 2 elements are, on the other hand, based on the additive nomenclature system.

It should be emphasized that the nomenclature described in this Chapter is confined to the precise description of the composition of a compound and the connectivity of atoms within a molecule or ion; it is often also important to specify the spatial relationships between the structural components of the molecule or ion (see Section IR-9.3). It is particularly true of organometallic chemistry that nomenclature should not attempt to convey details about the polarity of bonds, patterns of reactivity or methods of synthesis.

IR-10.2 NOMENCLATURE OF ORGANOMETALLIC COMPOUNDS OF THE TRANSITION ELEMENTS

IR-10.2.1 Concepts and conventions

The (additive) nomenclature of coordination complexes, the general definitions and rules of which are given in Sections IR-9.1 and IR-9.2, provides the basis for the system presented here for naming organometallic compounds of the transition elements. The general concepts of coordination chemistry can be applied to organometallic compounds but need to be expanded to deal with the additional modes of connectivity afforded by the interaction of metals with, for example, organic ligands containing unsaturated groupings, such as alkenes, alkynes and aromatic compounds. This section examines relevant concepts and conventions from coordination chemistry as they are applied to organometallic compounds, and indicates what new conventions need to be introduced in order to designate unambiguously the special bonding modes of organometallic compounds.

IR-10.2.1.1 *Coordination number*

The definition of coordination number as being equal to the number of σ-bonds between the ligands and the central atom (Section IR-9.1.2.6) also applies to ligands such as CN^-, CO, N_2 and PPh_3, where the bonding of a single ligating atom to a metal may involve a combination of σ- and π-components. The π-bond components are not considered in determining the coordination number, and so $[Ir(CO)Cl(PPh_3)_2]$, $[RhI_2(Me)(PPh_3)_2]$ and $[W(CO)_6]$ have coordination numbers of four, five and six, respectively.

However, this definition cannot be applied to the many organometallic compounds in which two or more adjacent atoms of a ligand interact with the central metal atom through what is often a combination of σ, π and δ bonding (the labels σ, π or δ referring to the symmetry of the orbital interactions between ligand and central atom).

For example, a ligand such as ethene, consisting of two ligating carbon atoms, nevertheless brings only one pair of electrons to the central atom. Likewise, ethyne, coordinating *via* both carbon atoms, can be thought to bring either one or two pairs of electrons to a single metal atom, depending on the type of coordination involved. Both ligands are normally regarded as monodentate. This changes when ethene or ethyne is considered to add oxidatively to a central metal atom; they are then considered to be bidentate chelating ligands which, on electron counting and dissection of the coordination entity to determine oxidation numbers, are assumed to take two pairs of electrons with them. This different view can be expressed by referring to compounds of such ligands as metallacyclopropanes or metallacyclopropenes rather than ethene or ethyne complexes.

IR-10.2.1.2 *Chelation*

The concept of chelation (Section IR-9.1.2.7) can again be applied strictly only to those organometallic complexes in which the donor atoms of a ligand are attached to the central metal atom through σ-bonds alone. Otherwise, ambiguities will result, as outlined above, even with a simple ligand such as ethene. Butadiene and benzene supply two and three pairs of electrons upon coordination and are therefore regarded as bi- and tridentate ligands, respectively. In stereochemistry, however, such ligands are often treated as if they were monodentate.

IR-10.2.1.3 *Specifying connectivity*

In the event of a ligand containing several different donor atoms, particularly when not all are used, the point or points of attachment to the metal are specified using the kappa (κ) convention (see Sections IR-9.2.4.1 and IR-9.2.4.2). In organometallic nomenclature the ligating carbon atoms are often sufficiently specified within the ligand name. However, use of the kappa notation becomes necessary to indicate the attachment of heteroatoms, and also to specify the particular points of attachment of a single ligand when bridging different metal centres in a polynuclear complex. The strength of the kappa convention is that its use completely avoids any ambiguities in describing the connectivities between a ligand and one or more metal centres Its use in organometallic nomenclature is discussed further in Section IR-10.2.3.3.

A complementary notation, the eta (η) convention, is used to specify the number ('hapticity') of *contiguous* ligating atoms that are involved in bonding to one or more metals. The need for this convention arises from the special nature of the bonding of unsaturated hydrocarbons to metals *via* their π-electrons, and it is used only when there are several contiguous atoms involved in the bond to the metal. The contiguous atoms of the π-coordinated ligand are often the same element, but they need not be, and they may also be atoms other than carbon. The eta convention is defined in Section IR-10.2.5.1, where its use is extensively illustrated. Even though all connectivity can be expressed by the kappa convention alone, the practice in organometallic nomenclature is that the eta convention should be used wherever there are contiguous ligating atoms. Complicated structures may require the use of both conventions (see Section IR-9.2.4.3).

Organic ligands with the ability to form more than one bond to a metal centre may be chelating (if bonding to a single metal), bridging (if bonding to more than one metal), or sometimes even both chelating and bridging. The bridging bonding mode is indicated by the Greek letter μ (mu) prefixing the ligand name (Section IR-9.2.5.2). This convention is further exemplified for organometallic compounds in Sections IR-10.2.3.1 and IR-10.2.3.4.

IR-10.2.1.4 *Oxidation number and net charge*

The concept of oxidation number or state (see also Sections IR-4.6.1, IR-5.4.2.2 and IR-9.1.2.8) is sometimes difficult to apply to organometallic compounds. This is especially true when it cannot be determined whether complexation by a ligand is better regarded as a Lewis-acid or Lewis-base association or as an oxidative addition. Thus, for nomenclature purposes it is only the net charge on the coordination entity that is important, and formal oxidation numbers will not be assigned to the central atoms of the organometallic complexes in the following sections. The reader is referred to standard textbooks on organometallic chemistry for discussion of the assignment of oxidation number in such compounds.

IR-10.2.2 Compounds with one metal–carbon single bond

In naming organometallic compounds the usual rules for naming ligands in coordination entities are applied if the ligand coordinates *via* an atom other than carbon (Section IR-9.2.2.3). Thus, the ligand $MeCOO^-$ is named acetato, Me_2As^- is named dimethylarsanido, and PPh_3 is named triphenylphosphane.

If an organic ligand coordinating *via* one carbon atom is regarded as an anion formed by the removal of one hydron from that atom, the ligand name is formed by replacing the ending 'ide' of the anion name by 'ido'.

Examples:

1. CH_3^- methanido
2. $CH_3CH_2^-$ ethanido
3. $(CH_2{=}CHCH_2)^-$ prop-2-en-1-ido
4. $C_6H_5^-$ benzenido
5. $(C_5H_5)^-$ cyclopentadienido

Although strictly speaking ambiguous, the anion name cyclopentadienide is acceptable as a short form of cyclopenta-2,4-dien-1-ide (and consequently the ligand name cyclopentadienido).

The compound $[TiCl_3Me]$ would be called trichlorido(methanido)titanium using the above type of ligand name.

The alternative for naming an organic ligand attached *via* a single carbon atom is to regard it as a substituent group, its name being derived from a parent hydride from which one hydrogen atom has been removed. This designation is somewhat arbitrary as such ligands in organometallic chemistry are generally treated as anions when deducing oxidation states, although the bonding in reality may be highly covalent. However, it has a long tradition in organic and organometallic chemistry, and its major advantage is that names used in common practice for organic groups can be applied unchanged.

There are two methods for constructing substituent group names from parent hydride names:

(a) The suffix 'yl' replaces the ending 'ane' of the parent hydride name. If the parent hydride is a chain, the atom with the free valence is understood to terminate the chain. In all cases that atom has the locant '1' (which is omitted from the name). This method is employed for saturated acyclic and monocyclic hydrocarbon substituent groups and for the mononuclear parent hydrides of silicon, germanium, tin and lead.

Examples:

6. CH_3-	methyl
7. CH_3CH_2-	ethyl
8. $C_6H_{11}-$	cyclohexyl
9. $CH_3CH_2CH_2CH_2-$	butyl
10. $CH_3CH_2CH_2C(Me)H-$	1-methylbutyl
11. Me_3Si-	trimethylsilyl

The compound $[TiCl_3Me]$ would be called trichlorido(methyl)titanium by this method.

(b) In a more general method, the suffix 'yl' is added to the name of the parent hydride with elision of the terminal 'e', if present. The atom with the free valence is given a number as low as is consistent with the established numbering of the parent hydride. The locant number, including '1', must always be cited. (See Section P-29 of Ref. 3 for a more complete discussion of substituent group names.)

Examples:

12. $CH_3CH_2CH_2C(Me)H-$	pentan-2-yl (*cf.* Example 10 above)
13. $CH_2{=}CHCH_2-$	prop-2-en-1-yl

In fused polycyclic hydrocarbons as well as in heterocyclic systems, special numbering schemes are adopted (see Section P-25 of Ref. 3).

Examples:

14. naphthalen-2-yl

15. H 1*H*-inden-1-yl

16. 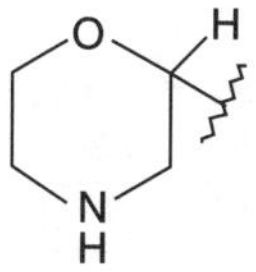 morpholin-2-yl

Table IR-10.1 gives the names used for ligands forming a single bond to a metal, and this is followed by examples illustrating the naming of compounds containing one metal–carbon single bond. In this Table (as well as in Tables IR-10.2 and IR-10.4) the organic ligands are listed both as anions and as neutral species. Acceptable alternative names are given in the final column.

Table IR-10.1 *Names for ligands forming a metal–carbon single bond (or bond to other group 14 element)*

Ligand formula	*Systematic name as anionic ligand*	*Systematic name as neutral ligand*	*Acceptable alternative name*
CH_3-	methanido	methyl	
CH_3CH_2-	ethanido	ethyl	
$CH_3CH_2CH_2-$	propan-1-ido	propyl	
$(CH_3)_2CH-$	propan-2-ido	propan-2-yl or 1-methylethyl	isopropyl
$CH_2{=}CHCH_2-$	prop-2-en-1-ido	prop-2-en-1-yl	allyl
$CH_3CH_2CH_2CH_2-$	butan-1-ido	butyl	
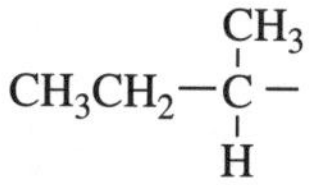	butan-2-ido	butan-2-yl or 1-methylpropyl	*sec*-butyl
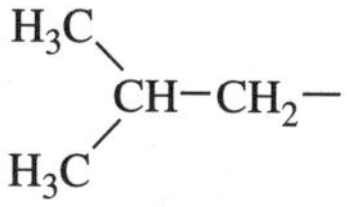	2-methylpropan-1-ido	2-methylpropyl	isobutyl
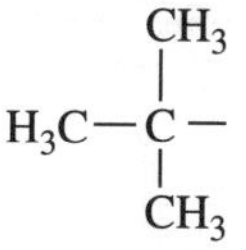	2-methylpropan-2-ido	2-methylpropan-2-yl or 1,1-dimethylethyl	*tert*-butyl
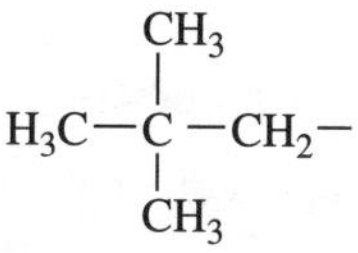	2,2-dimethylpropan-1-ido	2,2-dimethylpropyl	
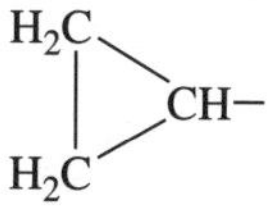	cyclopropanido	cyclopropyl	
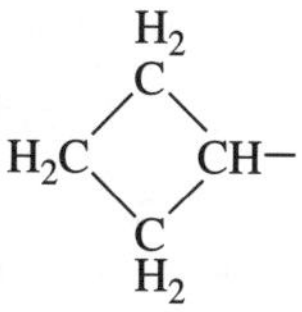	cyclobutanido	cyclobutyl	

Table IR-10.1 *Continued*

Ligand formula	*Systematic name as anionic ligand*	*Systematic name as neutral ligand*	*Acceptable alternative name*
C_5H_5-	cyclopenta-2,4-dien-1-ido	cyclopenta-2,4-dien-1-yl	cyclopentadienyl
C_6H_5-	benzenido	phenyl	
$C_6H_5CH_2-$	phenylmethanido	phenylmethyl	benzyl
$H_3C-C(=O)-$	1-oxoethan-1-ido	ethanoyl[a]	acetyl[a]
$C_2H_5-C(=O)-$	1-oxopropan-1-ido	propanoyl[a]	propionyl[a]
$C_3H_7-C(=O)-$	1-oxobutan-1-ido	butanoyl[a]	butyryl[a]
$C_6H_5-C(=O)-$	oxo(phenyl)methanido	benzenecarbonyl[a]	benzoyl[a]
$H_2C{=}CH-$	ethenido	ethenyl	vinyl
$HC{\equiv}C-$	ethynido	ethynyl	
H_3Si-	silanido	silyl	
H_3Ge-	germanido	germyl	
H_3Sn-	stannanido	stannyl	
H_3Pb-	plumbanido	plumbyl	

[a] These acyl names are preferred to 1-oxoethyl, *etc.*

Examples:

17. $[OsEt(NH_3)_5]Cl$ pentaammine(ethyl)osmium(1+) chloride
18. $Li[CuMe_2]$ lithium dimethylcuprate(1−)
19. CrR_4 (R = 1)

tetrakis(bicyclo[2.2.1]heptan-1-yl)chromium

20. $[Pt\{C(O)Me\}Me(PEt_3)_2]$ acetyl(methyl)bis(triethylphosphane)platinum

21.

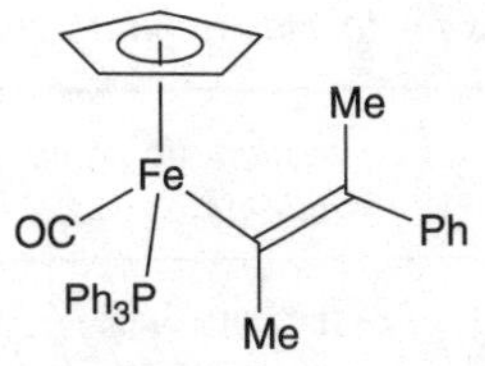

carbonyl(η^5-cyclopentadienyl)[(*E*)-3-phenylbut-2-en-2-yl](triphenylphosphane)iron

(The η term used here is explained in Section IR-10.2.5.1.)

22.

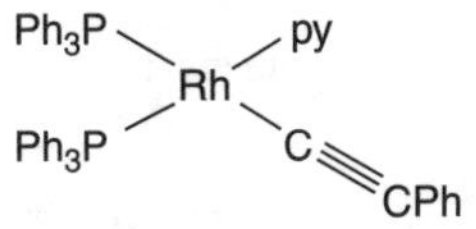

(phenylethynyl)(pyridine)bis(triphenylphosphane)rhodium

23.

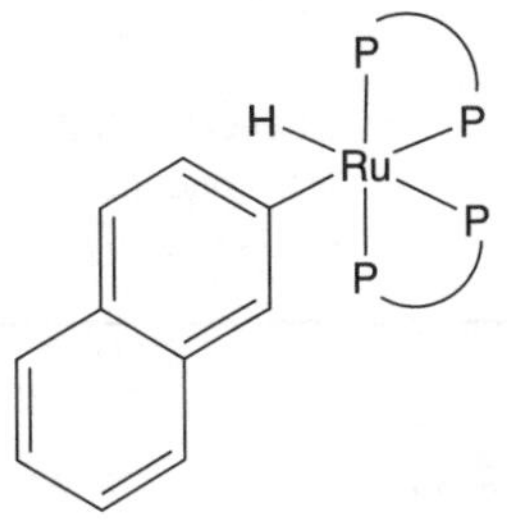

bis[ethane-1,2-diylbis(dimethylphosphane-κP)]hydrido(naphthalen-2-yl)ruthenium

P⌒P = $Me_2PCH_2CH_2PMe_2$ = ethane-1,2-diylbis(dimethylphosphane)

IR-10.2.3 Compounds with several metal–carbon single bonds from one ligand

When an organic ligand forms more than one metal–carbon single bond (to one or more metal atoms), the ligand name may be derived from the name of the parent hydrocarbon from which the appropriate number of hydrogen atoms have been removed. In the systematic substitutive name, the suffix 'diyl' or 'triyl' is attached to the name of the parent hydrocarbon if two or three hydrogen atoms, respectively, are replaced by one or more metal atoms. There is no removal of the terminal 'e'. The locant '1' is assigned so as to create the longest chain of carbon atoms, and the direction of numbering is chosen to give the lowest possible locants to side chains or substituents. The locant number(s) must always be cited, except for ligands derived from methane.

Alternatively, when considering these ligands as anions, the endings 'diido' and 'triido' should be used. This nomenclature also applies to hypervalent coordination modes, *e.g.* for bridging methyl groups. Typical ligands forming two or three metal–carbon single bonds are listed in Table IR-10.2.

Table IR-10.2 *Names for ligands forming several metal–carbon single bonds*

Ligand formula	*Systematic name as anionic ligand*	*Systematic name as neutral ligand*	*Acceptable alternative name*
$-CH_2-$	methanediido	methanediyl	methylene
$-CH_2CH_2-$	ethane-1,2-diido	ethane-1,2-diyl	ethylene
$-CH_2CH_2CH_2-$	propane-1,3-diido	propane-1,3-diyl	
$-CH_2CH_2CH_2CH_2-$	butane-1,4-diido	butane-1,4-diyl	
$HC\equiv$ (HC with three single bonds)	methanetriido	methanetriyl	
$CH_3CH<$	ethane-1,1-diido	ethane-1,1-diyl	
$CH_3C\equiv$ (CH₃C with three single bonds)	ethane-1,1,1-triido	ethane-1,1,1-triyl	
$-CH{=}CH-$	ethene-1,2-diido	ethene-1,2-diyl	
$H_2C{=}C<$	ethene-1,1-diido	ethene-1,1-diyl	
$-C\equiv C-$	ethyne-1,2-diido	ethyne-1,2-diyl	
$-C_6H_4-$	benzenediido (-1,2-diido, *etc.*)	benzenediyl (-1,2-diyl, *etc.*)	phenylene (1,2-, *etc.*)

IR-10.2.3.1 *The mu (μ) convention*

Organic ligands forming more than one metal–carbon bond can be either chelating, if coordinating to one metal atom, or bridging, if coordinating to two or more metal atoms. A bridging bonding mode is indicated by the Greek letter μ (Sections IR-9.2.5.2 and IR-10.2.3.4).

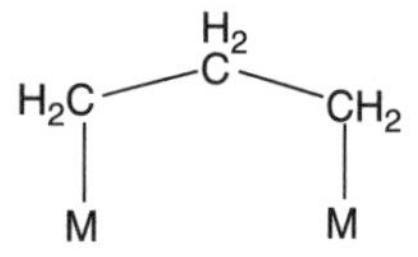

μ-propane-1,3-diyl (bridging)

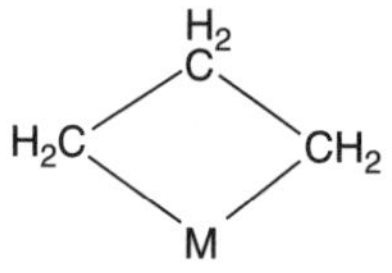

propane-1,3-diyl (chelating)

The number of metal atoms connected by a bridging ligand is indicated by a right subscript, μ_n, where $n \geq 2$, though the bridging index 2 is not normally indicated.

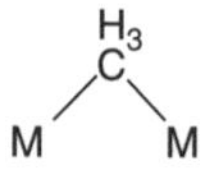

μ-methyl

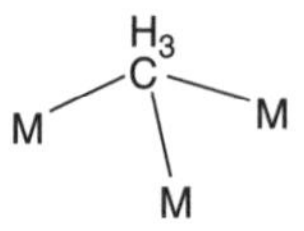

μ_3-methyl

The name methylene for CH_2 can only be used in connection with a bridging bonding mode (μ-methylene), whereas a CH_2 ligand bonding to one metal only has a metal–carbon double bond and should be named as methylidene (see Section IR-10.2.4).

μ-methylene methylidene

Likewise, the ligand HC will have at least three different bonding modes: bridging three metals (μ_3-methanetriyl), bridging two metals (μ-methanylylidene) and coordinating to one metal (methylidyne, see Section IR-10.2.4).

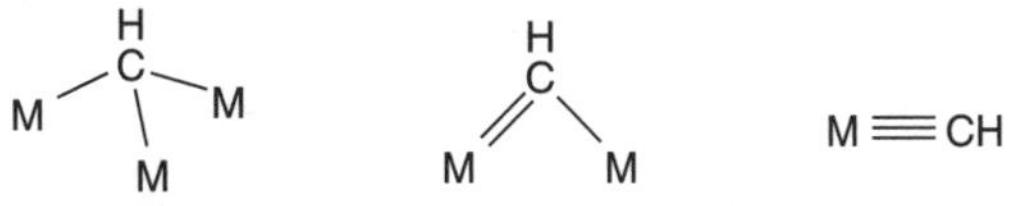

μ_3-methanetriyl μ-methanylylidene methylidyne

In a bridging mode the ligand CH_2CH_2 should be called μ-ethane-1,2-diyl, while the same ligand coordinating through both carbon atoms to a single metal centre should be called η^2-ethene (see Section IR-10.2.5).

μ-ethane-1,2-diyl η^2-ethene

A similar situation arises with CHCH which, when bridging with the carbon atoms individually bonded to each of two metals, should be called μ-ethene-1,2-diyl or, when the metal-carbon bonds are double, μ-ethanediylidene (see Section IR-10.2.4). The same ligand coordinating through both carbon atoms to both metal centres should be called μ-ethyne; when coordinated through both carbons to one metal it is named η^2-ethyne (see Section IR-10.2.5).

μ-ethene-1,2-diyl μ-ethanediylidene

μ-η^2:η^2-ethyne η^2-ethyne

IR-10.2.3.2 *Chelating ligands*

Where a chelating ligand is formed by removing two or more hydrogen atoms from a parent compound, the atoms with free valencies, understood to form the bonds to the central atoms, are indicated by using the appropriate ligand name (such as propane-1,3-diyl), *cf.* Section IR-10.2.3. This is demonstrated in Examples 1–3 below. Note that an alternative nomenclature for such metallacycles is currently being developed.

Examples:

1.

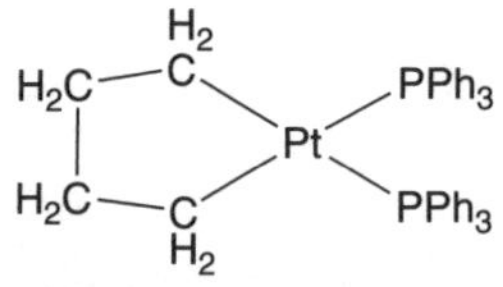

(butane-1,4-diyl)bis(triphenylphosphane)platinum

2.

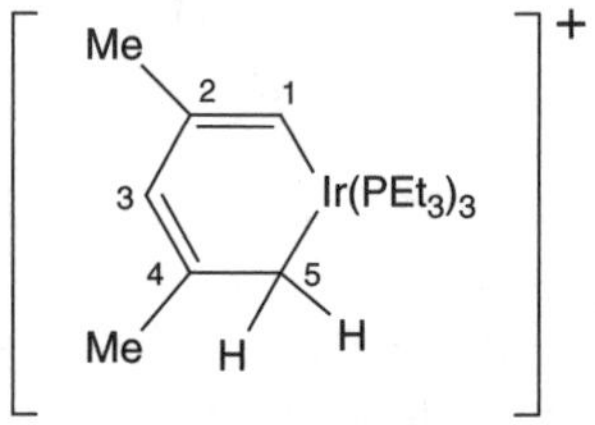

(2,4-dimethylpenta-1,3-diene-1,5-diyl)tris(triethylphosphane)iridium(1+)

3.

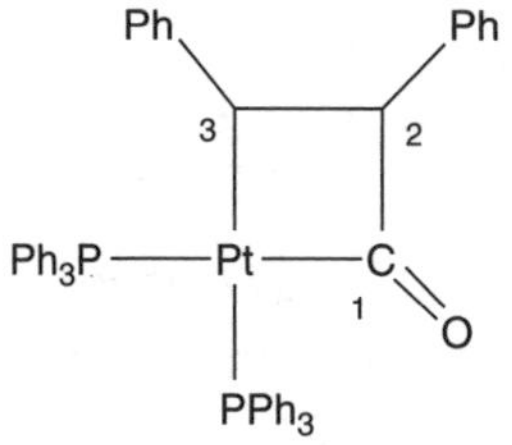

(1-oxo-2,3-diphenylpropane-1,3-diyl)bis(triphenylphosphane)platinum

IR-10.2.3.3 *The kappa (κ) convention*

Chelate rings that contain a coordinate (dative) bond from a heteroatom in addition to a carbon attachment should be named using the κ convention. In this convention (see Section IR-9.2.4.2) the coordinating atoms of a polydentate ligand bonding to a metal centre are indicated by the Greek letter kappa, κ, preceding the italicized element symbol of each ligating atom. A right superscript numeral may be added to the symbol κ to indicate the number of identical bonds from a type of ligating atom to the central atom(s); non-equivalent ligating atoms should each be indicated by an italicized element symbol preceded by κ.

In simple cases one or more superscript primes on the element symbol may be used to differentiate between donor atoms of the same element. Otherwise a right superscript numeral corresponding to the conventional numbering of the atoms in the ligand is used to define unambiguously the identity of the ligating atom. These symbols are placed after that portion of the ligand name which represents the particular functionality, substituent group, ring or chain in which the ligating atom is found.

Often it is only necessary for the coordinating heteroatom to be specified using the κ convention, the ligating carbon atom being adequately specified by the appropriate substitutive suffix. For illustrative purposes only, an arrow is used in the examples that follow to indicate a coordinate bond in the chelate ring. In Example 1 the κC^1 specification is included for clarity but is not strictly necessary as the bonding from carbon atom number 1 is implied by the name 'phenyl'.

Examples:

1.

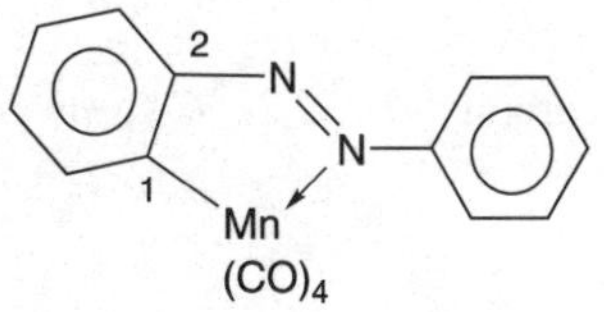

tetracarbonyl[2-(2-phenyldiazen-1-yl-κN^2)phenyl-κC^1]manganese

2.

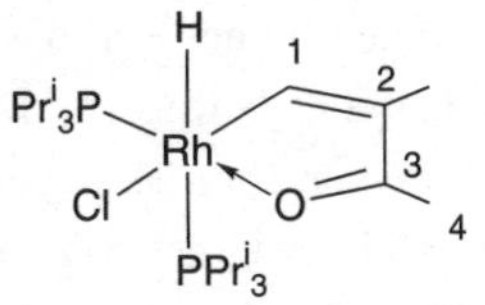

chloridohydrido(2-methyl-3-oxo-κ*O*-but-1-en-1-yl)bis(triisopropylphosphane)rhodium

IR-10.2.3.4 *Bridging ligands*

A bridging ligand is indicated by the Greek letter μ (mu) prefixing the ligand name (see Sections IR-9.2.5.2 and IR-10.2.3.1). Bridging ligands are listed in alphabetical order along with the other ligands, but in names a bridging ligand is cited before a corresponding non-bridging ligand, and multiple bridging is listed in decreasing order of complexity, *e.g.* μ_3 bridging before μ_2 bridging.

Example:

1.

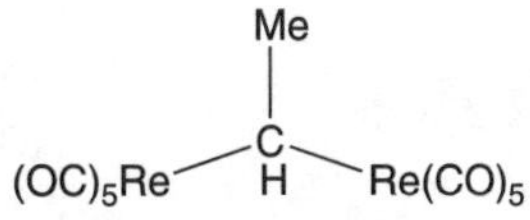

(μ-ethane-1,1-diyl)bis(pentacarbonylrhenium)

The metal centres in heterodinuclear coordination entities are numbered and listed according to the element sequence given in Table VI*, the central atom arrived at last when traversing this table being numbered '1' and listed in the name first (see Section IR-9.2.5).

The numerical locants of the central atoms are used in conjunction with the κ notation to indicate the distribution of the ligating atoms. Such locants are placed before the κ symbol which, as before, may be followed by a right superscript numeral to denote the number of equivalent bonds to the central atom specified by the locant (see Section IR-9.2.5.5). Thus, decacarbonyl-1κ^5*C*,2κ^5*C* indicates that the carbon atoms of five carbonyl ligands are bonded to central atom number 1 and another five to central atom number 2. In the names of bridging ligands, the κ terms indicating the bonding to each of the central atoms are separated by a colon, *e.g.* μ-propane-1,2-diyl-1κC^1:2κC^2.

* Tables numbered with a Roman numeral are collected together at the end of this book.

Example:

2.

decacarbonyl-1κ^5*C*,2κ^5*C*-(μ-propane-1,2-diyl-1κ*C*^{1}:2κ*C*^{2})rheniummanganese

IR-10.2.3.5 *Metal–metal bonding*

Metal–metal bonding is indicated by the italicized element symbols of the appropriate metal atoms, separated by an 'em' dash and enclosed in parentheses, placed after the list of central atom names and before the ionic charge. The element symbols are placed in the same order as the central atoms appear in the name, *i.e.* with the element met last in the sequence of Table VI given first. The number of such metal–metal bonds is indicated by an arabic numeral placed before the first element symbol and separated from it by a space. For the purpose of nomenclature, no distinction is made between different metal–metal bond orders.

Examples:

1.

(μ-ethane-1,2-diyl)bis(tetracarbonylosmium)(*Os*—*Os*)

2.

(μ$_3$-ethane-1,1,1-triyl)-*triangulo*-tris(tricarbonylcobalt)(3 *Co*—*Co*)

3.

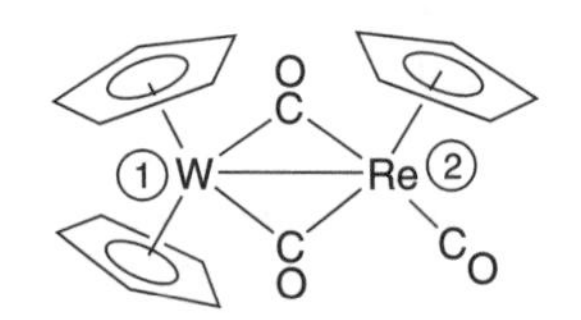

di-μ-carbonyl-carbonyl-2κ*C*-bis(1η^5-cyclopentadienyl)(2η^5-cyclopentadienyl)tungstenrhenium(*W*—*Re*)

The η terms involved here are explained in Section IR-10.2.5.1. For a more detailed discussion of dinuclear compounds and larger polynuclear clusters, with further examples, see Section IR-9.2.5.

IR-10.2.4 Compounds with metal–carbon multiple bonds

Ligands regarded as forming metal–carbon double or triple bonds may also be given substituent prefix names derived from the parent hydrides, the ligand names ending with 'ylidene' for a double bond and with 'ylidyne' for a triple bond. These suffixes are used according to two methods (see Section P-29 of Ref. 3).

(a) The suffix 'ylidene' or 'ylidyne' replaces the ending 'ane' of the parent hydride name. If the parent hydride is a chain, the atom with the free valencies is understood to terminate the chain. This atom has, in all cases, the locant '1' (which is omitted from the name). This method is used only for saturated acyclic and monocyclic hydrocarbon substituent groups and for the mononuclear parent hydrides of silicon, germanium, tin and lead. Note that the suffix 'ylene' should only be used in conjunction with μ to designate bridging $-CH_2-$ (methylene) or $-C_6H_4-$ (phenylene) (see Section IR-10.2.3.1).

(b) In a more general method, the suffix 'ylidene' or 'ylidyne' is added to the name of the parent hydride with elision of the teminal 'e', if present. The atom with the free valence is given a number as low as is consistent with the established numbering of the parent hydride. For ligand names with the suffix 'ylidene', this locant must always be cited, except if it is the only locant in the name and there is no ambiguity.

Example:

1. EtCH= propylidene [method (a)]
   $Me_2C=$ propan-2-ylidene [method (b)]

Note that in numbering a ligand that has several points of attachment, the longest chain of carbon atoms is chosen as the parent chain before assigning the lowest possible locant to the atom with the free valence. In a metallacycle, the direction of numbering is chosen so as to give the lowest possible locants to side chains or substituents. Once again, special numbering schemes apply to heterocyclic and polycyclic systems (see Sections P-25 and P-29 of Ref. 3).

If a ligand forms one or more metal–carbon single bonds as well as metal–carbon multiple bonds, the order of endings is 'yl', 'ylidene', 'ylidyne'. Method (b) should then be used to give the lowest possible set of locants for the free valencies. If a choice remains, lower numbers are selected for the 'yl' positions before the 'ylidene' positions and then for any side chains or substituents.

Example:

2. $CH_3-CH_2-\overset{|}{C}=$ propan-1-yl-1-ylidene

Typical ligands forming a metal–carbon double or triple bond are listed in Table IR-10.3, and this is followed by examples illustrating the naming of compounds containing one or more metal–carbon multiple bonds. The η term in Example 5 is explained in Section IR-10.2.5.1.

Note that the anion names given in Table IR-10.2 (methanediido, ethane-1,1-diido, *etc.*) may also be used for these ligands, but it is then not possible to communicate the concept of the carbon–metal bond as being a double or triple bond.

Table IR-10.3 *Names for ligands forming metal–carbon multiple bonds*

Ligand formula	*Systematic name*	*Acceptable alternative name*
$H_2C{=}$	methylidene	
$MeCH{=}$	ethylidene	
$H_2C{=}C{=}$	ethenylidene	vinylidene
$H_2C{=}HC{-}HC{=}$	prop-2-en-1-ylidene	allylidene
$H_2C{=}C{=}C{=}$	propa-1,2-dien-1-ylidene	allenylidene
$(H_3C)_2C{=}$	propan-2-ylidene	isopropylidene
$H_3C{-}C(CH_3)_2{-}CH{=}$	2,2-dimethylpropylidene	
$(CH_2)_2C{=}$ (cyclopropyl ring)	cyclopropylidene	
$(CH_2)_3C{=}$ (cyclobutyl ring)	cyclobutylidene	
$C_4H_4C{=}$ (cyclopentadienyl ring)	cyclopenta-2,4-dien-1-ylidene	
$PhHC{=}$	phenylmethylidene	benzylidene
$HC(=)(-)$	methanylylidene	
$HC{\equiv}$	methylidyne	
$MeC{\equiv}$	ethylidyne	
$EtC{\equiv}$	propylidyne	
$H_3C{-}C(CH_3)_2{-}C{\equiv}$	2,2-dimethylpropylidyne	
$PhC{\equiv}$	phenylmethylidyne	benzylidyne

Examples:

3.

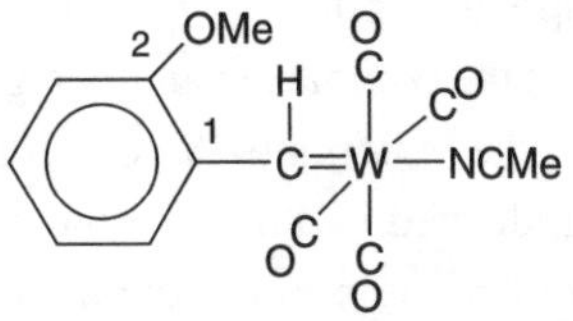

(acetonitrile)tetracarbonyl[(2-methoxyphenyl)methylidene]tungsten

4.

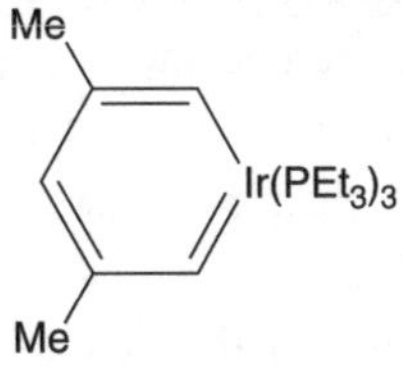

(2,4-dimethylpenta-1,3-dien-1-yl-5-ylidene)tris(triethylphosphane)iridium

5.

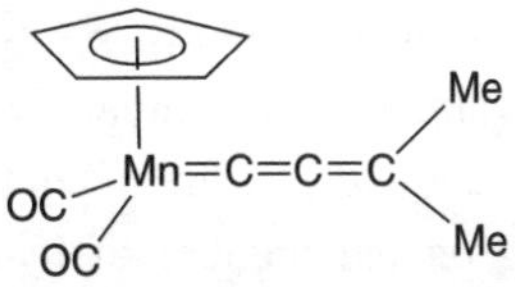

dicarbonyl(η^5-cyclopentadienyl)(3-methylbuta-1,2-dien-1-ylidene)manganese

6.

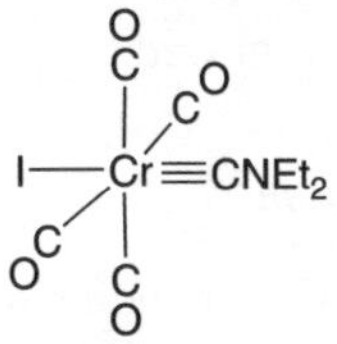

tetracarbonyl[(diethylamino)methylidyne]iodidochromium

7.

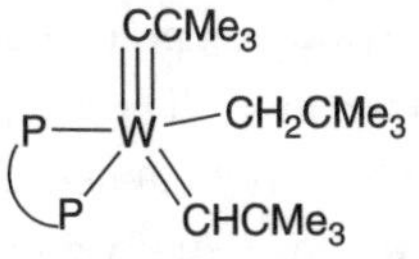

(2,2-dimethylpropyl)(2,2-dimethylpropylidene)(2,2-dimethylpropylidyne)⊂
[ethane-1,2-diylbis(dimethylphosphane-κ*P*)]tungsten*

P⁀P = $Me_2PCH_2CH_2PMe_2$ = ethane-1,2-diylbis(dimethylphosphane)

IR-10.2.5 Compounds with bonds to unsaturated molecules or groups

Since the discovery of Zeise's salt, $K[Pt(\eta^2\text{-}C_2H_4)Cl_3]$, the first organometallic complex of a transition element, and particularly since the first reported synthesis of ferrocene, $[Fe(\eta^5\text{-}C_5H_5)_2]$, the number and variety of organometallic compounds with unsaturated organic ligands has increased enormously.

Complexes containing ligands which coordinate to a central atom with at least two adjacent atoms in a 'side-on' fashion require a special nomenclature. These ligands normally contain

*The symbol '⊂' is used to divide the name, necessitated by the line break. In the absence of the line break this symbol is omitted. Note that all *hyphens* are true parts of the name.

groups that coordinate *via* the π-electrons of their multiple bonds, such as alkenes, alkynes and aromatic compounds, but they may also be carbon-free entities containing bonds between heteroelements; the complexes are then generally referred to as 'π-complexes'. However, the exact nature of the bonding (σ, π, δ) is often uncertain. The atoms bonded to the metal atom are therefore indicated in a manner independent of theoretical implications. Thus, the use of the prefixes σ and π is not recommended in nomenclature; these symbols refer to the symmetry of orbitals and their interactions, which are irrelevant for nomenclature purposes.

Ligands such as alkenes, alkynes, nitriles and diazenes, and others such as allyl (C_3H_5), butadiene (C_4H_6), cyclopentadienyl (C_5H_5), cycloheptatrienyl (C_7H_7) and cyclooctatetraene (C_8H_8), may be formally regarded as anionic, neutral (or sometimes cationic). The structures of, and bonding in, their complexes may also be complicated or ill-defined. Names for such ligands are therefore chosen that indicate stoichiometric composition and are derived in a similar way to those for the ligands discussed in preceding Sections.

Ligands considered as neutral molecules are given a name according to the rules of Ref. 3, including the special nomenclature and numbering applied to fused polycyclic or unsaturated heterocyclic ligands (see Section P-25 of Ref. 3).

Ligands regarded as substituent groups derived by removing hydrogen atoms from (substituted) parent hydrides are given the substituent names ending in 'yl', 'diyl', 'ylidene', *etc.*, depending on the number of hydrogen atoms removed, again following Ref. 3 (in particular Section P-29). Ligands regarded as anions obtained by removing hydrons from (substituted) parent hydrides are given the endings 'ido', 'diido', *etc.*, depending on the number of hydrons removed.

IR-10.2.5.1 *The eta (η) convention*

The special nature of the bonding of unsaturated hydrocarbons to metals *via* their π-electrons has led to the development of the 'hapto' nomenclature to designate unambiguously the unique bonding modes of the compounds so formed.[4] (See also Section IR-9.2.4.3.) The Greek symbol η (eta) provides a topological description by indicating the connectivity between the ligand and the central atom. The number of contiguous atoms in the ligand coordinated to the metal is indicated by a right superscript numeral, *e.g.* η^3 ('eta three' or 'trihapto'), η^4 ('eta four' or 'tetrahapto'), η^5 ('eta five' or 'pentahapto'), *etc.* The symbol η is added as a prefix to the ligand name, or to that portion of the ligand name most appropriate to indicate the connectivity, as in cyclopenta-2,4-dien-1-yl-η^2-ethene *versus* vinyl-η^5-cyclopentadienyl:

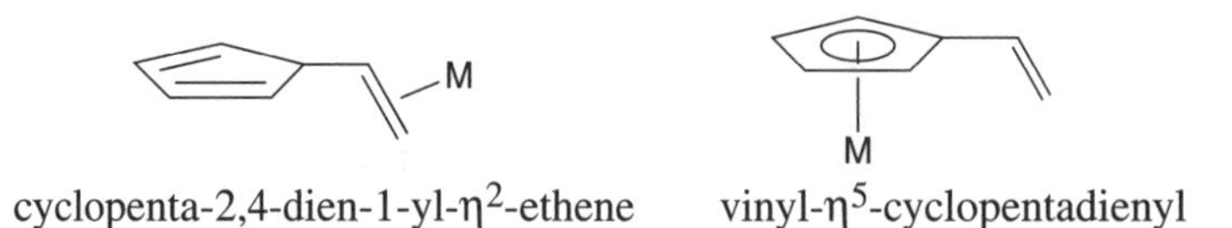

cyclopenta-2,4-dien-1-yl-η^2-ethene   vinyl-η^5-cyclopentadienyl

The ligand name η^5-cyclopentadienyl, although strictly speaking ambiguous, is acceptable as a short form of η^5-cyclopenta-2,4-dien-1-yl, due to common usage.

These ligand names are enclosed in parentheses in the full name of a complex. Note the importance of making rigorous use of enclosing marks, *etc.* to distinguish the above bonding modes from the other four cases below. Note also that when cyclopenta-2,4-dien-1-yl coordinates at the carbon with the free valence, a κ term is added for explicit indication of that bonding. In general, this is necessary with names of unsaturated ligands which may participate in several types of bonding (see Example 17 below, where the ligand name ends in 'yl', but the bonding is described using an η term placed elsewhere in the name, and Example 24, where the C^1 atoms in the cyclopentadienyl ligands are involved in binding to both central atoms).

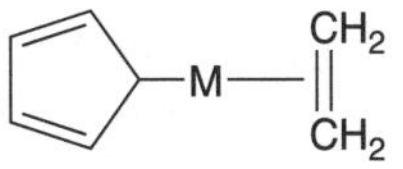

(cyclopenta-2,4-dien-1-yl-κC^1)(η^2-ethene) (η^5-cyclopentadienyl)(η^2-ethene)

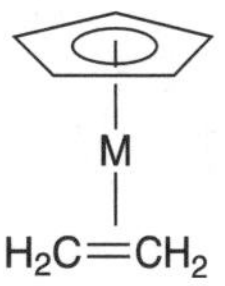

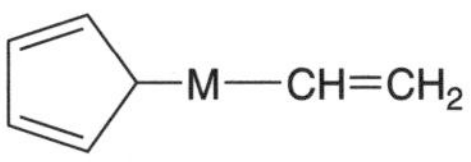

(cyclopenta-2,4-dien-1-yl-κC^1)(vinyl) (η^5-cyclopentadienyl)(vinyl)

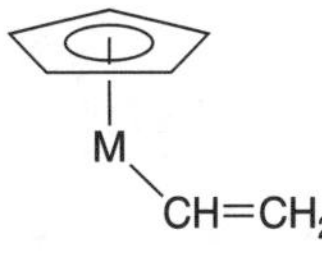

Complexes of unsaturated systems incorporating heteroatoms may be designated in the same manner if both the carbon atoms and adjacent heteroatoms are coordinated. Names for typical unsaturated molecules and groups acting as ligands are listed in Table IR-10.4, and this is followed by examples illustrating the naming of compounds containing such ligands. Note that when using the η prefixes, shorthand forms of anion and substituent group names are acceptable, *e.g.* η^5-cyclohexadienido instead of η^5-cyclohexa-2,4-dien-1-ido and η^5-cyclohexadienyl instead of η^5-cyclohexa-2,4-dien-1-yl.

Table IR-10.4 *Ligand names for unsaturated molecules and groups*

Ligand [a]	*Systematic name as anionic ligand*	*Systematic name as neutral ligand*	*Acceptable alternative name*
	η^3-propenido	η^3-propenyl	η^3-allyl
	η^3-(*Z*)-butenido	η^3-(*Z*)-butenyl	
	η^3-2-methylpropenido	η^3-2-methylpropenyl	η^3-2-methylallyl
	η^4-2-methylidenepropane-1,3-diido	η^4-2-methylidenepropane-1,3-diyl	
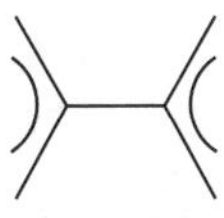	η^3,η^3-2,3-dimethylidenebutane-1,4-diido	η^3,η^3-2,3-dimethylidenebutane-1,4-diyl	η^3,η^3-2,2′-biallyl
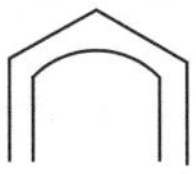	η^5-(*Z,Z*)-pentadienido	η^5-(*Z,Z*)-pentadienyl	
	η^5-cyclopentadienido	η^5-cyclopentadienyl	

Table IR-10.4 *Continued*

Ligand [a]	*Systematic name as anionic ligand*	*Systematic name as neutral ligand*	*Acceptable alternative name*
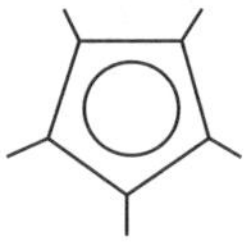	pentamethyl-η^5-cyclopentadienido	pentamethyl-η^5-cyclopentadienyl	
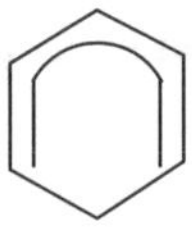	η^5-cyclohexadienido	η^5-cyclohexadienyl	
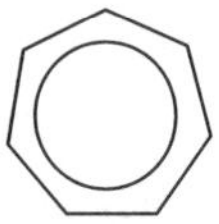	η^7-cycloheptatrienido	η^7-cycloheptatrienyl[b]	
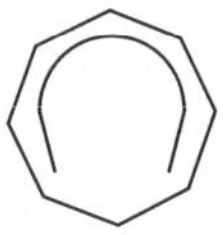	η^7-cyclooctatrienido	η^7-cyclooctatrienyl[c]	
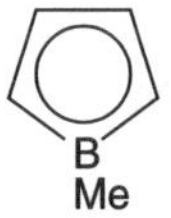 B Me		1-methyl-η^5-1*H*-borole	
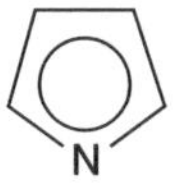 N	η^5-azacyclopentadienido	η^5-azacyclopentadienyl	η^5-1*H*-pyrrolyl
 P	η^5-phosphacyclopentadienido	η^5-phosphacyclopentadienyl	η^5-1*H*-phospholyl
 As	η^5-arsacyclopentadienido	η^5-arsacyclopentadienyl	η^5-1*H*-arsolyl
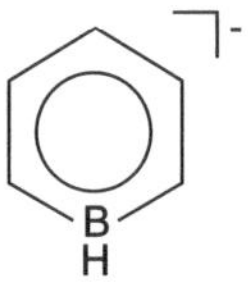 B H, -	η^6-borinin-1-uido		η^6-boranuidabenzene[d]
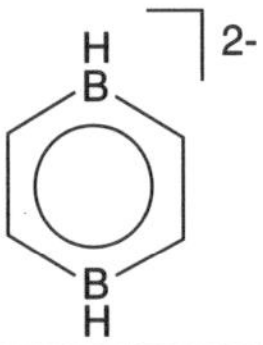 H B, B H, 2-	η^6-1,4-diborinine-1,4-diuido		η^6-1,4-diboranuidabenzene[e]

[a] The ligands are drawn as if complexed to a metal, *i.e.* these are depictions of bonded entities, not free ligands. The arcs used in these and later examples indicate delocalization (by analogy with the circle in benzene).
[b] The name η^7-tropyl has been used previously but is no longer acceptable.
[c] The name η^7-homotropyl has been used previously but is no longer acceptable.
[d] The name η^6-boratabenzene has been used previously but is no longer acceptable.
[e] The name η^6-1,4-diboratabenzene has been used previously but is no longer acceptable.

Examples:

1.

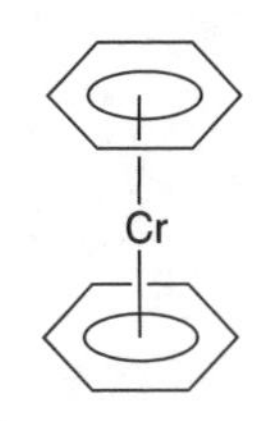

bis(η^6-benzene)chromium

2.

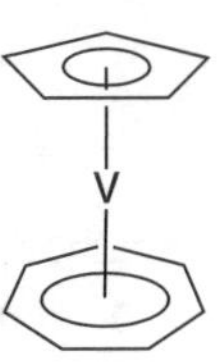

(η^7-cycloheptatrienyl)(η^5-cyclopentadienyl)vanadium

3.

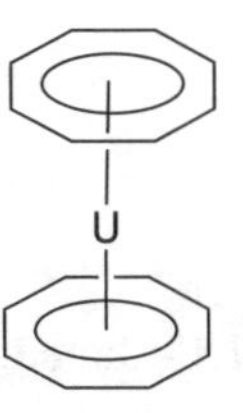

bis(η^8-cyclooctatetraene)uranium (*cf.* Section IR-10.2.6)

4.

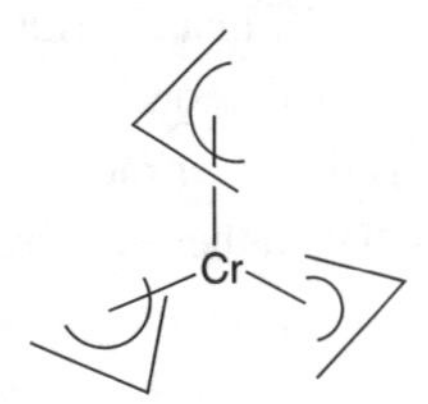

tris(η^3-allyl)chromium

5.

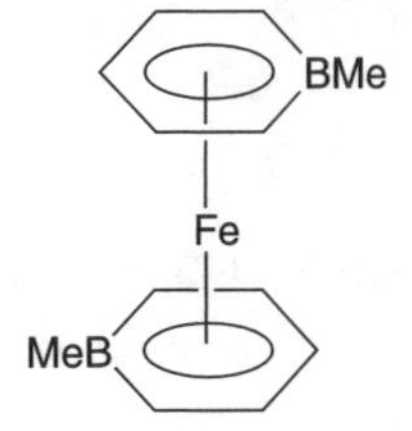

bis(η^6-1-methyl-1-boranuidabenzene)iron

6.

dicarbonyl(η^2-formaldehyde)bis(triphenylphosphane)osmium

7.

(η^2-carbon dioxide)bis(triethylphosphane)nickel

8.

tricarbonyl{*N*, *N*-dimethyl-1-[2-(diphenylphosphanyl)-
η^6-phenyl]ethane-1-amine}chromium

9.

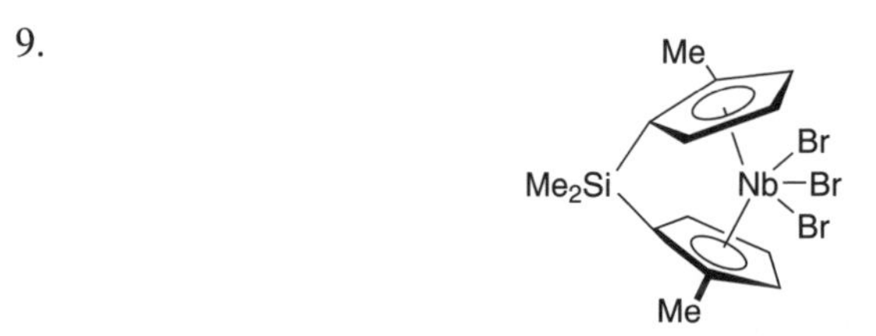

tribromido[1,1'-(dimethylsilanediyl)bis(2-methyl-η^5-cyclopentadienyl)]niobium

If not all unsaturated atoms of a ligand are involved in bonding, if a ligand can adopt several bonding modes, or if a ligand bridges several metal atoms, the locants of the ligating atoms appear in a numerical sequence before the symbol η, which is preceded by a hyphen. Extended coordination over more than two contiguous carbon atoms should be indicated by, for example, (1–4-η) rather than by (1,2,3,4-η). The locants and the symbol η are enclosed in parentheses. No superscript on the symbol η is then necessary.

Examples:

10.

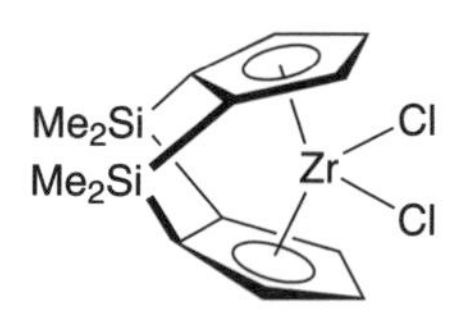

dichlorido[(1–3,3a,8a:4a,5–7,7a-η)-4,4,8,8-tetramethyl-1,4,5,8-tetrahydro-
4,8-disila-*s*-indacene-1,5-diyl]zirconium

11.

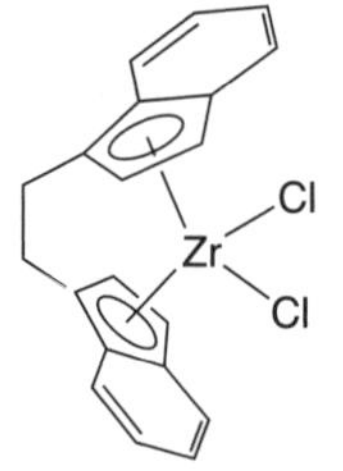

dichlorido[1,1′-(ethane-1,2-diyl)bis(1–3,3a,7a-η-1*H*-inden-1-yl)]zirconium

12.

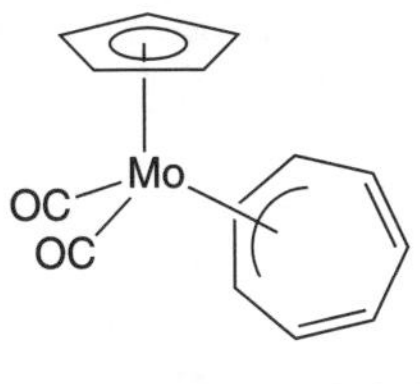

dicarbonyl[(1–3-η)-cyclohepta-2,4,6-trien-1-yl](η⁵-cyclopentadienyl)molybdenum

13.

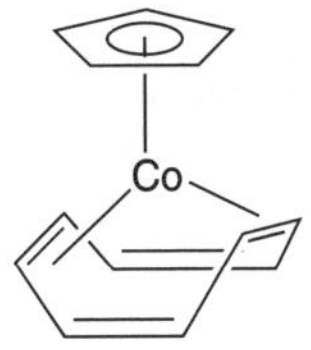

[(1,2,5,6-η)-cyclooctatetraene](η⁵-cyclopentadienyl)cobalt

14.

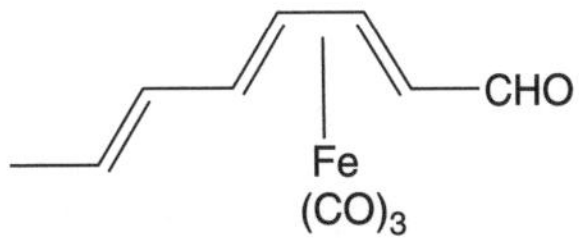

tricarbonyl[(2–5-η)-(*E,E,E*)-octa-2,4,6-trienal]iron

15.

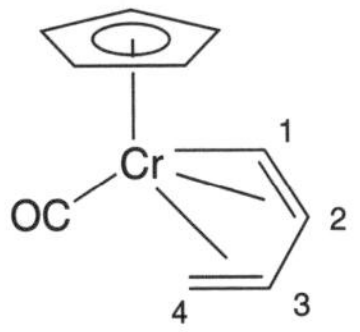

(η⁴-buta-1,3-dien-1-yl-κ*C*¹)carbonyl(η⁵-cyclopentadienyl)chromium

16.

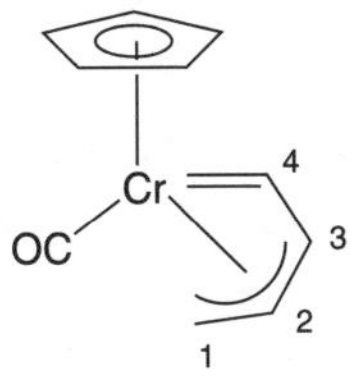

[(1–3-η)-but-2-en-1-yl-4-ylidene-κ*C*⁴]carbonyl(η⁵-cyclopentadienyl)chromium

17.

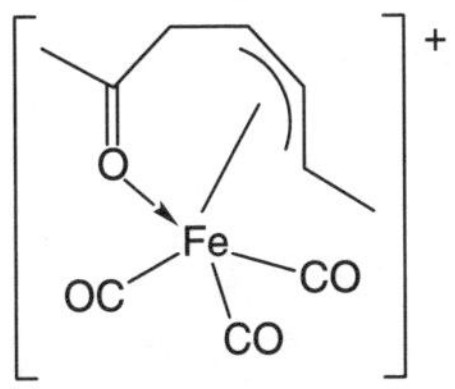

tricarbonyl[6-oxo-κ*O*-(2–4-η)-hept-3-en-2-yl]iron(1+)

As indicated in the previous example, the η symbol can, if necessary, be combined with the κ symbol (see Section IR-10.2.3.3). The symbol η then precedes the ligand name while the κ symbol is either placed at the end of the ligand name or, for more complicated structures, after that portion of the ligand name which denotes the particular function in which the ligating atom is found.

Examples:

18.

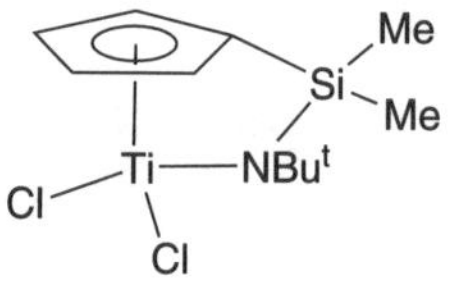

[*N*-*tert*-butyl(η^5-cyclopentadienyl)dimethylsilanaminido-κN]dichloridotitanium

19.

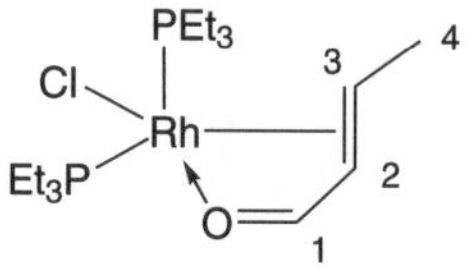

[(*E*)-η^2-but-2-enal-κO]chloridobis(triethylphosphane)rhodium

The symbol η^1 is not used. For a cyclopentadienyl ligand bonded by only one σ-bond one uses cyclopenta-2,4-dien-1-yl or cyclopenta-2,4-dien-1-yl-κC^1.

Example:

20.

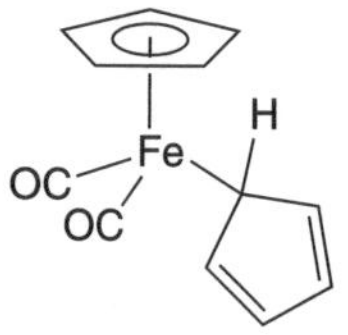

dicarbonyl(η^5-cyclopentadienyl)(cyclopenta-2,4-dien-1-yl-κC^1)iron

If an unsaturated hydrocarbon serves as a bridging ligand, the prefix μ (see Sections IR-10.2.3.1 and IR-10.2.3.4) is combined with both η and κ, where necessary. The colon is used to separate the locants of the bridging ligand which indicate binding to different metal atoms. The metal atoms are numbered according to the rules given in Section IR-9.2.5.6, and their numbers are placed before the η and κ symbols with no hyphens. If ligand locants are also specified, these are separated from the η symbol by a hyphen and the whole expression is enclosed in parentheses, as in 1(2–4-η).

Examples:

21.

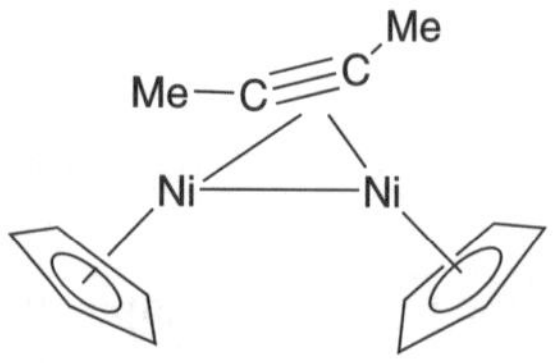

(μ-η^2:η^2-but-2-yne)bis[(η^5-cyclopentadienyl)nickel](*Ni*—*Ni*)

22.

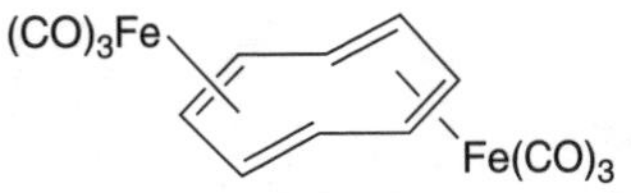

trans-[μ-(1–4-η:5–8-η)-cyclooctatetraene]bis(tricarbonyliron)

23.

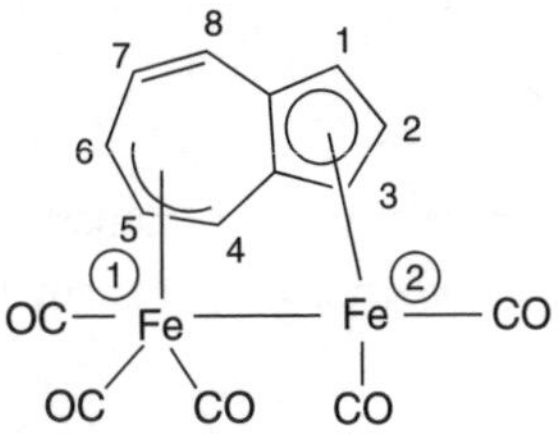

{μ-[2(1–3,3a,8a-η):1(4–6-η)]azulene}(pentacarbonyl-1κ^3*C*,2κ^2*C*)diiron(*Fe—Fe*)

24.

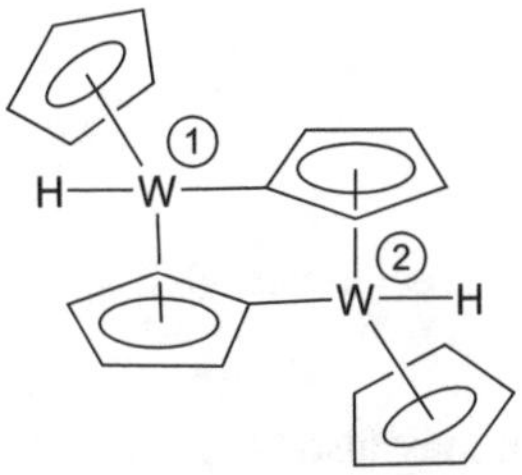

(μ-1η^5-cyclopenta-2,4-diene-1,1-diyl-2κ*C*)(μ-2η^5-cyclopenta-2,4-diene-1,1-diyl-1κ*C*)bis[(η^5-cyclopentadienyl)hydridotungsten]

25.

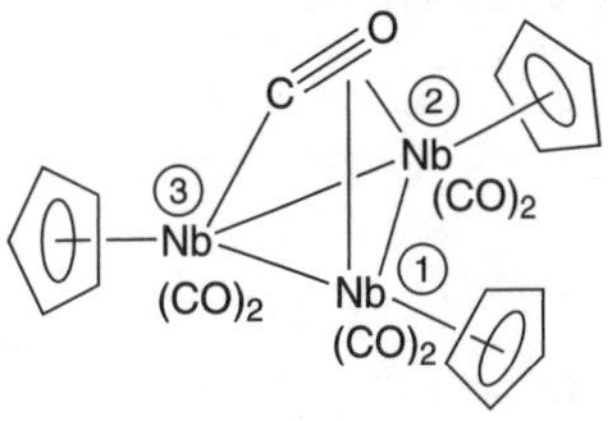

μ_3-1η^2:2η^2-carbonyl-3κ*C*-*triangulo*-tris[dicarbonyl(η^5-cyclopentadienyl)niobium](3 *Nb—Nb*)

26.

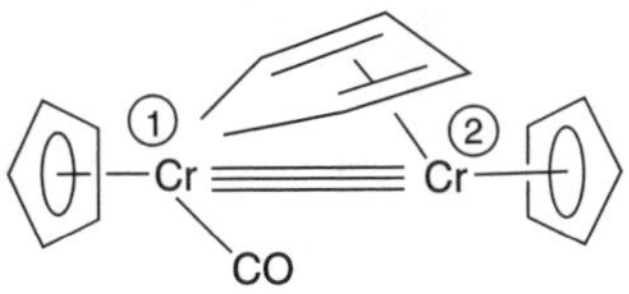

(μ-2η^4-buta-1,3-diene-1,4-diyl-1$\kappa^2 C^1$,C^4)carbonyl-1κ*C*-bis[(η^5-cyclopentadienyl)chromium](*Cr—Cr*)

The eta convention can also be extended to π-coordinated ligands containing no carbon atoms, such as cyclotriborazane and pentaphosphole ligands.

Examples:

27.

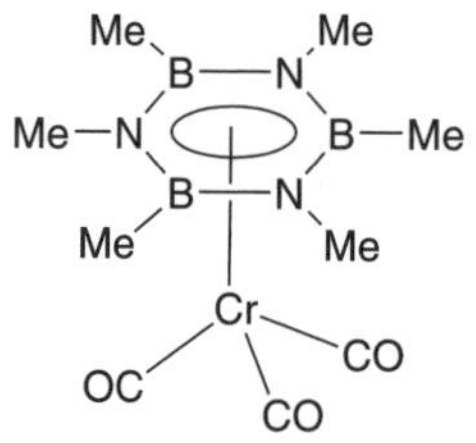

tricarbonyl(η^{6}-hexamethyl-1,3,5,2,4,6-triazatriborinane)chromium, or tricarbonyl(η^{6}-hexamethylcyclotriborazane)chromium

28.

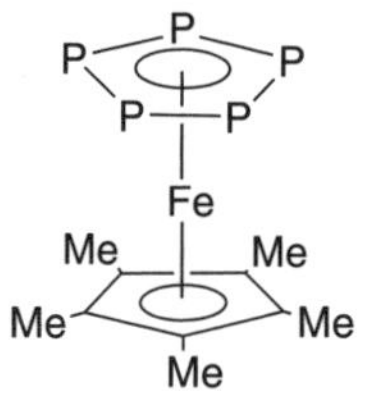

(pentamethyl-η^{5}-cyclopentadienyl)(η^{5}-pentaphospholyl)iron

This convention may also be used for ligands in which σ-bonds are coordinated in a side-on fashion, such as the H-H bond in complexes of dihydrogen (*i.e.* η^{2}-H_2)[5] or the saturated C-H bonds in 'agostic' interactions.[6] The η symbol and locants for agostic interactions are placed separately from other locants at the end of the ligand name. In Example 30 the agostic bond is denoted by a half arrow.

Examples:

29.

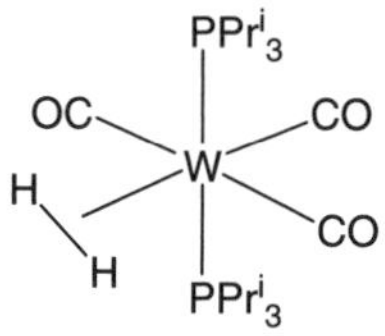

tricarbonyl(η^{2}-dihydrogen)bis(triisopropylphosphane)tungsten

30.

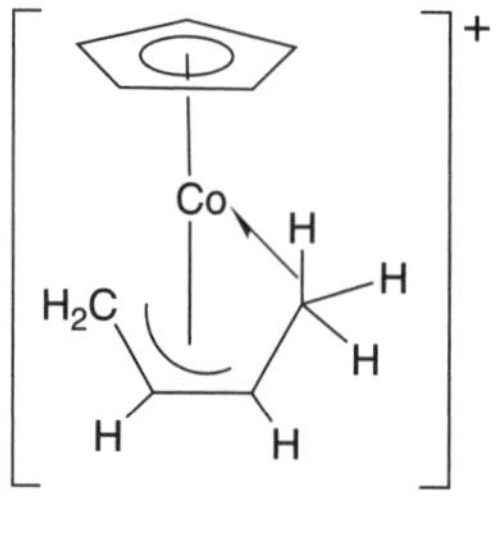

[(1–3-η)-but-2-en-1yl-η^{2}-C^4,H^4](η^{5}-cyclopentadienyl)cobalt(1+)

31.

(η^2,η^2-cycloocta-1,5-diene)(η^6-phenyltriphenylborato)rhodium, or
[(1,2,5,6-η)-cycloocta-1,5-diene)](η^6-phenyltriphenylboranuido)rhodium

IR-10.2.6 Metallocene nomenclature

The first transition element compound containing only carbocyclic rings as ligands was bis(η^5-cyclopentadienyl)iron, $[Fe(\eta^5\text{-}C_5H_5)_2]$, which has a 'sandwich' structure with two parallel η^5- or π-bonded rings. The recognition that this compound was amenable to electrophilic substitution, similar to the aromatic behaviour of benzene, led to the suggestion of the non-systematic name 'ferrocene' and to similar names for other 'metallocenes'.

Examples:

1. $[V(\eta^5\text{-}C_5H_5)_2]$ vanadocene
2. $[Cr(\eta^5\text{-}C_5H_5)_2]$ chromocene
3. $[Co(\eta^5\text{-}C_5H_5)_2]$ cobaltocene
4. $[Ni(\eta^5\text{-}C_5H_5)_2]$ nickelocene
5. $[Ru(\eta^5\text{-}C_5H_5)_2]$ ruthenocene
6. $[Os(\eta^5\text{-}C_5H_5)_2]$ osmocene

Metallocene derivatives may be named either by the standard organic suffix (functional) nomenclature or by prefix nomenclature. The organic functional suffix system is described in Section P-33 of Ref. 3. Metallocene substituent group names have endings 'ocenyl', 'ocenediyl', 'ocenetriyl', *etc.*

Examples:

7.

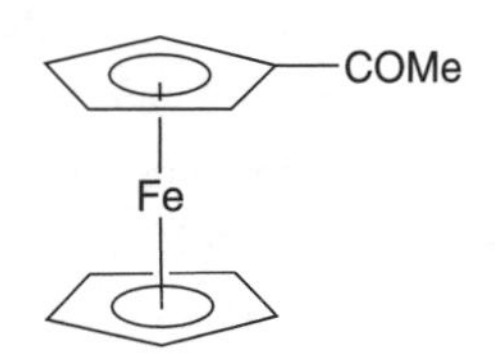

acetylferrocene, or 1-ferrocenylethan-1-one

8.

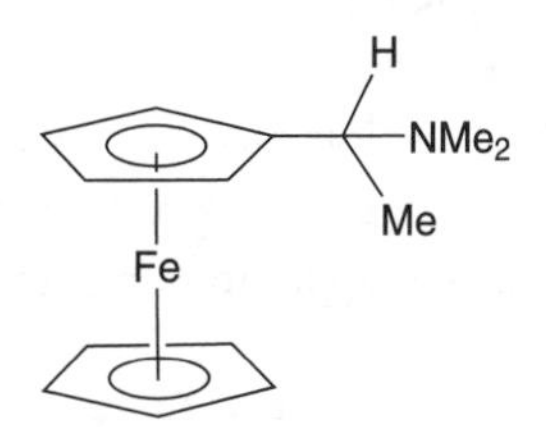

1-[1-(dimethylamino)ethyl]ferrocene, or
1-ferrocenyl-*N*,*N*-dimethylethan-1-amine

Substituents on the equivalent cyclopentadienyl rings of the metallocene entity are given the lowest possible numerical locants in the usual manner. The first ring is numbered 1–5 and the second ring 1′–5′ (see Examples 9 and 10).

Examples:

9.

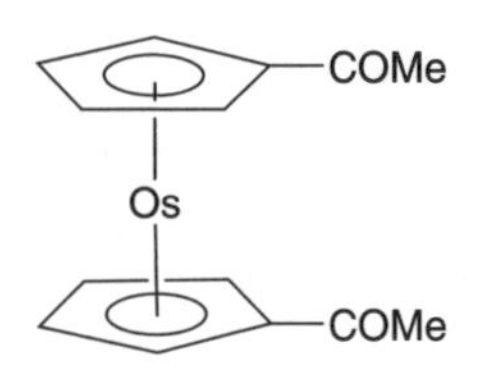

1,1′-diacetylosmocene, or 1,1′-(osmocene-1,1′-diyl)bis(ethan-1-one)

10.

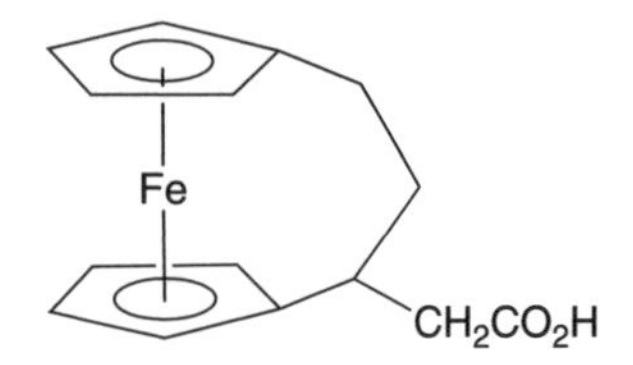

1,1′-(4-carboxybutane-1,3-diyl)ferrocene, or
3,5-(ferrocene-1,1′-diyl)pentanoic acid

11. $[Ru(\eta^5\text{-}C_5Me_5)_2]$
decamethylruthenocene, or
bis(pentamethyl-η^5-cyclopentadienyl)ruthenium

12. $[Cr(\eta^5\text{-}C_5Me_4Et)_2]$
1,1′-diethyloctamethylchromocene, or
bis(1-ethyl-2,3,4,5-tetramethyl-η^5-cyclopentadienyl)chromium

13. $[Co(\eta^5\text{-}C_5H_4PPh_2)_2]$
1,1′-bis(diphenylphosphanyl)cobaltocene, or
(cobaltocene-1,1′-diyl)bis(diphenylphosphane)

Metallocene nomenclature does not, however, apply to all transition elements. For example, there are at least two isomers with the empirical formula $C_{10}H_{10}Ti$ but neither has the regular sandwich structure analogous to that of ferrocene, and so neither should be named 'titanocene'. Similarly, 'manganocene' is a misnomer for $[Mn(\eta^5\text{-}C_5H_5)_2]$ since it has a chain structure in the solid state, with no individual sandwich entities. However, decamethylmanganocene, $[Mn(\eta^5\text{-}C_5Me_5)_2]$, has a normal sandwich structure, as does decamethylrhenocene, $[Re(\eta^5\text{-}C_5Me_5)_2]$. With increasing atomic number, the occurrence of the classic ferrocene-type bis(η^5-cyclopentadienyl) sandwich structure becomes rare.

The name-ending 'ocene' should therefore be confined to discrete molecules of the form bis(η^5-cyclopentadienyl)metal (and ring-substituted analogues), where the cyclopentadienyl rings are essentially parallel, and the metal is in the d-block [*i.e.* the terminology does not apply to compounds of the s- or p-block elements such as $Ba(C_5H_5)_2$ or $Sn(C_5H_5)_2$].

The oxidized species have been referred to as metallocenium(n+) salts, although it should be noted that in this case the ending 'ium' does not carry the usual meaning it has in substitutive nomenclature, *i.e.* the addition of a hydron to a neutral parent compound.

To avoid this ambiguity, the name bis(η^5-cyclopentadienyl)iron(1+), for example, is strongly preferred to ferrocenium(1+) for $[Fe(\eta^5\text{-}C_5H_5)_2]^+$. The same comment applies to substituted derivatives.

Examples:

14. $[Co(\eta^5\text{-}C_5H_5)_2][PF_6]$
    bis(η^5-cyclopentadienyl)cobalt(1+) hexafluoridophosphate
15. $[Co(\eta^5\text{-}C_5H_5)(\eta^5\text{-}C_5H_4COMe)][BF_4]$
    (acetyl-η^5-cyclopentadienyl)(η^5-cyclopentadienyl)cobalt(1+) tetrafluoridoborate

The oxidized form of osmocene is dinuclear in the solid state, with a long Os–Os bond, so should not in any case be named using the 'ocenium' nomenclature. However, $[Os(\eta^5\text{-}C_5Me_5)_2]^+$ has a mononuclear sandwich structure and may be described as the decamethylosmocenium(1+) ion, although bis(pentamethyl-η^5-cyclopentadienyl)osmium(1+) is strongly preferred.

In strong protic acid media, ferrocene is hydronated to $[Fe(\eta^5\text{-}C_5H_5)_2H]^+$. To avoid ambiguities, this should be named by the additive procedure, *i.e.* bis(η^5-cyclopentadienyl)hydridoiron(1+).

Transition element complexes derived from ligands with additional rings fused to the cyclopentadienyl rings are also known. The names of these complexes are derived from the retained common or semisystematic names of the hydrocarbon ligands, *e.g.* 1*H*-inden-1-yl (C_9H_7), fluoren-9-yl ($C_{13}H_9$), and azulene ($C_{10}H_8$). Thus, $[Fe(\eta^5\text{-}C_9H_7)_2]$ is named bis(η^5-indenyl)iron or, more specifically, bis[(1–3,3a,7a-η)-1*H*-inden-1-yl]iron. To avoid possible ambiguities, the use of fusion nomenclature, such as 'benzoferrocene', is strongly discouraged.

Many compounds have ligands in addition to two η^5-cyclopentadienyl rings. They are often referred to as metallocene di(ligand) species, *e.g.* $[Ti(\eta^5\text{-}C_5H_5)_2Cl_2]$ is frequently named 'titanocene dichloride'. This practice is discouraged since metallocene nomenclature applies only to compounds in which the two rings are parallel. Thus, $[Ti(\eta^5\text{-}C_5H_5)_2Cl_2]$ is named dichloridobis(η^5-cyclopentadienyl)titanium, and $[W(\eta^5\text{-}C_5H_5)_2H_2]$, $[Ti(CO)_2(\eta^5\text{-}C_5H_5)_2]$ and $[Zr(\eta^5\text{-}C_5H_5)_2Me_2]$ should be named bis(η^5-cyclopentadienyl)dihydridotungsten, dicarbonylbis(η^5-cyclopentadienyl)titanium and bis(η^5-cyclopentadienyl)dimethylzirconium, respectively.

The bis(cyclooctatetraene) compound $[U(\eta^8\text{-}C_8H_8)_2]$ has sometimes been described as 'uranocene'. Related species are obtained from zirconium, $[Zr(\eta^8\text{-}C_8H_8)_2]$, and the lanthanoids, *e.g.* $[Ce(\eta^8\text{-}C_8H_8)_2]^-$. In such complexes, the carbocyclic rings are parallel and there are certain similarities to ferrocene in the molecular orbital descriptions of their bonding. However, some lanthanoids also form metal(II) cyclopentadienyl complexes, such as $[Sm(\eta^5\text{-}C_5Me_5)_2]$. Extension of the 'ocene' nomenclature to $[U(\eta^8\text{-}C_8H_8)_2]$ and similar compounds can therefore lead to confusion and is strongly discouraged.

Furthermore, the cyclooctatetraene ring can also function as an η^4-ligand, as in $[Ti(\eta^4\text{-}C_8H_8)(\eta^8\text{-}C_8H_8)]$. Compounds of cyclooctatetraene should therefore be named using standard organometallic nomenclature, *e.g.* bis(η^8-cyclooctatetraene)uranium and [(1–4-η)-cyclooctatetraene](η^8-cyclooctatetraene)titanium. The ligand $C_8H_8^{2-}$ is occasionally referred to as 'cyclooctatetraenyl'. This name is incorrect as it can only be used for the (as yet hypothetical) ligand C_8H_7.

IR-10.3 NOMENCLATURE OF ORGANOMETALLIC COMPOUNDS OF THE MAIN GROUP ELEMENTS

IR-10.3.1 Introduction

The nomenclature of organometallic compounds of the main group elements is an area of current and ongoing development. This section briefly describes key aspects of the naming of such compounds, leaving a full treatment of the subject to a future IUPAC project. Detailed information on the nomenclature of organic compounds containing the elements of groups 13–16 may be found in Sections P-68 and P-69 of Ref. 3.

In principle, all organometallic compounds, whether of the transition or main group elements, can be given names based on the additive system of nomenclature that is applied to coordination compounds, provided the constitution of the compound is known. Examples of such names were given in Sections IR-7.2 and IR-7.3. In addition, compounds of elements such as boron, silicon, arsenic and selenium are often considered to be organometallic, and are commonly named by notionally substituting the hydrogen atoms of the parent hydride with the appropriate substituent groups.

If a choice must be made, it is recommended here that organometallic compounds derived from the elements of groups 13–16 be named by a substitutive process, while those derived from the elements of groups 1 and 2 be named using the additive system of nomenclature or in some cases just compositional nomenclature if less structural information is to be conveyed. Where an organometallic compound contains two or more central atoms (which may be associated with different nomenclature systems according to the above recommendation), a choice must again be made to provide the basis of the name. A general rule is recommended in Section IR-10.4.

IR-10.3.2 Organometallic compounds of groups 1 and 2

Organometallic compounds of the elements of groups 1 and 2 with a defined coordination structure are named according to the additive system of nomenclature, the general definitions and rules of which are given in Chapter IR-7 and Sections IR-9.1 and IR-9.2. Thus, prefixes denoting the organic groups and any other ligands are placed in alphabetical order before the name of the metal. These prefixes may adopt either the additive 'ido', 'diido', *etc.* endings or, in the case of hydrocarbyl groups, the substitutive 'yl', 'diyl', *etc.* endings (see Sections IR-10.2.2 and IR-10.2.3). The latter practice allows names in common usage for organic groups to be applied unchanged. The presence of a hydrogen atom attached to the metal centre must always be indicated (by the prefix 'hydrido') and the name of a cyclic compound with the central atom in the ring may be formed using appropriate locants of a divalent 'diido' or 'diyl' group to indicate chelate-type bonding to the metal, as in Example 5 below.

Many organometallic compounds of groups 1 and 2 exist in associated molecular form (as aggregates) or contain structural solvent, or both. However, their names are often based solely on the stoichiometric compositions of the compounds, unless it is specifically desired to draw attention to the extent of aggregation or the nature of any structural solvent, or both (see Example 3 below). In the examples below, note how the different types of name reflect the different structural content implied by the formulae shown. As usual, the formulae enclosed in square brackets designate coordination entities.

Note that metallocene terminology (Section IR-10.2.6) is not recommended for bis(cyclopentadienyl) compounds of the main group metals (see Examples 6 and 7).

Examples:

1. [BeEtH]
   ethylhydridoberyllium, or ethanidohydridoberyllium
2. $Na(CHCH_2)$
   sodium ethenide (compositional name)

   $Na–CH{=}CH_2$, or $[Na(CH{=}CH_2)]$
   ethenidosodium, ethenylsodium, or vinylsodium
3. $[\{Li(OEt_2)(\mu_3\text{-}Ph)\}_4]$
   tetrakis[(ethoxyethane)(μ_3-phenyl)lithium], or
   tetrakis[(μ_3-benzenido)(ethoxyethane)lithium]
4. $2Na^+(Ph_2CCPh_2)^{2-}$
   disodium 1,1,2,2-tetraphenylethane-1,2-diide (compositional name)

   $Ph_2C(Na)-C(Na)Ph_2$
   (μ-1,1,2,2-tetraphenylethane-1,2-diyl)disodium, or
   (μ-1,1,2,2-tetraphenylethane-1,2-diido-κ^2C^1,C^2)disodium
5.

   Mg 4 1 3 2 $CH_2CH_2CH{=}CMe_2$

   [2-(4-methylpent-3-en-1-yl)but-2-ene-1,4-diyl]magnesium, or
   [2-(4-methylpent-3-en-1-yl)but-2-ene-1,4-diido-κ^2C^1,C^4]magnesium
6. $[Mg(\eta^5\text{-}C_5H_5)_2]$
   bis(η^5-cyclopentadienyl)magnesium, or
   bis(η^5-cyclopentadienido)magnesium
7. $[PPh_4][Li(\eta^5\text{-}C_5H_5)_2]$
   tetraphenylphosphanium bis(η^5-cyclopentadienyl)lithate(1–), or
   tetraphenylphosphanium bis(η^5-cyclopentadienido)lithate(1–)
8. LiMe — lithium methanide (compositional name)
   [LiMe] — methyllithium
   $[(LiMe)_4]$ — tetra-μ_3-methyl-tetralithium
   $(LiMe)_n$ — poly(methyllithium)

9. MgIMe
   magnesium iodide methanide (compositional name)

   [MgI(Me)]
   iodido(methanido)magnesium (additive name of coordination type)

   [MgMe]I
   methylmagnesium iodide (compositional name with formally electropositive component named using additive nomenclature)

   $[MgI(Me)]_n$
   poly[iodido(methanido)magnesium], or poly[iodido(methyl)magnesium]

IR-10.3.3 Organometallic compounds of groups 13–16

Organometallic compounds of the elements of groups 13–16 are named according to the substitutive system of nomenclature, dealt with in Chapter IR-6. Thus, the name of the parent hydride (formed in accordance with the rules of Section IR-6.2) is modified by a prefix for each substituent replacing a hydrogen atom of the parent hydride. The prefix should be in appropriate substituent form (chloro, methyl, sulfanylidene, *etc.*) and not in ligand form (chlorido, methanido, sulfido, *etc.*).

Where there is more than one kind of substituent, the prefixes are cited in alphabetical order before the name of the parent hydride, parentheses being used to avoid ambiguity, and multiplicative prefixes being used as necessary. Non-standard bonding numbers are indicated using the λ-convention (see Section IR-6.2.2.2). An overview of the rules for naming substituted derivatives of parent hydrides is given in Section IR-6.3, while a detailed exposition may be found in Ref. 3.

Examples:

1. AlH_2Me methylalumane
2. $AlEt_3$ triethylalumane
3. $Me_2CHCH_2CH_2In(H)CH_2CH_2CHMe_2$
   bis(3-methylbutyl)indigane
4. $Sb(CH{=}CH_2)_3$ triethenylstibane, or trivinylstibane
5. $SbMe_5$ pentamethyl-λ^5-stibane
6. PhSb=SbPh diphenyldistibene
7. $GeCl_2Me_2$ dichlorodimethylgermane
8. $GeMe(SMe)_3$ methyltris(methylsulfanyl)germane
9. BiI_2Ph diiodo(phenyl)bismuthane
10. $Et_3PbPbEt_3$ hexaethyldiplumbane
11. $SnMe_2$ dimethyl-λ^2-stannane
12. $BrSnH_2SnCl_2SnH_2(CH_2CH_2CH_3)$
    1-bromo-2,2-dichloro-3-propyltristannane
13. $Me_3SnCH_2CH_2C{\equiv}CSnMe_3$
    but-1-yne-1,4-diylbis(trimethylstannane)

In the presence of one or more characteristic groups that may be expressed using one or more suffixes (–NH_2, –OH, –COOH, *etc.*), the name of the parent hydride carrying the highest-ranking such group is modified by the suffix, and other substituents are then denoted by prefixes as described in Section IR-6.3.1. If acting as a substituent, the group 13–16 parent hydride name in question is modified by changing the ending 'ane' to 'anyl' (or 'yl' for the group 14 elements), 'anediyl', *etc.*

Examples:

14. $(EtO)_3GeCH_2CH_2COOMe$
    methyl 3-(triethoxygermyl)propanoate
15. $H_2As(CH_2)_4SO_2Cl$
    4-arsanylbutane-1-sulfonyl chloride
16. $OCHCH_2CH_2GeMe_2GeMe_2CH_2CH_2CHO$
    3,3′-(1,1,2,2-tetramethyldigermane-1,2-diyl)dipropanal
17. $SiMe_3NH_2$ trimethylsilanamine

Sometimes it may be necessary or preferable to consider a parent hydride in which several (four or more) skeletal carbon atoms of a hydrocarbon have been replaced by main group elements. In this method of skeletal replacement the heteroatoms are designated by the 'a' terms of replacement nomenclature (Table X) cited in the order given by Table VI and preceded by the appropriate locant(s). The rules for locant numbering are specified in Section IR-6.2.4.1 and this nomenclature is fully described in Sections P-21.2 and P-22.2 of Ref. 3.

Examples:

18. $Me\overset{2}{Si}H_2\overset{3}{C}H_2\overset{4}{C}H_2\overset{5}{Si}H_2\overset{6}{C}H_2\overset{7}{C}H_2\overset{8}{Si}H_2\overset{9}{C}H_2\overset{10}{C}H_2\overset{11}{Si}H_2Me$
    2,5,8,11-tetrasiladodecane
19. $Me\overset{2}{Si}H_2\overset{3}{O}\overset{4}{P}(H)\overset{5}{O}CH_2Me$
    3,5-dioxa-4-phospha-2-silaheptane
20. $H\overset{1}{S}\overset{}{C}H{=}\overset{2}{N}\overset{3}{O}\overset{4}{C}H_2\overset{5}{Se}\overset{6}{C}H_2\overset{7}{O}\overset{8}{N}HMe$
    3,7-dioxa-5-selena-2,8-diazanon-1-ene-1-thiol
21.

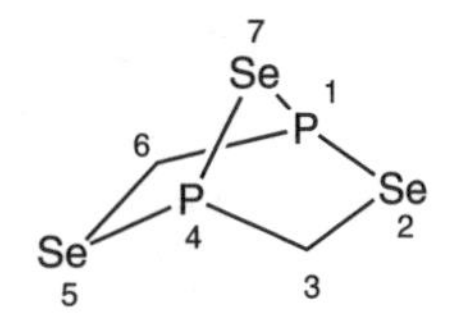

2,5,7-triselena-1,4-diphosphabicyclo[2.2.1]heptane

When elements from groups 13–16 replace carbon atoms in monocyclic systems, the resulting structures may be named using the extended Hantzsch–Widman procedures. This nomenclature is fully described in Section IR-6.2.4.3 and in Section P-22.2 of Ref. 3 and will not be elaborated further here.

Sections P-68 and P-69 of Ref. 3 offer a more comprehensive treatment of the nomenclature of organic compounds containing the elements of groups 13–16.

IR-10.4 ORDERING OF CENTRAL ATOMS IN POLYNUCLEAR ORGANOMETALLIC COMPOUNDS

When an organometallic compound contains two or more different metal atoms, a choice must be made to provide the basis of the name. It is thus convenient to classify the possible central atoms as belonging to either (i) the elements of groups 1–12 (whose compounds are named according to the additive system of nomenclature) or (ii) the elements of groups 13–16 (whose compounds are named according to the substitutive system).

IR-10.4.1 Central atoms from groups 1–12 only

If both or all potential central atoms belong to class (i), then the compound is named additively using the methodology described in Section IR-9.2.5, including the rules given there for ordering the central atoms. Ferrocenyllithium (ferrocenyl, see Section IR-10.2.6) could thus be systematically named:

(2η^5-cyclopentadienyl)(2η^5-cyclopenta-2,4-dien-1-yl-1κC^1)lithiumiron,

a name which also illustrates the use of the κ and η conventions. Further examples in which both or all central atoms belong to class (i) are given in Sections IR-10.2.3.4, IR-10.2.3.5 and IR-10.2.5.1.

IR-10.4.2 Central atoms from both groups 1–12 and groups 13–16

If at least one possible central atom belongs to class (i) and one or more others to class (ii), then the compound is named additively using the metal atom(s) of class (i) as central atom(s). The remaining atoms of the complex are named as ligands by rules already presented (Sections IR-9.1, IR-9.2 and IR-10.2.1 to IR-10.2.5).

Examples:

1. $[Li(GePh_3)]$ (triphenylgermyl)lithium
2. $(Me_3Si)_3CMgC(SiMe_3)_3$
   bis[tris(trimethylsilyl)methyl]magnesium
3. $[Mo(CO)_5(=Sn\{CH(SiMe_3)_2\}_2)]$
   {bis[bis(trimethylsilyl)methyl]-λ^2-stannylidene}pentacarbonylmolybdenum
4.

Ph_2Sb–(4)C_6H_4(1)–HgPh

[4-(diphenylstibanyl)phenyl](phenyl)mercury

5.

(phenylstibanediyl)bis[dicarbonyl(η^5-cyclopentadienyl)manganese]

IR-10.4.3 Central atoms from groups 13–16 only

If the possible central atoms are both or all from class (ii), then the compound is named substitutively as described in Section IR-10.3.3 (and in more detail in Section IR-6.3). The parent hydride is chosen on the basis of the following element order ('>' meaning 'chosen before', *cf.* Section P-41 of Ref. 3):

$$N > P > As > Sb > Bi > Si > Ge > Sn > Pb > B > Al > Ga > In > Tl > S > Se > Te > C$$

Thus, for a compound containing both arsenic and lead, the parent hydride would be selected as AsH_3, rather than PbH_4, the lead atom then appearing in the name as a prefixed substituent, often with its own substituent groups.

Examples:

1. $As(PbEt_3)_3$ tris(triethylplumbyl)arsane
2. (4-stibanylphenyl)arsane
3. 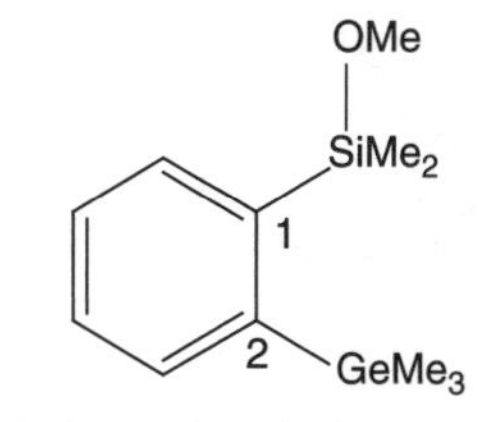

   methoxydimethyl[2-(trimethylgermyl)phenyl]silane
4. $Et_3PbCH_2CH_2CH_2BiPh_2$
   diphenyl[3-(triethylplumbyl)propyl]bismuthane
5. $SiClH_2Sn(Me){=}Sn(Me)SiClH_2$
   Si,Si'-(1,2-dimethyldistannene-1,2-diyl)bis(chlorosilane)

IR-10.5 REFERENCES

1. *Nomenclature of Inorganic Chemistry, IUPAC Recommendations 1990*, ed. G.J. Leigh, Blackwell Scientific Publications, Oxford, 1990.

2. Nomenclature of Organometallic Compounds of the Transition Elements, A. Salzer, *Pure Appl. Chem.*, **71**, 1557–1585 (1999).
3. *Nomenclature of Organic Chemistry, IUPAC Recommendations*, eds. W.H. Powell and H. Favre, Royal Society of Chemistry, in preparation.
4. F.A. Cotton, *J. Am. Chem. Soc.*, **90**, 6230–6232 (1968).
5. D.J. Heinekey and W.J. Oldham, Jr., *Chem. Rev.*, **93**, 913–926 (1993).
6. M. Brookhart, M.L.H. Green and L.-L. Wong, *Prog. Inorg. Chem.*, **36**, 1–124 (1988).

IR-11 Solids

CONTENTS

IR-11.1 INTRODUCTION

IR-11.1.1 General

This chapter deals with some aspects of terminology, nomenclature and notation for solids. However, in cases where detailed structural information is to be conveyed, fully systematic names can be difficult to construct. An attempt to deal with this problem is described in Ref. 1.

IR-11.1.2 Stoichiometric and non-stoichiometric phases

In binary and multi-component systems, intermediate crystalline phases (stable or metastable) may occur. Thermodynamically, the composition of any such phase is variable. In some cases, such as sodium chloride, the possible variation in composition is very small. Such phases are called stoichiometric. However, in other phases appreciable variations in composition can occur, for example in wustite (nominally FeO).

These are called non-stoichiometric phases. In general, it is possible to define an ideal composition relative to which the variations occur. This composition, called the stoichiometric composition, is usually that in which the ratio of the numbers of different atoms corresponds to the ratio of the numbers of normal crystallographic sites of different types in the ideal (ordered) crystal.

This concept can be used even when the stoichiometric composition is not included in the homogeneity range of the phase. The term 'non-stoichiometric' does not mean phases with complex formulae, but those with *variable composition*, for which the term *solid mixture* is an alternative. Formerly, the term *solid solution* was used, but this term is recommended to apply only in the following sense.[2–4] *Mixture* is used to describe a solid phase containing more than one substance, when all substances are treated in the same way. *Solution* is used to describe a liquid or solid phase containing more than one substance when, for convenience, one (or more) of the substances, called the *solvent*, is treated differently from the other substances, called *solutes*. For non-stoichiometric phases, each atom or group of atoms that contributes to the non-stoichiometry is treated equally, so the term *mixture* is appropriate.

IR-11.2 NAMES OF SOLID PHASES

IR-11.2.1 General

Names for stoichiometric phases, such as NaCl, are derived simply as in Chapter IR-5, whereas formulae are derived as presented in Chapter IR-4. Although NaCl in the solid state consists of an infinite network of units, $(NaCl)_{\infty}$, the compound is named sodium chloride and represented symbolically as NaCl.

However, for non-stoichiometric phases and solid solutions, formulae are preferable to names, since strictly systematic names tend to be inconveniently cumbersome. They should be used only when unavoidable (*e.g.* for indexing) and should be constructed in the following style.

Examples:

1. iron(II) sulfide (iron deficient)
2. molybdenum dicarbide (carbon excess)

IR-11.2.2 Mineral names

Mineral names should be used only to designate actual minerals and not to define chemical composition. Thus, the name calcite refers to a particular mineral (contrasted with other minerals of similar composition) and is not a term for the chemical compound the composition of which is properly expressed by the name calcium carbonate.

A mineral name may, however, be used to indicate the structure type. Where possible, a name that refers to a general group should replace a more specific name. For example, large numbers of minerals have been named that are all spinels, but which have widely differing atomic constituents. In this case, the generic name '*spinel* type' should be used rather than the more specific names chromite, magnetite, *etc*. The generic mineral name, printed in italics, should be accompanied by a representative chemical formula. This is particularly important for *zeolite* types.[5]

Examples:

1. $FeCr_2O_4$ (*spinel* type)
2. $BaTiO_3$ (*perovskite* type)

IR-11.3 CHEMICAL COMPOSITION

IR-11.3.1 Approximate formulae

The formula used in any given case depends upon how much information is to be conveyed. A general notation, which can be used even when the mechanism of the variation in composition is unknown, is to put the sign ~ (read as *circa*, or 'approximately') before the formula.

Examples:

1. ~FeS
2. ~CuZn

If it is desirable to give more information, one of the notations described below may be used.

IR-11.3.2 Phases with variable composition

For a phase where the variation in composition is caused solely or partially by isovalent substitution, the symbols of the atoms or groups that replace each other may be separated by a comma and placed together between parentheses. If possible, the formula is written so that the limits of the homogeneity range, when one or the other of the two atoms or groups is lacking, are represented.

Examples:

1. (Cu,Ni) denotes the complete range of compositions from pure Cu to pure Ni.
2. K(Br,Cl) comprises the range from pure KBr to pure KCl.

Phases for which substitution also results in vacant positions are denoted in the same way.

Examples:

3. $(Li_2,Mg)Cl_2$ denotes a solid mixture intermediate in composition between LiCl and $MgCl_2$.
4. $(Al_2,Mg_3)Al_6O_{12}$ represents a solid mixture intermediate in composition between $MgAl_2O_4$ ($= Mg_3Al_6O_{12}$) and Al_2O_3 (*spinel* type) ($= Al_2Al_6O_{12}$).

In general, however, a notation in which there are variables which define composition should be used. The ranges of the variables can also be indicated. Thus, a phase involving substitution of atom A for B is written $A_{m+x}B_{n-x}C_p$ $(0 \leq x \leq n)$. The commas and parentheses called for above are not then required.

Examples:

5. Cu_xNi_{1-x} $(0 \leq x \leq 1)$ is equivalent to (Cu,Ni).
6. KBr_xCl_{1-x} $(0 \leq x \leq 1)$ is equivalent to K(Br,Cl).
7. $Li_{2-2x}Mg_xCl_2$ $(0 \leq x \leq 1)$ is equivalent to $(Li_2,Mg)Cl_2$ but shows explicitly that one vacant cation position appears for every 2 Li^+ replaced by Mg^{2+}.
8. $Co_{1-x}O$ indicates that there are vacant cation sites; for $x = 0$ the formula corresponds to the stoichiometric composition CoO.
9. $Ca_xZr_{1-x}O_{2-x}$ indicates that Zr has been partly replaced by Ca, resulting in vacant anion sites; for $x = 0$ the formula corresponds to the stoichiometric composition ZrO_2.

If the variable x is limited this may be shown by using δ or ε instead of x. A specific composition or composition range can be indicated by stating the actual value of the variable x (or δ, or ε). This value can be written in parentheses after the general formula. However, the value of the variable may also be introduced in the formula itself. This notation can be used both for substitutional and for interstitial solid solutions.[6]

Examples:

10. $Fe_{3x}Li_{4-x}Ti_{2(1-x)}O_6$ $(x = 0.35)$, or $Fe_{1.05}Li_{3.65}Ti_{1.30}O_6$
11. $LaNi_5H_x$ $(0 < x < 6.7)$
12. $Al_4Th_8H_{15.4}$
13. $Ni_{1-\delta}O$

IR-11.4 POINT DEFECT (KRÖGER–VINK) NOTATION

IR-11.4.1 General

As well as the chemical composition, information about point defects, site symmetry, and site occupancy can be given by using additional symbols. These symbols may also be used to write quasi-chemical equilibria between point defects.[6]

IR-11.4.2 Indication of site occupancy

In a formula, the main symbols indicate the species present at a certain site, defined with respect to empty space. This will generally be the symbol of an element. If a site is vacant this is denoted by the italicized symbol *V*. (In certain contexts other symbols, such as a square box, □, are used for vacancies, but the use of italicized *V* is preferred, the element vanadium being written with the upright letter V).

The site and its occupancy in a structure of ideal composition are represented by right lower indexes. The first index indicates the type of site, and the second index (if used), separated from the first by a comma, indicates the number of atoms on this site. Thus, an atom A on a site normally occupied by A in the ideal structure is expressed by $\mathrm{A_A}$; an atom A on a site normally occupied by B is expressed $\mathrm{A_B}$; and $\mathrm{M_{M,1-\mathit{x}}N_{M,\mathit{x}}M_{N,\mathit{x}}N_{N,1-\mathit{x}}}$ stands for a disordered alloy, where the ideal composition is $\mathrm{M_MN_N}$ with all M atoms on one type of crystallographic site and all N atoms on a second type of crystallographic site. An alternative description is $\mathrm{(M_{1-\mathit{x}}N_{\mathit{x}})_M(M_{\mathit{x}}N_{1-\mathit{x}})_N}$. A species occupying an interstitial site (*i.e.* a site which is unoccupied in the ideal structure) is indicated by the subscript 'i'.

Examples:

1. $\mathrm{Mg_{Mg,2-\mathit{x}}Sn_{Mg,\mathit{x}}Mg_{Sn,\mathit{x}}Sn_{Sn,1-\mathit{x}}}$ shows that in $\mathrm{Mg_2Sn}$ some of the Mg atoms are present on Sn sites and *vice versa*.
2. $\mathrm{(Bi_{2-\mathit{x}}Te_{\mathit{x}})_{Bi}(Bi_{\mathit{x}}Te_{3-\mathit{x}})_{Te}}$ shows that in $\mathrm{Bi_2Te_3}$ some of the Bi atoms are present on Te sites and *vice versa*.
3. $\mathrm{Na_{Na,1-\mathit{x}}\mathit{V}_{Na,\mathit{x}}Cl_{Cl,1-\mathit{x}}\mathit{V}_{Cl,\mathit{x}}}$ shows that *x* Na and *x* Cl sites in NaCl are vacant, giving Schottky defects.
4. $\mathrm{Ca_{Ca,1}F_{F,2-\mathit{x}}\mathit{V}_{F,\mathit{x}}F_{i,\mathit{x}}}$ shows that in $\mathrm{CaF_2}$, *x* F sites are vacant, while *x* F ions are situated on interstitial sites, creating Frenkel defects.
5. $\mathrm{(Ca_{0.15}Zr_{0.85})_{Zr}(O_{1.85}\mathit{V}_{0.15})_O}$, or $\mathrm{Ca_{Zr,0.15}Zr_{Zr,0.85}O_{O,1.85}\mathit{V}_{O,0.15}}$, shows that in CaO-stabilized $\mathrm{ZrO_2}$, 0.85 of the Zr sites are occupied by Zr, 0.15 of the Zr sites are occupied by Ca, and that, of the two oxygen sites, 1.85 sites are occupied by oxygen ions, leaving 0.15 sites vacant.
6. $\mathrm{V_{V,1}C_{C,0.8}\mathit{V}_{C,0.2}}$ shows that 0.2 C-sites are vacant in vanadium carbide, VC.

The defect symbols can be used in writing quasi-chemical reactions.

Examples:

7. $\mathrm{Na_{Na} \rightarrow \mathit{V}_{Na} + Na(g)}$ indicates the evaporation of a Na atom, leaving behind a sodium vacancy in the lattice.
8. $\mathrm{0.5Cl_2(g) + \mathit{V}_{Cl} \rightarrow Cl_{Cl}}$ indicates the incorporation of a chlorine atom, from a dichlorine molecule, on a vacant chlorine site in the lattice.

IR-11.4.3 Indication of crystallographic sites

Crystallographic sites can be distinguished by subscripts, *e.g.* tet, oct and dod, denoting tetrahedrally, octahedrally and dodecahedrally coordinated sites, respectively. The use of subscripts such as a, b, . . . , which are not self-explanatory, is not approved. In some cases, such as oxides and sulfides, the number of subscripts can be reduced by defining specific symbols to indicate site symmetries, *e.g.* ( ) for tetrahedral sites, [ ] for octahedral sites, { } for dodecahedral sites. To avoid confusion, such enclosing marks should be restricted to cases where they are not being used to express multiplication. The meaning of the symbols should be clearly stated in the text.

Examples:

1. $Mg_{tet}Al_{oct,2}O_4$ or $(Mg)[Al_2]O_4$ denotes a normal spinel.
2. $Fe_{tet}Fe_{oct}Ni_{oct}O_4$ or $(Fe)[FeNi]O_4$ denotes $NiFe_2O_4$ (*inverse spinel* type).

IR-11.4.4 Indication of charges

Charges are indicated by a right upper index. When formal charges are given, the usual convention holds: one unit of positive charge is indicated by a superscript +, n units of positive charge by a superscript $n+$, one unit of negative charge by a superscript −, n units of negative charge by a superscript $n-$. Thus A^{n+} denotes n units of formal positive charge on an atom of symbol A. In defect chemistry, charges are defined preferably with respect to the ideal unperturbed crystal. In this case, they are called *effective charges.* One unit of positive effective charge is shown by a superscript dot, $^{\bullet}$, (not to be confused with the radical dot described in Section IR-4.6.2) and one unit of negative effective charge by a prime, $'$; n units of effective charge are indicated by superscript $^{n\bullet}$ or $^{n\prime}$. The use of double dots $^{\bullet\bullet}$ or double primes $''$ in the case of two effective charges is also allowed. Thus $A^{2\bullet}$ and $A^{\bullet\bullet}$ indicate that an atom of symbol A has two units of effective positive charge. Sites that have no effective charge relative to the unperturbed lattice may be indicated explicitly by a superscript cross, *i.e.* '^{x}'.

Examples:

1. $Li_{Li,1-2x}Mg^{\bullet}_{Li,x}V'_{Li,x}Cl_{Cl}$ and $Li^{x}_{Li,1-2x}Mg^{\bullet}_{Li,x}V'_{Li,x}Cl^{x}_{Cl}$ are equivalent expressions for a substitutional solid solution of $MgCl_2$ in LiCl.
2. $Y_{Y,1-2x}Zr^{\bullet}_{Y,2x}O''_{i,x}O_3$ and $Y^{x}_{Y,1-2x}Zr^{\bullet}_{Y,2x}O''_{i,x}O^{x}_3$ are equivalent expressions for an interstitial solid solution of ZrO_2 in Y_2O_3.
3. $Ag_{Ag,1-x}V'_{Ag,x}Ag^{\bullet}_{i,x}Cl_{Cl}$ indicates that a fraction x of the Ag^+ ions is removed from the Ag sites to interstitial sites, leaving the silver site vacant.

Formal charges may be preferred in cases where the unperturbed crystal contains an element in more than one oxidation state.

Examples:

4. $La^{2+}_{La,1-3x}La^{3+}_{La,2+2x}V_{La,x}(S^{2-})_4$ $(0<x<1/3)$
5. $Cu^{+}_{Cu,2-x}Fe^{3+}_{Cu,x}Tl^{+}_{Tl}Se^{2-}_{Se,1+2x}Se^{-}_{Se,1-2x}$ $(0<x<1/2)$ shows that Fe^{3+} partly replaces Cu^+ in $Cu^+{}_2Tl^+Se^{2-}Se^-$.

Free electrons are denoted by e', free holes by $h^{\bullet}$. As crystals are macroscopically neutral bodies, the sums of the formal charges and of the effective charges must be zero.

Key aspects of the Kröger–Vink point defect notation are summarized in Table IR-11.1.

Table IR-11.1 *Examples*[a] *of defect notation in* $M^{2+}(X^{-})_2$ *containing a foreign ion Q*

interstitial M^{2+} ion	$M_i^{\bullet\bullet}$	M atom vacancy	V_M^x
interstitial X^- ion	X_i'	X atom vacancy	V_X^x
M^{2+} ion vacancy	V_M''	normal M^{2+} ion	M_M^x
X^- ion vacancy	$V_X^{\bullet}$	normal X^- ion	X_X^x
interstitial M atom	M_i^x	Q^{3+} ion at M^{2+} site	$Q_M^{\bullet}$
interstitial X atom	X_i^x	Q^{2+} ion at M^{2+} site	Q_M^x
interstitial M^+ ion	$M_i^{\bullet}$	Q^+ ion at M^{2+} site	Q_M'
M^+ ion vacancy	V_M'	free electron	e'
		free hole	$h^{\bullet}$

[a] Consider an ionic compound $M^{2+}(X^{-})_2$. The formal charge on M is 2+, the formal charge on X is 1−. If an atom X is removed, one negative unit of charge remains on the vacant X site. The vacancy is neutral with respect to the ideal MX_2 lattice and is therefore indicated by V_X or V_X^x. If the electron is also removed from this site, the resultant vacancy is effectively positive, *i.e.* $V_X^{\bullet}$. Similarly, removal of an M atom leaves V_M, removal of an M^+ ion leaves V_M', removal of an M^{2+} ion leaves V_M''. If an impurity with a formal charge of three positive units Q^{3+} is substituted on the M^{2+} site, its effective charge is one positive unit. Therefore it is indicated by $Q_M^{\bullet}$.

IR-11.4.5 Defect clusters and use of quasi-chemical equations

Pairs or more complicated clusters of defects can be present in a solid. Such a defect cluster is indicated between parentheses. The effective charge of the cluster is indicated as an upper right index.

Examples:

1. $(Ca_K^{\bullet}V_K')^x$ denotes a neutral defect pair in a solid solution, for example of $CaCl_2$ in KCl.
2. $(V_{Pb}''V_{Cl}^{\bullet})'$ or $(V_{Pb}V_{Cl})'$ indicates a charged vacancy pair in $PbCl_2$.

Quasi-chemical reactions may be written for the formation of such defect clusters.

Examples:

3. $Cr_{Mg}^{\bullet} + V_{Mg}'' \rightarrow (Cr_{Mg}V_{Mg})'$ describes the association reaction of a Cr^{3+} impurity in MgO with magnesium vacancies.
4. $2Cr_{Mg}^{\bullet} + V_{Mg}'' \rightarrow (Cr_{Mg}V_{Mg}Cr_{Mg})^x$ gives another possible association reaction in the system of Example 3.
5. $Gd_{Ca}^{\bullet} + F_i' \rightarrow (Gd_{Ca}F_i)^x$ describes the formation of a dipole between a Gd^{3+} impurity and a fluorine interstitial in CaF_2.

IR-11.5 PHASE NOMENCLATURE

IR-11.5.1 Introduction

The use of the Pearson notation[7] (see also Section IR-3.4.4) is recommended for the designation of the structures of metals and solid solutions in binary and more complex systems. The use of Greek letters, which do not convey the necessary information, and of the *Strukturbericht* designations, which are not self-explanatory, is not acceptable.

IR-11.5.2 Recommended notation

The Pearson symbol consists of three parts: first, a lower-case italic letter (*a, m, o, t, h, c*) designating the crystal system; second, an italic capital letter (*P, S, F, I, R*) designating the lattice setting and, finally, a number designating the number of atoms or ions in the conventional unit cell. Table IR-3.1 summarizes the system.

Examples:

1. Cu, symbol (*cF*4), indicates copper of cubic symmetry, with face-centred lattice, containing 4 atoms per unit cell.
2. NaCl, symbol (*cF*8), indicates a cubic face-centred lattice with 8 ions per unit cell.
3. CuS(*hP*12), indicates a hexagonal primitive lattice with 12 ions per unit cell.

If required, the Pearson symbol can be followed by the space group and a prototype formula.

Example:

4. $CaMg_{0.5}Ag_{1.5}$(*hP*12, $P6_3/mmc$) ($MgZn_2$ type).

IR-11.6 NON-STOICHIOMETRIC PHASES

IR-11.6.1 Introduction

Several special problems of nomenclature for non-stoichiometric phases have arisen with the improvements in the precision with which their structures can be determined. Thus, there are references to homologous series, non-commensurate and semi-commensurate structures, Vernier structures, crystallographic shear phases, Wadsley defects, chemical twinned phases, infinitely adaptive phases and modulated structures. Many of the phases that fall into these classes have no observable composition ranges although they have complex structures and formulae; an example is $Mo_{17}O_{47}$. These phases, despite their complex formulae, are essentially stoichiometric and possession of a complex formula must not be taken as an indication of a non-stoichiometric compound (*cf.* Section IR-11.1.2).

IR-11.6.2 Modulated structures

Modulated structures possess two or more periodicities in the same direction of space. If the ratio of these periodicities is a rational number, the structures are called *commensurate*;

if the ratio is irrational, the structures are called *non-commensurate* or *incommensurate*. Commensurately modulated structures exist in many stoichiometric and non-stoichiometric compounds; they may be regarded as superstructures and be described by the usual rules. Non-commensurately modulated structures occur in several stoichiometric compounds (and some elements), usually in a limited temperature range, *e.g.* U, SiO_2, TaS_2, $NbSe_3$, $NaNO_2$, Na_2CO_3 and Rb_2ZnBr_4.

Many modulated structures can be regarded as being composed of two or more substructures. The substructure with the shortest periodicity often represents a simple *basic structure*, while the other periodicities cause modulations of the basic structure. The basic structure often remains unchanged within a certain composition range, while the other substructures take up the change in stoichiometry. If this change takes place continuously, a non-stoichiometric phase with a non-commensurate structure results. If the change occurs discontinuously, a series of (essentially stoichiometric) *homologous compounds* with commensurate structures (superstructures of the basic structure) may result or, in the intermediate case, a series of compounds with *semi-commensurate* or *Vernier* structures.

Examples:

1. Mn_nSi_{2n-m}
   The structure is of the *TiSi*$_2$ type which has two atom substructures, the Mn array being identical to that of the Ti array in $TiSi_2$ and the Si_2 array being identical to that of the Si_2 array in $TiSi_2$. Removal of Si leads to a composition Mn_nSi_{2n-m} in which the Mn array is completely unchanged. The Si atoms are arranged in rows and, as the Si content falls, the Si atoms in the rows spread out. In this case there will be a Vernier relationship between the Si atom rows and the static Mn positions which will change as the composition varies, giving rise to non-commensurate structures.

2. $YF_{2+x}O$
   The structure is of the *fluorite* type with extra sheets of atoms inserted into the parent YX_2 structure. When these are ordered, a homologous series of phases results. When they are disordered, there is a non-commensurate, non-stoichiometric phase, while partial ordering will give a Vernier or semi-commensurate effect. Other layer structures can be treated in the same way.

Misfit structures consist of two or more different, often mutually non-commensurate, units which are held together by electrostatic or other forces; no basic structure can be defined. The composition of compounds with misfit structures is determined by the ratio of the periodicities of their structural units and by electroneutrality.

Examples:

3. $Sr_{1-p}Cr_2S_{4-p}$ with $p = 0.29$, where chains of compositions Sr_3CrS_3 and $Sr_{3-x}S$ lie in tunnels of a framework of composition $Cr_{21}S_{36}$; the three units are mutually non-commensurate.

4. $LaCrS_3$, which is built from non-commensurate sheets of $(LaS)^+$ and $(CrS_2)^-$.

IR-11.6.3 Crystallographic shear structures

Crystallographic shear planes (*CS* planes) are planar faults in a crystal that separate two parts of the crystal which are displaced with respect to each other. The vector describing the displacement is called the crystallographic shear vector (*CS* vector). Each *CS* plane causes the composition of the crystal to change by a small increment because the sequence of crystal planes that produces the crystal matrix is changed at the *CS* plane. (From this it follows that the *CS* vector must be at an angle to the *CS* plane. If it were parallel to the plane, the succession of crystal planes would not be altered and no composition change would result. A planar boundary where the displacement vector is parallel to the plane is more properly called an *antiphase boundary*.)

Because each *CS* plane changes the composition of the crystal slightly, the overall composition of a crystal containing a population of *CS* planes will depend upon the number of *CS* planes present and their orientation. If the *CS* planes are disordered, the crystals will be non-stoichiometric, the stoichiometric variation being due to the *CS* plane 'defect'. If the *CS* planes are ordered into a parallel array, a stoichiometric phase with a complex formula results. In this case, a change in the separation of the *CS* planes in the ordered array will produce a new phase with a new composition. The series of phases produced by changes in the spacing between *CS* planes forms an *homologous* series. The general formula of a particular series will depend upon the type of *CS* plane in the array and the separation between the *CS* planes. A change in the *CS* plane may change the formula of the homologous series.

Examples:

1. Ti_nO_{2n-1}
   The parent structure is TiO_2 (*rutile* type). The *CS* planes are the (121) planes. Ordered arrays of *CS* planes can exist, producing an homologous series of oxides with formulae Ti_4O_7, Ti_5O_9, Ti_6O_{11}, Ti_7O_{13}, Ti_8O_{15} and Ti_9O_{17}. The series formula is Ti_nO_{2n-1}, with n between 4 and 9.
2. $(Mo,W)_nO_{3n-1}$
   The parent structure is WO_3. The *CS* planes are the (102) planes. Ordered arrays of *CS* planes can form, producing oxides with formulae Mo_8O_{23}, Mo_9O_{26}, $(Mo,W)_{10}O_{29}$, $(Mo,W)_{11}O_{32}$, $(Mo,W)_{12}O_{35}$, $(Mo,W)_{13}O_{38}$ and $(Mo,W)_{14}O_{41}$. The series formula is $(Mo,W)_nO_{3n-1}$, with n between 8 and 14.
3. W_nO_{3n-2}
   The parent structure is WO_3. The *CS* planes are the (103) planes. Ordered arrays of *CS* planes can form, producing oxides with formulae W_nO_{3n-2}, with n between approximately 16 and 25.

IR-11.6.4 Unit cell twinning or chemical twinning

This is a structure-building component in which two constituent parts of the structure are twin-related across the interface. The twin plane changes the composition of the host crystal by a definite amount (which may be zero). Ordered, closely spaced arrays of twin planes will lead to homologous series of phases. Disordered twin planes will lead to non-stoichiometric phases in which the twin planes serve as the defects. There is a close parallel between chemical twinning and crystallographic shear (see Section IR-11.6.3).

Example:

1. $(Bi,Pb)_nS_{n-4}$
   The parent structure is PbS which has the *cF*8 (*NaCl* type) structure. The twin planes are (311) with respect to the PbS unit cell. Two members of the homologous series are known, $Bi_8Pb_{24}S_{36}$ and $Bi_8Pb_{12}S_{24}$, but other members are found in the quaternary Ag-Bi-Pb-S system. The difference between compounds lies in the separation of the twin planes; each structure is built from slabs of PbS of varying thickness, alternate slabs being twinned across (311) with respect to the parent structure.

IR-11.6.5 Infinitely adaptive structures

In some systems it would appear that any composition can yield a fully ordered crystal structure over certain temperature and composition ranges. As the composition changes, so the structure changes to meet this need. The term *infinitely adaptive structures* has been applied to this group of substances.[8]

Examples:

1. Compounds in the Cr_2O_3-TiO_2 system between the composition ranges $(Cr,Ti)O_{2.93}$ and $(Cr,Ti)O_{2.90}$.
2. Compounds in the Nb_2O_5-WO_3 system with block-type structure between the composition limits Nb_2O_5 and $8WO_3{\cdot}9Nb_2O_5$ ($Nb_{18}W_8O_{69}$).

IR-11.6.6 Intercalation compounds

There are several materials in which a guest species is inserted into a host matrix. The process is called intercalation, and the product is called an *intercalation compound.* Common examples of intercalated materials are found in the clay silicates, layered dichalcogenides and electrode materials for lithium batteries; graphite intercalation is considered in detail in Ref. 9. Intercalated materials can be designated by conventional chemical formulae such as Li_xTaS_2 ($0<x<1$) or by host-guest designations, such as $TaS_2{:}xLi$ ($0<x<1$). If the stoichiometry is definite, ordinary compound designations may be used, *e.g.* $3TaS_2{\cdot}4N_2H_4$, $C_5H_5N{\cdot}2TiSe_2$ and KC_8.

Many intercalation compounds are layered structures and intercalation is a two-dimensional reaction. The term *insertion* is sometimes used for three-dimensional examples, as in the tungsten bronzes, *e.g.* Na_xWO_3, and the spinels, *e.g.* $Li_xMn_2O_4$, and also as a general term for a reaction involving the transfer of a guest atom, ion or molecule into a host crystal lattice[4] instead of intercalation. More specifically, *intercalation* is used for an insertion reaction that does not cause a major structural modification of the host.[4] If the structure of the host is modified significantly, for example by breaking of bonds, then the insertion can be referred to as *topochemical* or *topotactic.*[4]

IR-11.7 POLYMORPHISM

IR-11.7.1 Introduction

A number of chemical compounds and elements change their crystal structure with external conditions such as temperature and pressure. These various structures are termed polymorphic forms or modifications, and in the past have been designated using a number of labelling systems, including Greek letters and Roman numerals; the use of such non-systematic labels is not acceptable. A rational system based upon crystal structure should be used wherever possible (*cf.* Sections IR-3.4.4 and IR-4.2.5).

Polytypes and polytypoids can be regarded as a special form of polymorphism and are treated in more detail in Ref. 10.

IR-11.7.2 Use of crystal systems

Polymorphs are indicated by adding an italicized symbol denoting the crystal system after the name or formula. The symbols used are given in Table IR-3.1. For example, ZnS(*c*) corresponds to the zinc blende structure or sphalerite, and ZnS(*h*) to the wurtzite structure. Slightly distorted lattices may be indicated by using the *circa* sign ~. Thus, a slightly distorted cubic lattice would be expressed as (~*c*). In order to give more information, simple well-known structures should be designated by giving the type compound in parentheses whenever possible. For example, AuCd above 343 K should be designated AuCd (*CsCl* type) rather than AuCd(*c*).

Properties which strongly depend on lattice and point symmetries may require the addition of the space group to the crystal system abbreviation. For more details see Ref. 11.

IR-11.8 FINAL REMARKS

This Chapter deals with some basic notation and nomenclature of solid-state chemistry. In some areas, such as amorphous systems and glasses, the nomenclature needs further development. The reader is also referred to the work of the International Union of Crystallography.

IR-11.9 REFERENCES

1. Nomenclature of Inorganic Structure Types, J. Lima-de-Faria, E. Hellner, F. Liebau, E. Makovicky and E. Parthé, *Acta Crystallogr., Sect. A*, **46**, 1–11 (1990).
2. M.L. McGlashan, *Chemical Thermodynamics*, Academic Press, London, 1979, pp. 35–36.
3. *Quantities, Units and Symbols in Physical Chemistry*, Second Edn., eds. I. Mills, T. Cvitas, K. Homann, N. Kallay and K. Kuchitsu, Blackwell Scientific Publications, Oxford, 1993, p. 53. (The Green Book. The third edition is planned for publication in 2006.)
4. *Compendium of Chemical Terminology, IUPAC Recommendations*, Second Edn., eds. A.D. McNaught and A. Wilkinson, Blackwell Scientific Publications, Oxford, 1997. (The Gold Book.)
5. Chemical Nomenclature and Formulation of Compositions of Synthetic and Natural Zeolites, R.M. Barrer, *Pure Appl. Chem.*, **51**, 1091–1100 (1979).
6. F.A. Kröger and H.J. Vink, *Solid State Phys.*, **3**, 307–435 (1956).
7. W.B. Pearson, *A Handbook of Lattice Spacings and Structures of Metals and Alloys*, Vol. 2, Pergamon Press, Oxford, 1967, pp. 1–2. For tabulated lattice parameters and

data on elemental metals and semi-metals, see pp. 79–91. See also, P. Villars and L.D. Calvert, *Pearson's Handbook of Crystallographic Data for Intermetallic Phases*, Vols. 1–3, American Society for Metals, Metals Park, Ohio, USA, 1985.

8. J.S. Anderson. *J. Chem. Soc., Dalton Trans.*, 1107–1115 (1973).
9. Graphite Intercalation Compounds, Chapter II-6 in *Nomenclature of Inorganic Chemistry II, IUPAC Recommendations 2000*, eds. J.A. McCleverty and N.G. Connelly, Royal Society of Chemistry, 2001.
10. Nomenclature of Polytype Structures, A. Guinier, G.B. Bokij, K. Boll-Dornberger, J.M. Cowley, S. Durovic, H. Jagodzinski, P. Krishna, P.M. de Wolff, B.B. Zvyagin, D.E. Cox, P. Goodman, Th. Hahn, K. Kuchitsu and S.C. Abrahams, *Acta Crystallogr., Sect. A*, **40**, 399–404 (1984). See also, S.W. Bailey, V.A. Frank-Kamenetskii, S. Goldsztaub, A. Kato, A. Pabst, H. Schulz, H.F.W. Taylor, M. Fleischer and A.J.C. Wilson, *Acta Crystallogr., Sect. A*, **33**, 681–684 (1977).
11. Structural Phase Transition Nomenclature, J.-C. Tolédano, A.M. Glazer, Th. Hahn, E. Parthé, R.S. Roth, R.S. Berry, R. Metselaar and S.C. Abrahams, *Acta Crystallogr., Sect. A*, **54**, 1028–1033 (1998). Nomenclature of magnetic, incommensurate, composition-changed morphotropic, polytype, transient-structural and quasicrystalline phases undergoing phase transitions, J.-C. Tolédano, R.S. Berry, P.J. Brown, A.M. Glazer, R. Metselaar, D. Pandey, J.M. Perez-Mato, R.S. Roth and S.C. Abrahams, *Acta Crystallogr., Sect A*, **57**, 614–626 (2001), and erratum in *Acta Crystallogr., Sect. A*, **58**, 79 (2002).

TABLES

Table I *Names, symbols and atomic numbers of the elements (see also Section IR-3.1)*

Name	*Symbol*	*Atomic number*	*Name*	*Symbol*	*Atomic number*
actinium	Ac	89	indium	In	49
aluminium[a]	Al	13	iodine	I	53
americium	Am	95	iridium	Ir	77
antimony	Sb[b]	51	iron	Fe[g]	26
argon	Ar	18	krypton	Kr	36
arsenic	As	33	lanthanum	La	57
astatine	At	85	lawrencium	Lr	103
barium	Ba	56	lead	Pb[h]	82
berkelium	Bk	97	lithium	Li	3
beryllium	Be	4	lutetium	Lu	71
bismuth	Bi	83	magnesium	Mg	12
bohrium	Bh	107	manganese	Mn	25
boron	B	5	meitnerium	Mt	109
bromine	Br	35	mendelevium	Md	101
cadmium	Cd	48	mercury	Hg[i]	80
caesium[c]	Cs	55	molybdenum	Mo	42
calcium	Ca	20	neodymium	Nd	60
californium	Cf	98	neon	Ne	10
carbon	C	6	neptunium	Np	93
cerium	Ce	58	nickel	Ni	28
chlorine	Cl	17	niobium	Nb	41
chromium	Cr	24	nitrogen[j]	N	7
cobalt	Co	27	nobelium	No	102
copper	Cu[d]	29	osmium	Os	76
curium	Cm	96	oxygen	O	8
darmstadtium	Ds	110	palladium	Pd	46
dubnium	Db	105	phosphorus	P	15
dysprosium	Dy	66	platinum	Pt	78
einsteinium	Es	99	plutonium	Pu	94
erbium	Er	68	polonium	Po	84
europium	Eu	63	potassium	K[k]	19
fermium	Fm	100	praseodymium	Pr	59
fluorine	F	9	promethium	Pm	61
francium	Fr	87	protactinium	Pa	91
gadolinium	Gd	64	radium	Ra	88
gallium	Ga	31	radon	Rn	86
germanium	Ge	32	rhenium	Re	75
gold	Au[e]	79	rhodium	Rh	45
hafnium	Hf	72	roentgenium	Rg	111
hassium	Hs	108	rubidium	Rb	37
helium	He	2	ruthenium	Ru	44
holmium	Ho	67	rutherfordium	Rf	104
hydrogen	H[f]	1	samarium	Sm	62

Table I *Continued*

Name	*Symbol*	*Atomic number*	*Name*	*Symbol*	*Atomic number*
scandium	Sc	21	thorium	Th	90
seaborgium	Sg	106	thulium	Tm	69
selenium	Se	34	tin	Sn[o]	50
silicon	Si	14	titanium	Ti	22
silver	Ag[l]	47	tungsten	W[p]	74
sodium	Na[m]	11	uranium	U	92
strontium	Sr	38	vanadium	V	23
sulfur[n]	S	16	xenon	Xe	54
tantalum	Ta	73	ytterbium	Yb	70
technetium	Tc	43	yttrium	Y	39
tellurium	Te	52	zinc	Zn	30
terbium	Tb	65	zirconium	Zr	40
thallium	Tl	81			

[a] The alternative spelling 'aluminum' is commonly used.
[b] The element symbol Sb derives from the name stibium.
[c] The alternative spelling 'cesium' is commonly used.
[d] The element symbol Cu derives from the name cuprum.
[e] The element symbol Au derives from the name aurum.
[f] The hydrogen isotopes ^{2}H and ^{3}H are named deuterium and tritium, respectively, for which the symbols D and T may be used. However, ^{2}H and ^{3}H are preferred (see Section IR-3.3.2).
[g] The element symbol Fe derives from the name ferrum.
[h] The element symbol Pb derives from the name plumbum.
[i] The element symbol Hg derives from the name hydrargyrum.
[j] The name azote provides the root 'az' for nitrogen.
[k] The element symbol derives K from the name kalium.
[l] The element symbol Ag derives from the name argentum.
[m] The element symbol Na derives from the name natrium.
[n] The name theion provides the root 'thi' for sulfur.
[o] The element symbol Sn derives from the name stannum.
[p] The element symbol W derives from the name wolfram.

Table II *Temporary names and symbols for elements of atomic number greater than 111*[a]

Atomic number	*Name*[b]	*Symbol*
112	ununbium	Uub
113	ununtrium	Uut
114	ununquadium	Uuq
115	ununpentium	Uup
116	ununhexium	Uuh
117	ununseptium	Uus
118	ununoctium	Uuo
119	ununennium	Uue
120	unbinilium	Ubn
121	unbiunium	Ubu
130	untrinilium	Utn
140	unquadnilium	Uqn
150	unpentnilium	Upn
160	unhexnilium	Uhn
170	unseptnilium	Usn
180	unoctnilium	Uon
190	unennilium	Uen
200	binilnilium	Bnn
201	binilunium	Bnu
202	binilbium	Bnb
300	trinilnilium	Tnn
400	quadnilnilium	Qnn
500	pentnilnilium	Pnn
900	ennilnilium	Enn

[a] These names are used only when the permanent name has not yet been assigned by IUPAC (see Section IR-3.1.1).
[b] One may also write, for example, 'element 112'.

Table III *Suffixes and endings*[a]

a	Terminal vowel of prefixes indicating replacement of:
	carbon atoms by atoms of other elements in skeletal replacement nomenclature (Section IR-6.2.4.1) and Hantzsch–Widman nomenclature (Section IR-6.2.4.3), *e.g.* 'oxa', 'aza';
	boron atoms by atoms of other elements in boron hydride-based nomenclature (Section IR-6.2.4.4), *e.g.* 'carba', 'thia';
	heteroatoms by carbon atoms in natural product nomenclature (prefix 'carba').
	See Table X for 'a' prefixes for all elements.
ane	Ending of names of neutral saturated parent hydrides of elements of Groups 13–17, *e.g.* thallane, cubane, cyclohexane, cyclohexasilane, diphosphane, tellane, λ^4-tellane. *Cf.* Section IR-6.2.2 and Table IR-6.1.
	Last part of endings of a number of parent names of saturated heteromonocycles in Hantzsch–Widman nomenclature, *i.e.* of 'irane', 'etane', 'olane', 'ane', 'inane', 'epane', 'ocane', 'onane' and 'ecane' (see Section IR-6.2.4.3).
anide	Combined ending of names of anions resulting from the removal of a hydron from a parent hydride with an 'ane' name, formed by adding the suffix 'ide', *e.g.* methanide, CH_3^-. *Cf.* Section IR-6.4.4.
anium	Combined ending of names of cations resulting from the addition of a hydron to a parent structure with an 'ane' name, formed by adding the suffix 'ium', *e.g.* phosphanium, PH_4^+. *Cf.* Section IR-6.4.1.
ano	Ending resulting from the change of the 'ane' ending in names of parent hydrides to form prefixes denoting bridging divalent substituent groups, *e.g.* diazano, −HNNH−.
ate	General ending of additive names of anions, *e.g.* tetrahydridoaluminate(1−), $[AlH_4]^-$. *Cf.* Section IR-7.1.4 and Table X.
	Ending of names of anions and esters of inorganic oxoacids having the 'ic' ending in the acid name, *e.g.* nitrate, phosphonate, trimethyl phosphate, and of anions and esters of organic acids, *e.g.* acetate, methyl acetate, thiocyanate. See Tables IR-8.1 and IR-8.2 and Table IX for more examples of 'ate' anion names. See also 'inate', 'onate'.
ato	Ending of name of any anion with an 'ate' name (see above) acting as a ligand, *e.g.* tetrahydridoaluminato(1−), nitrato, acetato. *Cf.* Sections IR-7.1.3 and IR-9.2.2.3 and Table IX. See also 'inato', 'onato'.
	Ending of prefixes for certain anionic substituent groups, *e.g.* carboxylato, $-C(=O)O^-$; phosphato, $-O-P(=O)(O^-)_2$. See also 'onato'.
diene	See 'ene'.
diide	See 'ide'.
diido	See 'ido'.
diium	See 'ium'.
diyl	Combined suffix composed of the suffix 'yl' and the multiplicative prefix 'di', indicating the loss of two hydrogen atoms from a parent hydride resulting in a diradical, or a substituent group with two single bonds, if necessary accompanied by locants, *e.g.* hydrazine-1,2-diyl, $^{\bullet}HNNH^{\bullet}$ or −HNNH−; phosphanediyl, $HP<$. See also 'ylidene'.
diylium	See 'ylium'.
ecane	Ending of parent names of ten-membered saturated heteromonocycles in Hantzsch–Widman nomenclature, *cf.* Section IR-6.2.4.3.

Table III *Continued*

ecine	Ending of parent names of ten-membered heteromonocycles with the maximum number of non-cumulative double bonds in Hantzsch–Widman nomenclature, *cf.* Section IR-6.2.4.3.
ene	Ending of systematic names of acyclic and cyclic parent structures with double-bond unsaturation, replacing 'ane' in the name of the corresponding saturated parent hydride, and if necessary accompanied by locants and multiplicative prefixes specifying the locations and number of double bonds, *e.g.* diazene, triazene, pentasil-1-ene, cyclopenta-1,3-diene. *Cf.* Sections IR-6.2.2.3 and IR- 6.2.2.4.
	Ending of certain acceptable non-systematic names of unsaturated cyclic parent hydrides, *e.g.* benzene, azulene.
	See also 'irene', 'ocene'.
enide	Combined ending of names of anions resulting from the removal of a hydron from a parent hydride with an 'ene' name, formed by adding the suffix 'ide', *e.g.* diazenide, $HN{=}N^{-}$. *Cf.* Section IR-6.4.4.
enium	Combined ending of names of cations resulting from the addition of a hydron to a parent structure with an 'ene' name, formed by adding the suffix 'ium', *e.g.* diazenium. *Cf.* Section IR-6.4.1.
	Combined ending resulting from the addition of the suffix 'ium' to a metallocene name. This leads to ambiguous names, see Section IR-10.2.6.
eno	Ending resulting from the change of the 'ene' ending in names of cyclic mancude ring systems to 'eno' to form prefixes in fusion nomenclature. (See Section P-25.3 of the Blue Book[b].)
	Ending resulting from the change of the 'ene' ending in names of parent hydrides to form prefixes denoting bridging divalent substituent groups, *e.g.* diazeno, $-N{=}N-$.
epane	Ending of parent names of seven-membered saturated heteromonocycles in Hantzsch–Widman nomenclature, *cf.* Section IR-6.2.4.3.
epine	Ending of parent names of seven-membered heteromonocycles with the maximum number of non-cumulative double bonds in Hantzsch–Widman nomenclature, *cf.* Section IR-6.2.4.3.
etane	General ending of parent names of four-membered saturated heteromonocycles in Hantzsch–Widman nomenclature, *cf.* Section IR-6.2.4.3. See also 'etidine'.
ete	Ending of parent names of four-membered heteromonocycles with the maximum number of non-cumulative double bonds in Hantzsch–Widman nomenclature, *cf.* Section IR-6.2.4.3.
etidine	Ending of parent names of four-membered nitrogen-containing saturated heteromonocycles in Hantzsch–Widman nomenclature, *cf.* Section IR-6.2.4.3.
ic	Ending of names of many acids, both inorganic and organic, *e.g.* sulfuric acid, acetic acid, benzoic acid. For more examples, particularly of inorganic 'ic' acid names, see Tables IR-8.1 and IR-8.2, and Table IX. See also 'inic' and 'onic'.
	Ending formerly added to stems of element names to indicate a higher oxidation state, *e.g.* ferric chloride, cupric oxide, ceric sulfate. Such names are no longer acceptable.
ide	Ending of names of monoatomic and homopolyatomic anions, *e.g.* chloride, sulfide, disulfide(2−), triiodide(1−). *Cf.* Sections IR-5.3.3.2 and IR-5.3.3.3 and Table IX.
	Ending of names of formally electronegative homoatomic constituents in compositional names, *e.g.* disulfur dichloride. *Cf.* Section IR-5.4.

Table III *Continued*

	Ending of some acceptable non-systematic names of heteropolyatomic anions: cyanide, hydroxide.
	Suffix for names of anions formed by removal of one or more hydrons from a parent hydride, accompanied by locants and multiplicative prefixes as appropriate, *e.g.* hydrazinide, H_2NNH^-; hydrazine-1,2-diide, $^-HNNH^-$; disulfanediide, S_2^{2-}; methanide, CH_3^-.
ido	Ending of name of any anion with an 'ide' name (see above) acting as a ligand, *e.g.* chlorido, disulfido(2−) or disulfanediido, hydrazinido, hydrazine-1,2-diido, methanido. *Cf.* Sections IR-7.1.3 and IR-9.2.2.3 and Table IX.
	Ending of certain prefixes for anionic substituent groups, *e.g.* oxido for $-O^-$.
inane	Ending of parent names of six-membered saturated heteromonocycles in Hantzsch–Widman nomenclature, *cf.* Section IR-6.2.4.3.
inate	Ending of names of anions and esters of 'inic' oxoacids, *e.g.* borinate, phosphinate.
inato	Modification of the 'inate' ending of an anion name (see above) used when the anion acts as a ligand.
ine	Ending of the non-systematic, but still acceptable, parent name hydrazine (N_2H_4) and of the now obsolete names of other Group 15 hydrides, *e.g.* phosphine (PH_3).
	Ending of names of large heteromonocycles (more than 10 ring atoms) with the maximum number of non-cumulative double bonds for use in fusion nomenclature, *e.g.* 2*H*-1-oxa-4,8,11-triazacyclotetradecine.
	Last part of endings of a number of parent names in Hantzsch–Widman nomenclature of heteromonocycles, *i.e.* of 'irine', 'iridine', 'etidine', 'olidine', 'ine', 'inine', 'epine', 'ocine', 'onine' and 'ecine' (see Section IR-6.2.4.3).
	Ending of a number of parent names of nitrogeneous heterocyclic parent hydrides, *e.g.* pyridine, acridine.
inic	Ending of the parent names of acids of the types $H_2X(=O)(OH)$ (X=N, P, As, Sb), *e.g.* stibinic acid; $HX(=O)(OH)$ (X=S, Se, Te), *e.g.* sulfinic acid; and of borinic acid, H_2BOH.
inide	Combined ending of names of anions resulting from the removal of a hydron from a parent hydride with an 'ine' name, formed by adding the suffix 'ide', *e.g.* hydrazinide, H_2NNH^-. *Cf.* Section IR-6.4.4.
inine	Ending of parent names of six-membered heteromonocycles with the maximum number of non-cumulative double bonds in Hantzsch–Widman nomenclature, *cf.* Section IR-6.2.4.3.
inite	Ending of names of anions and esters of oxoacids with an 'inous' name, *e.g.* phosphinite, H_2PO^-, from phosphinous acid.
inito	Modification of the 'inite' ending of an anion name (see above) used when the anion acts as a ligand.
inium	Combined ending of names of cations resulting from the addition of a hydron to a parent structure with an 'ine' name, formed by adding the suffix 'ium', *e.g.* hydrazinium, pyridinium. *Cf.* Section IR-6.4.1.
ino	Ending of some non-systematic substituent group prefixes, *e.g.* amino, NH_2-; hydrazino, H_2NNH-.
	Ending resulting from the change of the 'ine' ending in names of cyclic mancude ring systems to 'ino' to form prefixes in fusion nomenclature. (See Section P-25.3 of the Blue Book[b].)

Table III *Continued*

inous	Ending of the parent names of acids of the types $H_2X(OH)$ (X = N, P, As, Sb), *e.g.* stibinous acid. See Table IR-8.1 for other such names.
inoyl	Ending of prefixes for substituent groups formed by removing all hydroxy groups from 'inic' acids (see above), *e.g.* phosphinoyl, $H_2P(O)$–; seleninoyl, HSe(O)–. (See Table IR-8.1 for phosphinic and seleninic acids.)
inyl	Ending of prefixes for the divalent substituent groups >X=O (sulfinyl, seleninyl and tellurinyl for X = S, Se and Te, respectively).
io	Ending of acceptable alternative prefixes for certain cationic substituent groups, *e.g.* ammonio for azaniumyl, pyridinio for pyridiniumyl (*cf.* Section IR-6.4.9).
	Now abandoned ending of prefixes for substituent groups consisting of a single atom, *e.g.* mercurio, –Hg–.
irane	General ending of parent names of three-membered saturated heteromonocycles in Hantzsch–Widman nomenclature, *cf.* Section IR-6.2.4.3. See also 'iridine'.
irene	General ending of parent names of three-membered heteromonocycles with the maximum number of non-cumulative double bonds (*i.e.* one double bond) in Hantzsch–Widman nomenclature, *cf.* Section IR-6.2.4.3. See also 'irine'.
iridine	Ending of parent names of three-membered nitrogen-containing saturated heteromonocycles in Hantzsch–Widman nomenclature, *cf.* Section IR-6.2.4.3.
irine	Ending of parent names of three-membered heteromonocycles with the maximum number of non-cumulative double bonds (*i.e.* one double bond) and N as the only heteroatom(s) in Hantzsch–Widman nomenclature, *cf.* Section IR-6.2.4.3.
ite	Ending of names of anions and esters of oxoacids having the 'ous' or the 'orous' ending in the acid name, *e.g.* hypochlorite (from hypochlorous acid), methyl sulfite (from sulfurous acid). *Cf.* Table IR-8.1. See also 'inite', 'onite'.
ito	Ending of name of any anion with an 'ite' name (see above) acting as a ligand, *e.g.* nitrito, sulfito. *Cf.* Sections IR-7.1.3 and IR-9.2.2.3 and Table IX. See also 'inito', 'onito'.
ium	Ending of names of many elements and their cations, *e.g.* helium, seaborgium, thallium(1+), and of the name of any new element (*cf.* Ref. 1 of Chapter IR-3).
	Suffix to indicate addition of hydrons to a parent hydride or other parent structure (see 'anium', 'enium', 'inium', 'onium', 'ynium'), accompanied by multiplying prefixes and locants as appropriate, *e.g.* hydrazinium, $H_2NNH_3^+$; hydrazine-1,2-diium, $^+H_3NNH_3^+$.
o	Terminal vowel indicating a negatively charged ligand; see 'ato', 'ido', 'ito'.
	Terminal vowel of prefixes for many inorganic and organic substituent groups, *e.g.* amino, chloro, oxido, sulfo, thiolato.
	Terminal vowel of prefixes for fusion components. (See Section P-25.3 of the Blue Book[b].) See also 'eno', 'ino'.
	Terminal vowel of infixes used in functional replacement nomenclature (Section IR-8.6) to indicate replacement of oxygen atoms and/or hydroxy groups, *e.g.* 'amido', 'nitrido', 'thio'.
ocane	Ending of parent names of eight-membered saturated heteromonocycles in Hantzsch–Widman nomenclature, *cf.* Section IR-6.2.4.3.
ocene	Ending of the names of certain bis(cyclopentadienyl)metal compounds, *e.g.* ferrocene. *Cf.* Section IR-10.2.6.

Table III *Continued*

ocine	Ending of parent names of eight-membered heteromonocycles with the maximum number of non-cumulative double bonds in Hantzsch–Widman nomenclature, *cf.* Section IR-6.2.4.3.
ol	Suffix specifying substitution af a hydrogen atom in a parent hydride for the group –OH, accompanied by locants and multiplicative prefixes if appropriate, *e.g.* silanol, SiH_3OH; trisilane-1,3-diol, $SiH_2(OH)SiH_2SiH_2OH$.
	Ending of corresponding suffixes 'thiol', 'selenol', 'tellurol' for –SH, –SeH and –TeH, respectively.
olane	General ending of parent names of five-membered saturated heteromonocycles in Hantzsch–Widman nomenclature, *cf.* IR-6.2.4.3. See also 'olidine'.
olate	Suffix specifying substitution of a hydrogen atom in a parent hydride for the substituent $-O^-$, accompanied by locants and multiplicative prefixes if appropriate, *e.g.* silanolate, SiH_3O^-; trisilane-1,3-diolate, $SiH_2(O^-)SiH_2SiH_2O^-$.
	Ending of corresponding suffixes 'thiolate', 'selenolate', 'tellurolate' for $-S^-$, $-Se^-$ and $-Te^-$, respectively.
olato	Modification of the suffix 'olate' used when the anion in question acts as a ligand.
ole	Ending of parent names of five-membered heteromonocycles with the maximum number of non-cumulative double bonds in Hantzsch–Widman nomenclature, *cf.* Section IR-6.2.4.3.
olidine	Ending of parent names of five-membered nitrogen-containing saturated heteromonocycles in Hantzsch–Widman nomenclature, *cf.* Section IR-6.2.4.3.
onane	Ending of parent names of nine-membered saturated heteromonocycles in Hantzsch–Widman nomenclature, *cf.* Section IR-6.2.4.3.
onate	Ending of names of anions and esters of 'onic' oxoacids, *e.g.* boronate, phosphonate, tetrathionate.
onato	Modification of the 'onate' ending of an anion used when the anion acts as a ligand.
	Ending of prefixes of certain anionic substituent groups, *e.g.* phosphonato, $-P(=O)(O^-)_2$; sulfonato, $-S(=O)_2(O^-)$.
one	Suffix specifying the substitution of two hydrogen atoms on the same skeletal atom in a parent hydride for the substituent =O, accompanied by locants and multiplicative prefixes as appropriate, *e.g.* phosphanone, HP=O; pentane-2,4-dione, $CH_3C(=O)CH_2C(=O)CH_3$.
	Ending of corresponding suffixes 'thione', 'selenone', 'tellurone' for =S, =Se and =Te, respectively.
onic	Ending of the parent names of acids of the types $HXO(OH)_2$ (X=N, P, As, Sb), *e.g.* stibonic acid; $HXO_2(OH)$ (X = S, Se, Te), *e.g.* sulfonic acid; and of boronic acid, $HB(OH)_2$. See Table IR-8.1.
	Ending of the parent names dithionic, trithionic, *etc.*, acids (see Table IR-8.1).
onine	Ending of parent names of nine-membered heteromonocycles with the maximum number of non-cumulative double bonds in Hantzsch–Widman nomenclature, *cf.* Section IR-6.2.4.3.
onite	Ending of names of anions and esters of 'onous' oxoacids, *e.g.* phosphonite, tetrathionite.
onito	Modification of the 'onite' ending of an anion name used when the anion acts as a ligand.

Table III *Continued*

onium	Ending of still acceptable non-systematic names of cations formed by hydron addition to a mononuclear parent hydride: ammonium, oxonium (see Section IR-6.4.1).
ono	Ending of prefixes for substituent groups formed from 'onic' acids by removal of a hydrogen atom, *e.g.* phosphono for $-P(=O)(OH)_2$. *Exception:* note that $-S(=O)_2OH$ is 'sulfo' rather than 'sulfono'.
onous	Ending of the parent names of acids of the types $HX(OH)_2$ (X = N, P, As, Sb), *e.g.* stibonous acid.
	Ending of the parent names dithionous, trithionous, *etc.*, acids (see Table IR-8.1).
onoyl	Ending of prefixes for substituent groups formed by removing all hydroxy groups from 'onic' acids, *e.g.* phosphonoyl, $HP(O){<}$; selenonoyl, $HSe(O)_2-$. (See Table IR-8.1 for phosphonic and selenonic acids.)
onyl	Ending of prefixes for the divalent substituent groups ${>}X(=O)_2$ (sulfonyl, selenonyl and telluronyl for X = S, Se and Te, respectively).
orane	Ending of the acceptable alternative names phosphorane for λ^5-phosphane (PH_5), arsorane for λ^5-arsane (AsH_5) and stiborane for λ^5-stibane (SbH_5).
oryl	Ending of prefixes for substituent groups formed by removing all hydroxy groups from 'oric' acids, *e.g.* phosphoryl, $P(O)\mathord{<}\!\!-$, from phosphoric acid.
ous	Ending of parent names of certain inorganic oxoacids, *e.g.* arsorous acid, seleninous acid. For more examples of 'ous' acid names, see Tables IR-8.1 and IR-8.2. See also 'inous', 'onous'.
	Ending formerly added to stems of element names to indicate a lower oxidation state, *e.g.* ferrous chloride, cuprous oxide, cerous hydroxide. Such names are no longer acceptable.
triene	See 'ene'.
triide	See 'ide'.
triium	See 'ium'.
triyl	Combined suffix composed of the suffix 'yl' and the multiplying prefix 'tri', indicating the loss of three hydrogen atoms from a parent hydride resulting in a triradical or a substituent group forming three single bonds, *e.g.* the substituent groups boranetriyl, $-B{<}$; trisilane-1,2,3-triyl, $-SiH_2\overset{\vert}{Si}HSiH_2-$; λ^5-phosphanetriyl, $H_2P\mathord{<}\!\!-$. (See also 'ylidyne' and 'ylylidene'.)
uide	Suffix specifying the addition of hydride to a parent structure, accompanied by locants and multiplicative prefixes if appropriate, *e.g.* tellanuide, TeH_3^-.
uido	Modification of the 'uide' suffix in an anion name used when the anion acts as a ligand.
y	Terminal vowel of prefixes for some substituent groups, *e.g.* carboxy, $-COOH$; hydroxy, $-OH$; oxy, $-O-$.
	Terminal vowel in prefixes used in specifying chain and ring atoms in additive nomenclature for inorganic chains and rings, *cf.* Section IR-7.4. These prefixes are given for all elements in Table X.
yl	Suffix to indicate removal of hydrogen atoms from a parent hydride to form radicals or substituent groups, accompanied by multiplicative prefixes and locants as appropriate, *e.g.* hydrazinyl, $H_2NNH^\bullet$ or H_2NNH-; hydrazine-1,2-diyl, $^\bullet HNNH^\bullet$ or $-HNNH-$. (See also 'diyl', 'ylene', 'ylidene', 'triyl', 'ylylidene', 'ylidyne'.)

Table III *Continued*

	Ending of certain non-systematic names of oxidometal cations, *e.g.* vanadyl for oxidovanadium(2+). These names are no longer acceptable.
ylene	Ending of a few still acceptable names for divalent substituent groups, meaning the same as 'diyl': methylene for methanediyl, $-CH_2-$; phenylene for benzenediyl, $-C_6H_4-$; (1,2-phenylene for benzene-1,2-diyl *etc.*).
ylidene	Suffix for names of divalent substituent groups formed by the loss of two hydrogen atoms from the same atom of a parent hydride and forming a double bond, *e.g.* azanylidene, HN= , and for names of corresponding diradicals. (See also 'diyl'.)
ylidyne	Suffix for names of trivalent substituent groups formed by the loss of three hydrogen atoms from the same atom of a parent hydride and forming a triple bond, *e.g.* phosphanylidyne, P≡. (See also 'ylylidene' and 'triyl'.)
ylium	Suffix for names of cations formed by the loss of hydride ions from parent hydrides, accompanied by locants and multiplicative prefixes as appropriate, *e.g.* azanylium, NH_2^+; disilane-1,2-diylium, $^+H_2SiSiH_2^+$.
ylylidene	Combined suffix ('yl' plus 'ylidene') for names of trivalent substituent groups formed by the loss of three hydrogen atoms from the same atom, forming a single bond and a double bond, *e.g.* azanylylidene, –N= . (See also 'ylidyne' and 'triyl'.)
yne	Ending of systematic names of acyclic and cyclic parent structures with triple-bond unsaturation, replacing 'ane' in the name of the corresponding saturated parent hydride, and if necessary accompanied by locants and multiplicative prefixes specifying the locations and number of triple bonds, *e.g.* diazyne (see 'ynium' for an application of this name), ethyne, penta-1,4-diyne.
ynide	Combined ending of names of anions resulting from the removal of a hydron from a parent hydride with an 'yne' name, formed by adding the suffix 'ide', *e.g.* ethynide, $CH{\equiv}C^-$. *Cf.* Section IR-6.4.4.
ynium	Combined ending of names of cations resulting from the addition of a hydron to a parent structure with an 'yne' name, formed by adding the suffix 'ium', *e.g.* diazynium ($N{\equiv}NH^+$). *Cf.* Section IR-6.4.1.

[a] The term 'suffix' is understood here to mean a name part added to a parent name in order to specify a modification of that parent, *e.g.* substitution of a hydrogen atom in a parent hydride by a characteristic group (suffixes such as 'carboxylic acid', 'thiol', *etc.*) or formation of a radical or substituent group by removal of one or more hydrogen atoms (suffixes such as 'yl', 'ylidene', *etc.*). The term 'ending' is used in a broader sense, but also to designate specifically the common last part (last syllable or last few syllables) of systematic names for members of classes of compounds (such as 'ane', 'ene', 'diene', 'yne', *etc.*, for parent hydrides, and 'onic acid', 'inic acid', *etc.*, for inorganic oxoacids).
[b] *Nomenclature of Organic Chemistry, IUPAC Recommendations*, eds. W.H. Powell and H. Favre, Royal Society of Chemistry, in preparation. (The Blue Book.)

Table IV *Multiplicative prefixes*

1	mono	21	henicosa
2	di[a] (bis[b])	22	docosa
3	tri (tris)	23	tricosa
4	tetra (tetrakis)	30	triaconta
5	penta (pentakis)	31	hentriaconta
6	hexa (hexakis)	35	pentatriaconta
7	hepta (heptakis)	40	tetraconta
8	octa (octakis)	48	octatetraconta
9	nona (nonakis)	50	pentaconta
10	deca (decakis)	52	dopentaconta
11	undeca	60	hexaconta
12	dodeca	70	heptaconta
13	trideca	80	octaconta
14	tetradeca	90	nonaconta
15	pentadeca	100	hecta
16	hexadeca	200	dicta
17	heptadeca	500	pentacta
18	octadeca	1000	kilia
19	nonadeca	2000	dilia
20	icosa		

[a] In the case of a ligand using two donor atoms, the term 'bidentate' rather than 'didentate' is recommended because of prevailing usage.

[b] The prefixes bis, tris, *etc.* (examples are given for 1–10 but continue throughout) are used with composite ligand names or in order to avoid ambiguity.

Table V *Geometrical and structural affixes*

Except for those denoted by Greek letters, geometrical and structural affixes are italicized. All are separated from the rest of the name by hyphens.

antiprismo	eight atoms bound into a regular antiprism
arachno	a boron structure intermediate between *nido* and *hypho* in degree of openness
asym	asymmetrical
catena	a chain structure; often used to designate linear polymeric substances
cis	two groups occupying adjacent positions in a coordination sphere
closo	a cage or closed structure, especially a boron skeleton that is a polyhedron having all faces triangular
cyclo	a ring structure. (Here, *cyclo* is used as a modifier indicating structure and hence is italicized. In organic nomenclature, 'cyclo' is considered to be part of the parent name since it changes the molecular formula. It is therefore not italicized.)
δ (delta)	denotes the absolute configuration of chelate ring conformations
Δ (delta)	a structural descriptor to designate deltahedra, or shows absolute configuration
dodecahedro	eight atoms bound into a dodecahedron with triangular faces
η (eta)	specifies the bonding of contiguous atoms of a ligand to a central atom
fac	three groups occupying the corners of the same face of an octahedron
hexahedro	eight atoms bound into a hexahedron (*e.g.* cube)
hexaprismo	twelve atoms bound into a hexagonal prism
hypho	an open structure, especially a boron skeleton, more closed than a *klado* structure but more open than an *arachno* structure
icosahedro	twelve atoms bound into an icosahedron with triangular faces
κ (kappa)	specifies the donor atoms in a ligand
klado	a very open polyboron structure
λ (lambda)	signifies, with its superscript, the bonding number, *i.e.* the sum of the number of skeletal bonds and the number of hydrogen atoms associated with an atom in a parent compound; denotes the absolute configuration of chelate ring conformations
Λ (lambda)	shows absolute configuration
mer	meridional; three groups occupying vertices of an octahedron so that one is *cis* to the other two which are themselves mutually *trans*
μ (mu)	signifies that a group so designated bridges two or more coordination centres
nido	a nest-like structure, especially a boron skeleton that is almost closed
octahedro	six atoms bound into an octahedron
pentaprismo	ten atoms bound into a pentagonal prism
quadro	four atoms bound into a quadrangle (*e.g.* square)
sym	symmetrical
tetrahedro	four atoms bound into a tetrahedron
trans	two groups occupying positions in a coordination sphere directly opposite each other
triangulo	three atoms bound into a triangle
triprismo	six atoms bound into a triangular prism

Table VI *Element sequence*

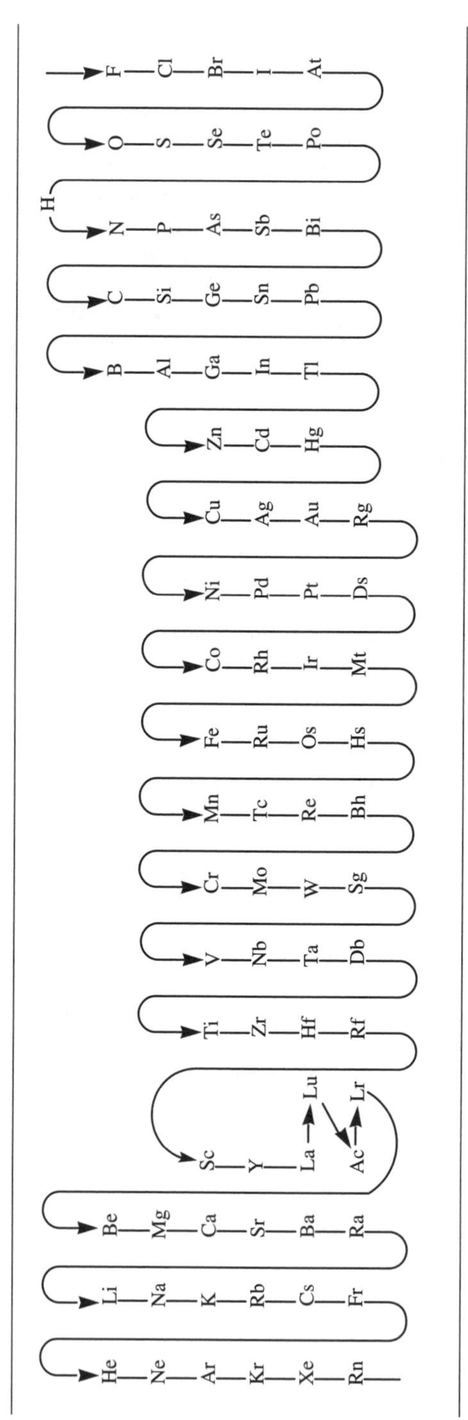

Table VII *Ligand abbreviations*

Guidelines for the construction and use of ligand abbreviations are given in Section IR-4.4.4 and their use in the formulae of coordination complexes is described in Section IR-9.2.3.4. Abbreviations are listed in alphabetical order but those beginning with a numeral are listed by the first letter of the abbreviation (*e.g.* 2,3,2-tet appears under the letter 't').

Structural formulae of selected ligands are shown in Table VIII (numbered according to the present Table).

Number and abbreviation	*Systematic name*	*Other name (from which abbreviation derived)*[a]
1. 4-abu	4-aminobutanoato	
2. Ac	acetyl	
3. acac	2,4-dioxopentan-3-ido	acetylacetonato
4. acacen	2,2′-[ethane-1,2-diylbis(azanylylidene)]bis(4-oxopentan-3-ido)	bis(acetylacetonato)ethylenediamine
5. ade	9*H*-purin-6-amine	adenine
6. ado	9-β-D-ribofuranosyl-9*H*-purin-6-amine	adenosine
7. adp	adenosine 5′-diphosphato(3−)	
8. aet	2-aminoethanethiolato	
9. ala	2-aminopropanoato	alaninato
10. ama	2-aminopropanedioato	aminomalonato
11. amp	adenosine 5′-phosphato(2−)	adenosine monophosphato
12. [9]aneN_3 (also tacn)	1,4,7-triazonane	
13. [12]aneN_4 (also cyclen)	1,4,7,10-tetraazacyclododecane	
14. [14]aneN_4 (also cyclam)	1,4,8,11-tetraazacyclotetradecane	
15. [18]aneP_4O_2	1,10-dioxa-4,7,13,16-tetraphosphacyclooctadecane	
16. [9]aneS_3	1,4,7-trithionane	
17. [12]aneS_4	1,4,7,10-tetrathiacyclododecane	
18. arg	2-amino-5-carbamimidamidopentanoato	argininato
19. asn	2,4-diamino-4-oxobutanoato	asparaginato
20. asp	2-aminobutanedioato	aspartato
21. atmp	[nitrilotris(methylene)]tris(phosphonato)	aminotris(methylenephosphonato)
22. atp	adenosine 5′-triphosphato(4−)	
23. 2,3-bdta	2,2′,2′′,2′′′-(butane-2,3-diyldinitrilo)tetraacetato	

Table VII *Continued*

Number and abbreviation	*Systematic name*	*Other name (from which abbreviation derived)*[a]
24. benzo-15-crown-5	2,3,5,6,8,9,11,12-octahydro-1,4,7,10,13-benzopentaoxacyclopentadecine	
25. big	bis(carbamimidoyl)azanido	biguanid-3-ido
26. biim	2,2′-bi(1*H*-imidazole)-1,1′-diido	2,2′-biimidazolato
27. binap	1,1′-binaphthalene-2,2′-diylbis(diphenylphosphane)	
28. bn	butane-2,3-diamine	
29. bpy	2,2′-bipyridine	
30. 4,4′-bpy	4,4′-bipyridine	
31. Bu	butyl	
32. bzac	1,3-dioxo-1-phenylbutan-2-ido	benzoylacetonato
33. bzim	1*H*-benzimidazol-1-ido	
34. Bz[b]	benzyl	
35. bztz	1,3-benzothiazole	
36. cat	benzene-1,2-diolato	catecholato
37. cbdca	cyclobutane-1,1-dicarboxylato	
38. cdta	2,2′,2″,2‴-(cyclohexane-1,2-diyldinitrilo)tetraacetato	
39. C_5H_4Me	methylcyclopentadienyl	
40. chxn (also dach)	cyclohexane-1,2-diamine	
41. cit	2-hydroxypropane-1,2,3-tricarboxylato	citrato
42. C_5Me_5[c]	pentamethylcyclopentadienyl	
43. cod	cycloocta-1,5-diene	
44. cot	cycloocta-1,3,5,7-tetraene	
45. Cp	cyclopentadienyl	
46. cptn	cyclopentane-1,2-diamine	
47. 18-crown-6	1,4,7,10,13,16-hexaoxacyclooctadecane	
48. crypt-211	4,7,13,18-tetraoxa-1,10-diazabicyclo[8.5.5]icosane	cryptand 211
49. crypt-222	4,7,13,16,21,24-hexaoxa-1,10-diazabicyclo[8.8.8]hexacosane	cryptand 222
50. Cy	cyclohexyl	

cyclam (see [14]aneN_4, No. 14)		
cyclen (see [12]aneN_4, No. 13)		
51. cys	2-amino-3-sulfanylpropanoato	cysteinato
52. cyt	4-aminopyrimidin-2(1*H*)-one	cytosine
53. dabco	1,4-diazabicyclo[2.2.2]octane	
dach (see chxn, No. 40)		diaminocyclohexane
54. dbm	1,3-dioxo-1,3-diphenylpropan-2-ido	dibenzoylmethanato
55. dea	2,2′-azanediyldi(ethan-1-olato)	diethanolaminato
56. depe	ethane-1,2-diylbis(diethylphosphane)	1,2-bis(diethylphosphino)ethane
57. diars	benzene-1,2-diylbis(dimethylarsane)	
58. dien	*N*-(2-aminoethyl)ethane-1,2-diamine	diethylenetriamine
59. [14]1,3-dieneN_4	1,4,8,11-tetraazacyclotetradeca-1,3-diene	
60. diop	[(2,2-dimethyl-1,3-dioxolane-4,5-diyl)bis(methylene)]bis(diphenylphosphane)	
61. diox	1,4-dioxane	
62. dipamp	ethane-1,2-diylbis[(2-methoxyphenyl)phenylphosphane]	'dimer of phenylanisylmethylphosphine'
63. dma	*N*,*N*-dimethylacetamide	dimethylacetamide
64. dme	1,2-dimethoxyethane	
65. dmf	*N*,*N*-dimethylformamide	
66. dmg	butane-2,3-diylidenebis(azanolato)	dimethylglyoximato
67. dmpe	ethane-1,2-diylbis(dimethylphosphane)	1,2-bis(dimethylphosphino)ethane
68. dmpm	methylenebis(dimethylphosphane)	bis(dimethylphosphino)methane
69. dmso	(methanesulfinyl)methane	dimethyl sulfoxide
70. dpm	2,2,6,6-tetramethyl-3,5-dioxoheptan-4-ido	dipivaloylmethanato
71. dppe	ethane-1,2-diylbis(diphenylphosphane)	1,2-bis(diphenylphosphino)ethane
72. dppf	1,1′-bis(diphenylphosphanyl)ferrocene	
73. dppm	methylenebis(diphenylphosphane)	bis(diphenylphosphino)methane
74. dppp	propane-1,3-diylbis(diphenylphosphane)	1,3-bis(diphenylphosphino)propane
75. dtmpa	(phosphonatomethyl)azanediylbis[ethane-2,1-diylnitrilobis(methylene)]tetrakis(phosphonato)	diethylenetriaminepentakis(methylenephosphonato)[d]
76. dtpa	2,2′,2′′2′′′-(carboxylatomethyl)azanediylbis[ethane-2,1-diylnitrilo]tetraacetato	diethylenetriaminepentaacetato
77. ea	2-amino(ethan-1-olato)	ethanolaminato

Table VII *Continued*

Number and abbreviation	*Systematic name*	*Other name (from which abbreviation derived)*[a]
78. edda	2,2′-[ethane-1,2-diylbis(azanediyl)]diacetato	ethylenediaminediacetato
79. edta	2,2′,2″,2‴-(ethane-1,2-diyldinitrilo)tetraacetato	ethylenediaminetetraacetato
80. edtmpa	ethane-1,2-diylbis[nitrilobis(methylene)]tetrakis(phosphonato)	ethylenediaminetetrakis(methylenephosphonato)[d]
81. egta	2,2′,2″,2‴-[ethane-1,2-diylbis(oxyethane-2,1-diylnitrilo)]tetraacetato	ethylene glycol-bis(2-aminoethyl)-*N*,*N*,*N*′,*N*′-tetraacetic acid
82. en	ethane-1,2-diamine	
83. Et	ethyl	
84. Et_2dtc	*N*,*N*-diethylcarbamodithioato	*N*,*N*-diethyldithiocarbamato
85. fod	6,6,7,7,8,8,8-heptafluoro-2,2-dimethyl-3,5-dioxooctan-4-ido	
86. fta	1,1,1-trifluoro-2,4-dioxopentan-3-ido	trifluoroacetylacetonato
87. gln	2,5-diamino-5-oxopentanoato	glutaminato
88. glu	2-aminopentanedioato	glutamato
89. gly	aminoacetato	glycinato
90. gua	2-amino-9*H*-purin-6(1*H*)-one	guanine
91. guo	2-amino-9-β-D-ribofuranosyl-9*H*-purin-6(1*H*)-one	guanosine
92. hdtmpa	hexane-1,6-diylbis[nitrilobis(methylene)]tetrakis(phosphonato)	hexamethylenediaminetetrakis(methylenephosphonato)[d]
93. hedp	1-hydroxyethane-1,1-diylbis(phosphonato)	1-hydroxyethane-1,1-diphosphonato
94. hfa	1,1,1,5,5,5-hexafluoropentane-2,4-dioxopentan-3-ido	hexafluoroacetylacetonato
95. his	2-amino-3-(imidazol-4-yl)propanoato	histidinato
96. hmpa	hexamethylphosphoric triamide	
97. hmta	1,3,5,7-tetraazatricyclo[3.3.1.1^{3,7}]decane	hexamethylenetetramine
98. ida	2,2′-azanediyldiacetato	iminodiacetato
99. ile	2-amino-3-methylpentanoato	isoleucinato
100. im	1*H*-imidazol-1-ido	
101. isn	pyridine-4-carboxamide	isonicotinamide

102. leu	2-amino-4-methylpentanoato	leucinato
103. lut	2,6-dimethylpyridine	lutidine
104. lys	2,6-diaminohexanoato	lysinato
105. mal	2-hydroxybutanedioato	malato
106. male	(*Z*)-butenedioato	maleato
107. malo	propanedioato	malonato
108. Me	methyl	
109. 2-Mepy	2-methylpyridine	
110. met	2-amino-4-(methylsulfanyl)butanoato	methioninato
111. mnt	1,2-dicyanoethene-1,2-dithiolato	maleonitriledithiolato
112. napy	1,8-naphthyridine	
113. nbd	bicyclo[2.2.1]hepta-2,5-diene	norbornadiene
114. nia	pyridine-3-carboxamide	nicotinamide
115. nmp	*N*-methylpyrrolidine	
116. nta	2,2′,2″-nitrilotriacetato	
117. oep	2,3,7,8,12,13,17,18-octaethylporphyrin-21,23-diido	
118. ox	ethanedioato	oxalato
119. pc	phthalocyanine-29,31-diido	
120. 1,2-pdta	2,2′,2″,2‴-(propane-1,2-diyldinitrilo)tetraacetato	1,2-propylenediaminetetraacetato
121. 1,3-pdta	2,2′,2″,2‴-(propane-1,3-diyldinitrilo)tetraacetato	1,3-propylenediaminetetraacetato
122. Ph	phenyl	
123. phe	2-amino-3-phenylpropanoato	phenylalaninato
124. phen	1,10-phenanthroline	
125. pip	piperidine	
126. pmdien	2,2′-(methylazanediyl)bis(*N*,*N*-dimethylethan-1-amine)	*N*,*N*,*N*′,*N*″,*N*″-pentamethyl‿diethylenetriamine[d]
127. pn	propane-1,2-diamine	
128. ppIX	2,18-bis(2-carboxyethyl)-3,7,12,17-tetramethyl-8,13-divinylporphyrin-21,23-diido	protoporphyrinato IX
129. pro	pyrrolidine-2-carboxylato	prolinato
130. ptn	pentane-2,4-diamine	
131. py	pyridine	

Table VII *Continued*

Number and abbreviation	*Systematic name*	*Other name (from which abbreviation derived)*[a]
132. pyz	pyrazine	
133. pz	1*H*-pyrazol-1-ido	
134. qdt	quinoxaline-2,3-dithiolato	
135. quin	quinolin-8-olato	
136. sal	2-hydroxybenzoato	salicylato
137. salan	2-[(phenylimino)methyl]phenolato	salicylideneanilinato
138. saldien	2,2′-[azanediylbis(ethane-2,1-diylazanylylidenemethanylylidene)]diphenolato	bis(salicylidene)diethylenetriaminato
139. salen	2,2′-[ethane-1,2-diylbis(azanylylidenemethanylylidene)]diphenolato	bis(salicylidene)ethylenediaminato
140. salgly	*N*-(2-oxidobenzylidene)glycinato	salicylideneglycinato
141. salpn	2,2′-[propane-1,2-diylbis(azanylylidenemethanylylidene)]diphenolato	bis(salicylidene)propylenediaminato
142. saltn	2,2′-[propane-1,3-diylbis(azanylylidenemethanylylidene)]diphenolato	bis(salicylidene)trimethylenediaminato
143. sdta[e]	2,2′,2″,2‴-[(1,2-diphenylethane-1,2-diyl)dinitrilo]tetraacetato	stilbenediaminetetraacetato
144. sep[f]	1,3,6,8,10,13,16,19-octaazabicyclo[6.6.6]icosane	
145. ser	2-amino-3-hydroxypropanoato	serinato
146. stien[e]	1,2-diphenylethane-1,2-diamine	stilbenediamine
tacn (see [9]aneN_3, No. 12)		1,4,7-triazacyclononane
147. tap	propane-1,2,3-triamine	1,2,3-triaminopropane
148. tart	2,3-dihydroxybutanedioato	tartrato
149. tcne	ethenetetracarbonitrile	tetracyanoethylene
150. tcnq	2,2′-(cyclohexa-2,5-diene-1,4-diylidene)di(propanedinitrile)	tetracyanoquinodimethane
151. tdt	4-methylbenzene-1,2-dithiolato	
152. tea	2,2′,2″-nitrilotri(ethan-1-olato)	triethanolaminato
153. terpy	2,2′:6′,2″-terpyridine	terpyridine
154. 2,3,2-tet	*N*,*N*′-bis(2-aminoethyl)propane-1,3-diamine	1,4,8,11-tetraazaundecane
155. 3,3,3-tet	*N*,*N*′-bis(3-aminopropyl)propane-1,3-diamine	1,5,9,13-tetraazatridecane
156. tetren	*N*,*N*′-(azanediyldiethane-2,1-diyl)di(ethane-1,2-diamine)	tetraethylenepentamine

157. tfa	trifluoroacetato	
158. thf	oxolane	tetrahydrofuran
159. thiox	1,4-oxathiane	thioxane
160. thr	2-amino-3-hydroxybutanoato	threoninato
161. tht	thiolane	tetrahydrothiophene
162. thy	5-methylpyrimidine-2,4(1*H*,3*H*)-dione	thymine
163. tmen	*N*,*N*,*N*′,*N*′-tetramethylethane-1,2-diamine	
164. tmp	5,10,15,20-tetrakis(2,4,6-trimethylphenyl)porphyrin-21,23-diido	5,10,15,20-tetramesitylporphyrin-21,23-diido
165. tn	propane-1,3-diamine	trimethylenediamine
166. Tol (*o*-, *m*- or *p*-)	2-, 3- or 4-methylphenyl	tolyl (*o*-, *m*- or *p*-)
167. Tp	hydridotris(pyrazolido-*N*)borato(1−), or tris(1*H*-pyrazol-1-yl)boranuido	hydrotris(pyrazolyl)borato
168. Tp′[g]	tris(3,5-dimethylpyrazolido-*N*)hydridoborato(1−)	hydrotris(3,5-dimethylpyrazolyl)borato
169. tpp	5,10,15,20-tetraphenylporphyrin-21,23-diido	
170. tren	*N*,*N*-bis(2-aminoethyl)ethane-1,2-diamine	tris(2-aminoethyl)amine
171. trien	*N*,*N*′-bis(2-aminoethyl)ethane-l,2-diamine	triethylenetetramine
172. triphos[h]	[(phenylphosphanediyl)bis(ethane-2,1-diyl)]bis(diphenylphosphane)	
173. tris	2-amino-2-(hydroxymethyl)propane-1,3-diol	tris(hydroxymethyl)aminomethane
174. trp	2-amino-3-(1*H*-indol-3-yl)propanoato	tryptophanato
175. tsalen	2,2′-[ethane-1,2-diylbis(azanylylidenemethanylylidene)]dibenzenethiolato	bis(thiosalicylidene)ethylenediaminato
176. ttfa	4,4,4-trifluoro-1,3-dioxo-1-(2-thienyl)butan-2-ido	thenoyltrifluoroacetonato
177. ttha	2,2′,2″,2‴-(ethane-1,2-diylbis{[(carboxylatomethyl)azanediyl]ethane-2,1-diylnitrilo})tetraacetato	triethylenetetraminehexaacetato
178. ttp	5,10,15,20-tetrakis(4-methylphenyl)porphyrin-21,23-diido	5,10,15,20-tetra-*p*-tolylporphyrin-21,23-diido
179. tu	thiourea	
180. tyr	2-amino-3-(4-hydroxyphenyl)propanoato	tyrosinato
181. tz	1,3-thiazole	thiazole
182. ura	pyrimidine-2,4(1*H*,3*H*)-dione	uracil
183. val	2-amino-3-methylbutanoato	valinato

[a] Many of these names are no longer acceptable.
[b] The abbreviation Bz has often been used previously for ‘benzoyl’, and Bzl has been used for ‘benzyl’. Use of the alternatives, PhCO and $PhCH_2$, is therefore preferable.
[c] The use of the abbreviation Cp* for pentamethylcyclopentadienyl is discouraged. It can lead to confusion because the asterisk,*, is also used to represent an excited state, an optically active substance, *etc.*

[d] The symbol '⁀' is used to divide the name, necessitated by the line break. In the absence of the line break this symbol is omitted. Note that all *hyphens* are true parts of the name.
[e] The abbreviation derives from the non-systematic name stilbenediamine which incorrectly implies the presence of a C=C double bond in the ligand.
[f] The abbreviation derives from the non-systematic name sepulchrate which incorrectly implies that the ligand is anionic.
[g] The use of Tp′ is preferred to Tp* for the reasons given in footnote c. A general procedure for abbreviating substituted hydridotris(pyrazolido-*N*)borate ligands has been proposed [see S. Trofimenko, *Chem. Rev.*, **93**, 943–980 (1993)]. For example, Tp′ becomes Tp^{Me_2}, the superscript denoting the methyl groups at the 3- and 5-positions of the pyrazole rings.
[h] The abbreviation triphos should not be used for the four-phosphorus ligand $PhP(CH_2PPh_2)_3$.

Table VIII *Structural formulae of selected ligands (numbered according to Table VII)*

1 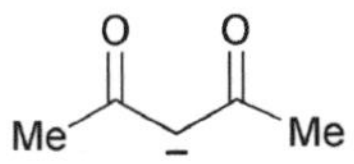 3 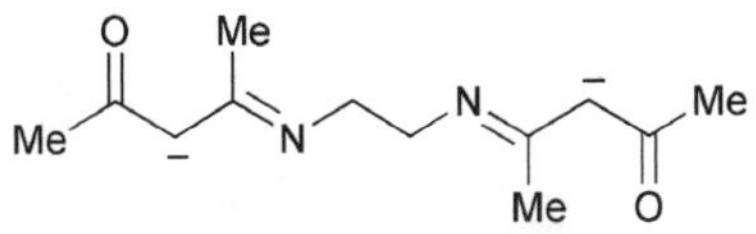 4

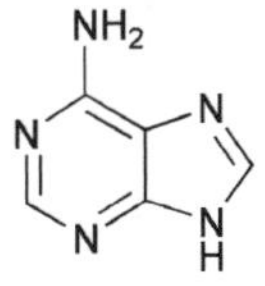

5 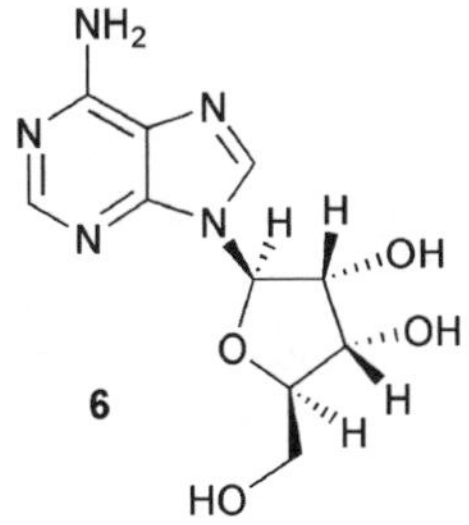 6 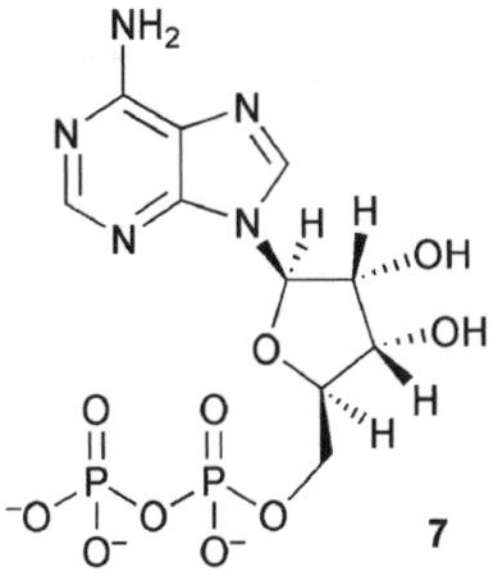 7

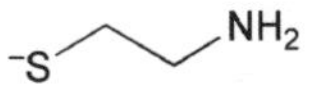

8 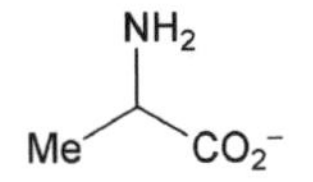 9 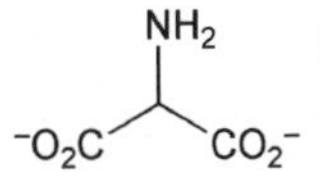 10 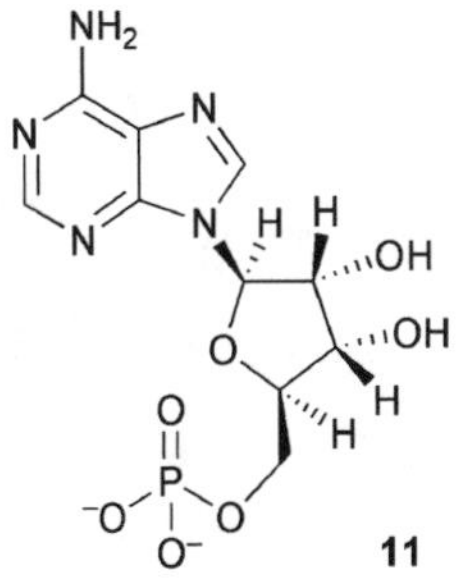 11

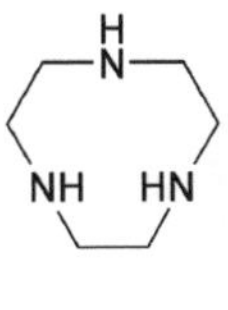

12 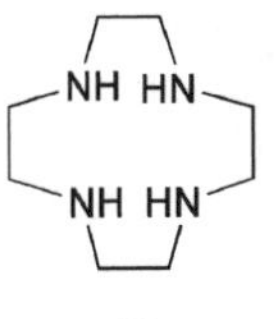 13 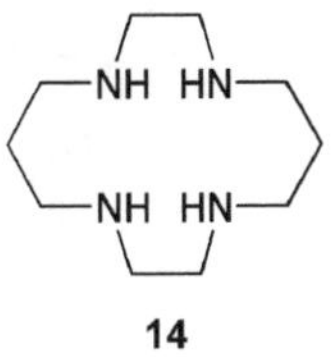 14

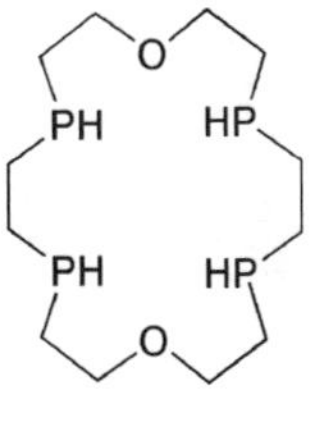

15 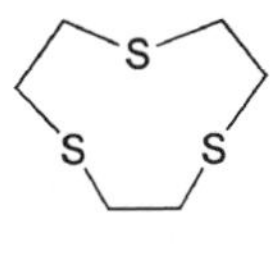 16 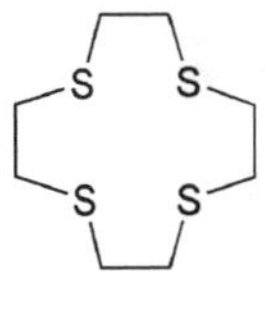 17

Table VIII *Continued*

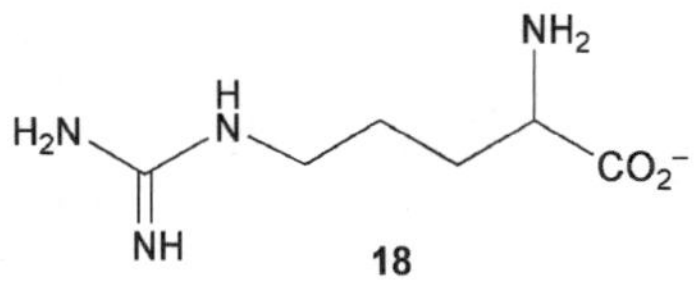

18

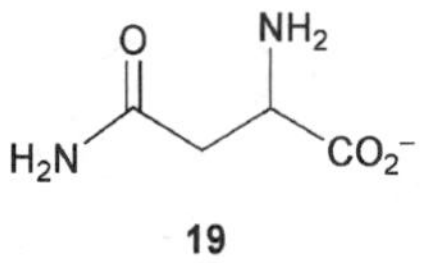

19

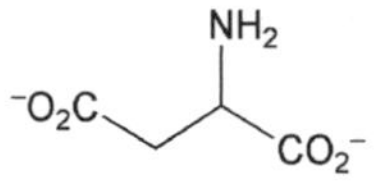

20

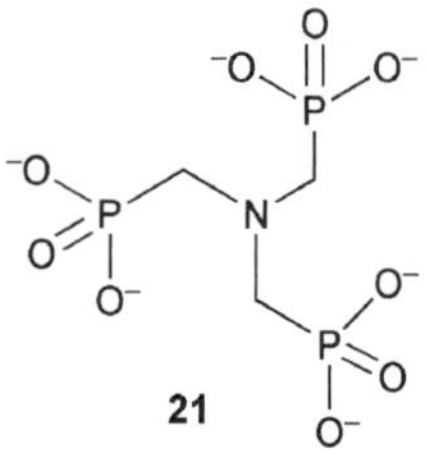

21

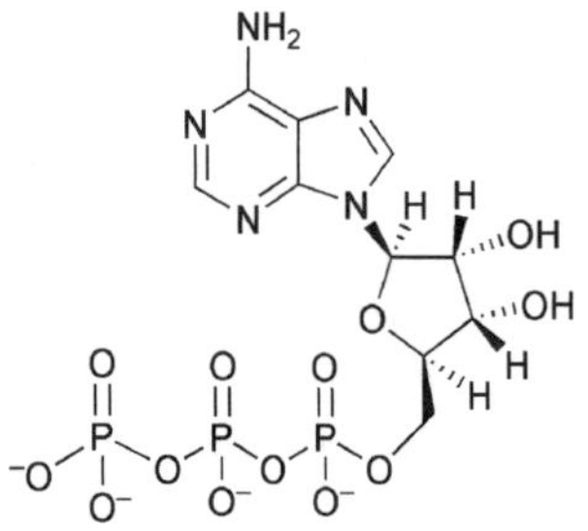

22

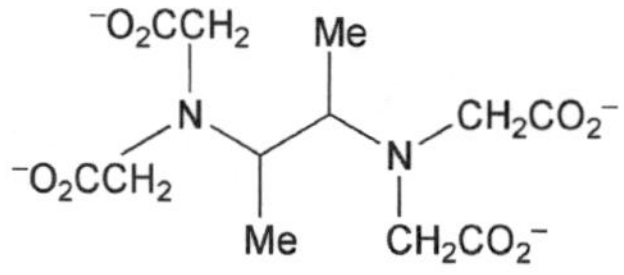

23

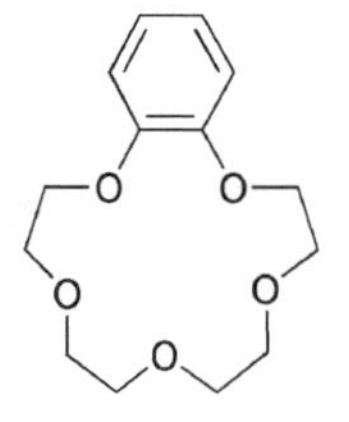

24

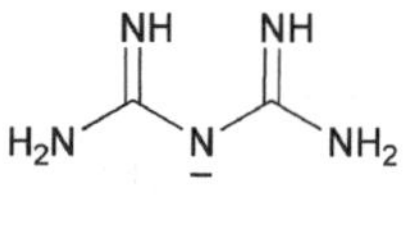

25

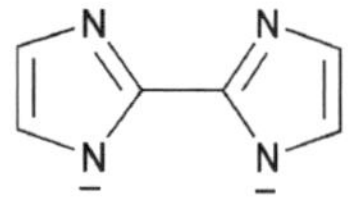

26

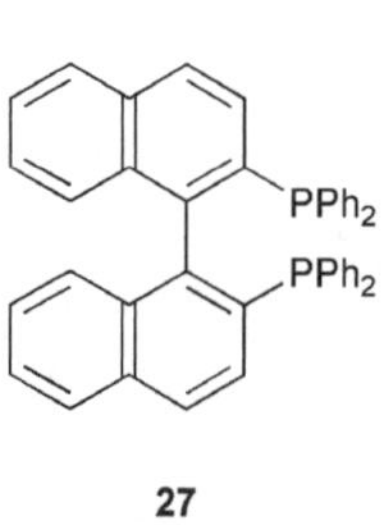

27

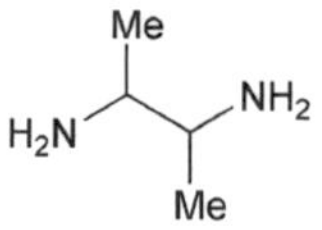

28

Table VIII *Continued*

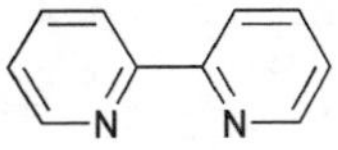

29 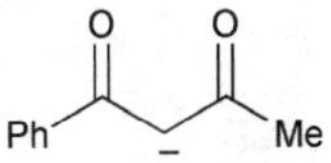 32 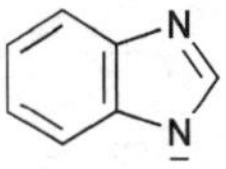 33 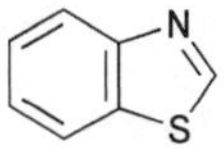 35

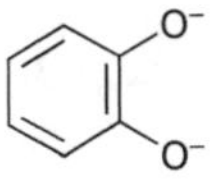

36 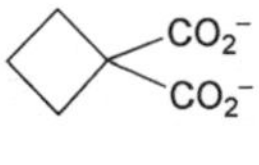 37 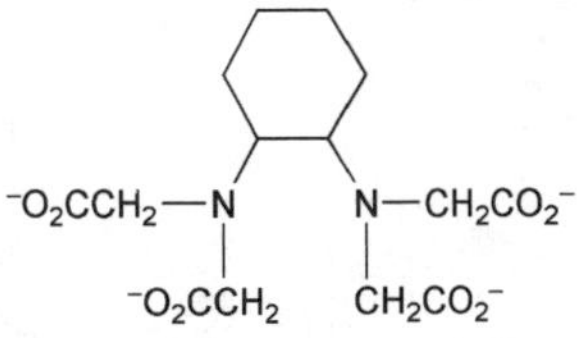 38 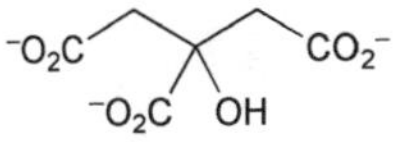 41

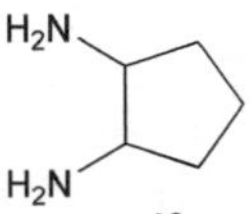

46 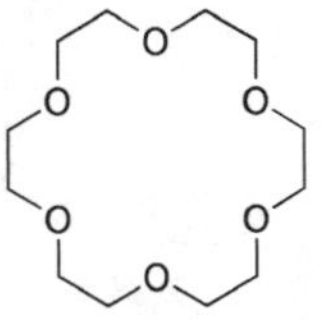 47 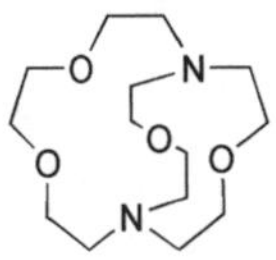 48

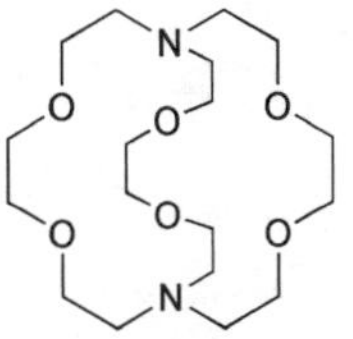

49 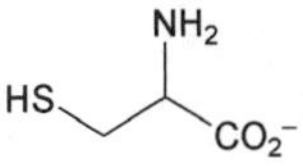 51 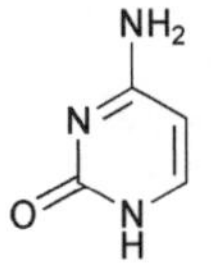 52 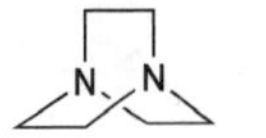 53

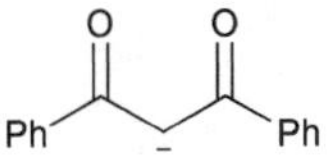

54 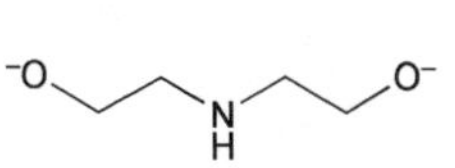 55 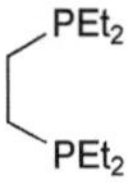 56 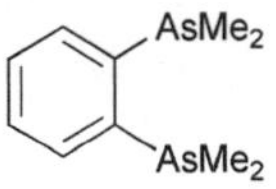 57

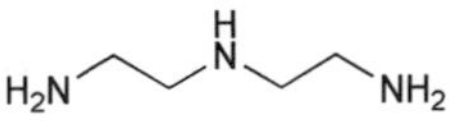

58 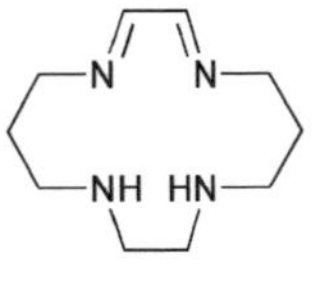 59 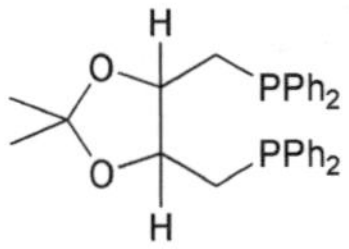 60

Table VIII *Continued*

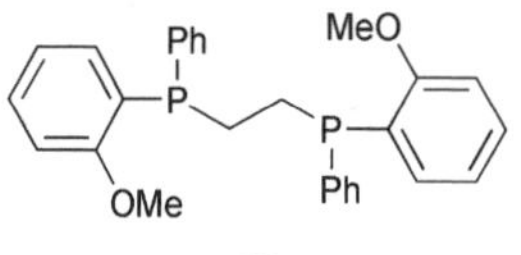

62

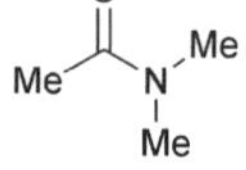

63

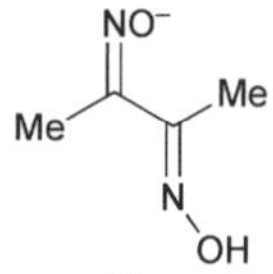

66

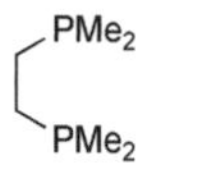

67

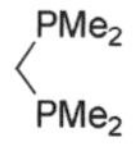

68

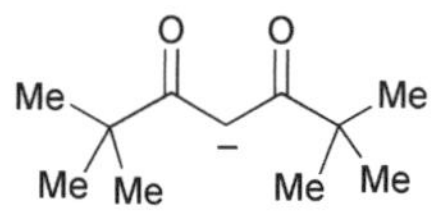

70

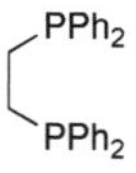

71

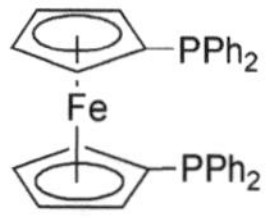

72

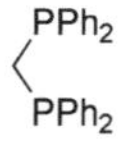

73

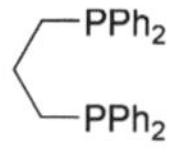

74

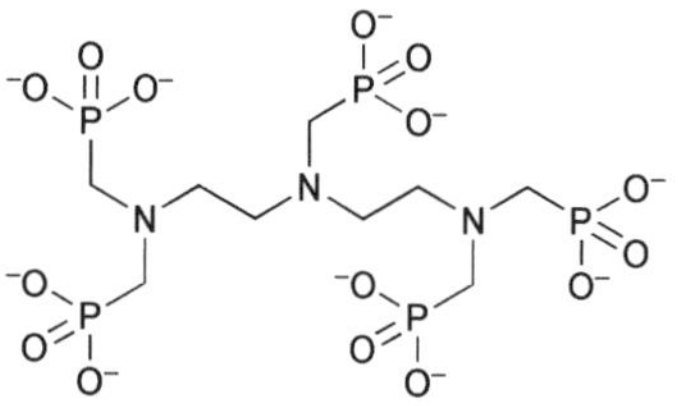

75

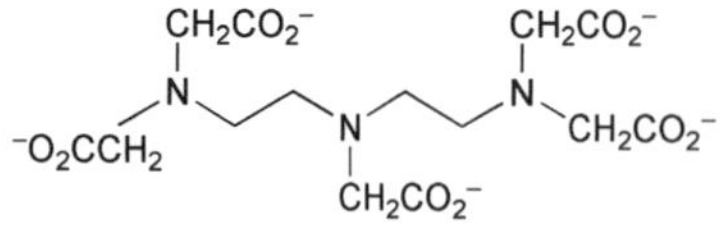

76

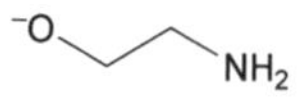

77

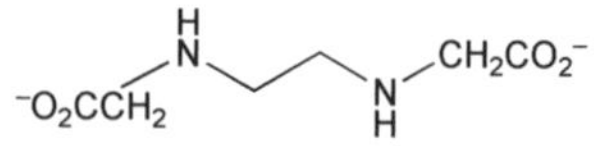

78

Table VIII *Continued*

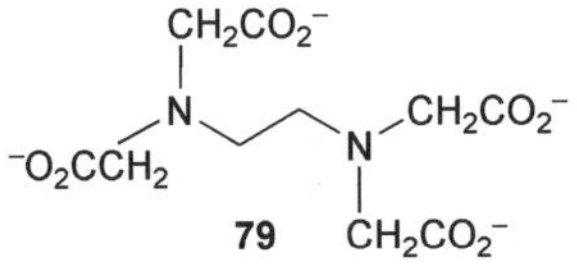

$CH_2CO_2^-$ N N $CH_2CO_2^-$ $^-O_2CCH_2$ $CH_2CO_2^-$

79

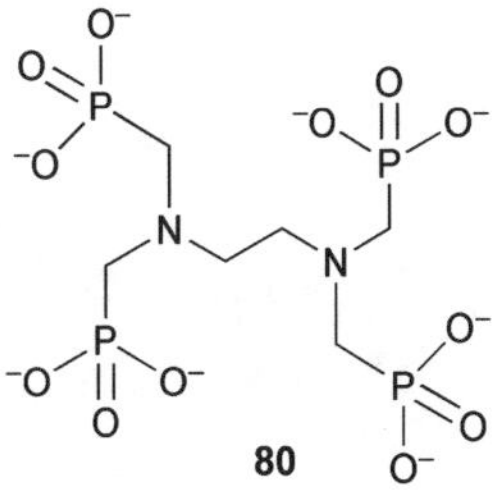

80

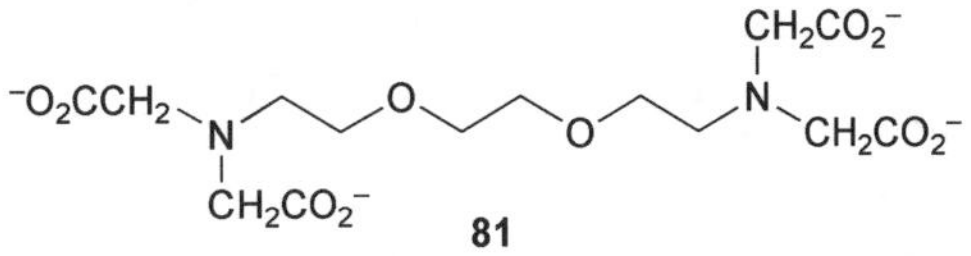

$^-O_2CCH_2$ N O O N $CH_2CO_2^-$ $CH_2CO_2^-$ $CH_2CO_2^-$

81

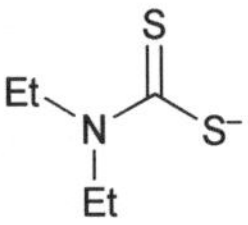

Et N S S^- Et

84

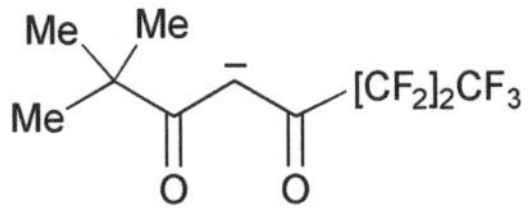

Me Me Me O O $[CF_2]_2CF_3$

85

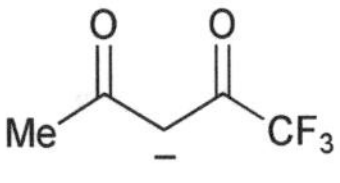

Me O O CF_3

86

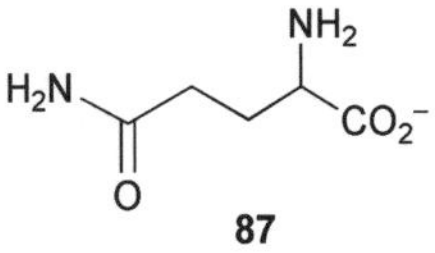

H_2N O NH_2 CO_2^-

87

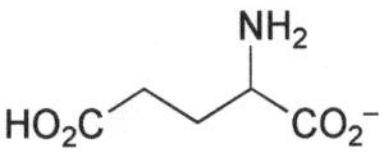

HO_2C NH_2 CO_2^-

88

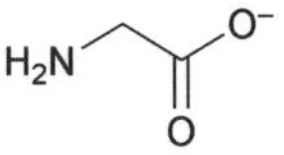

H_2N O O^-

89

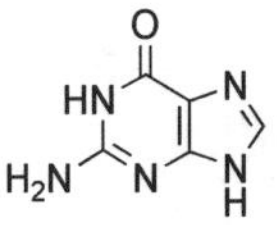

HN O N H_2N N N H

90

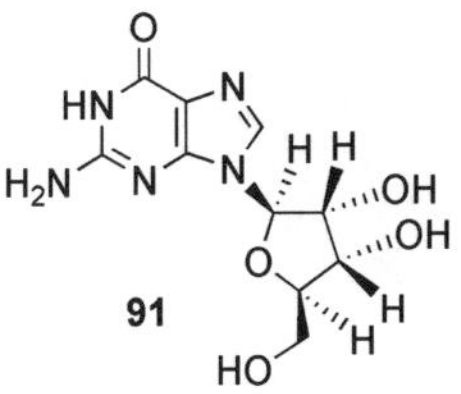

HN O N H_2N N N H H OH OH O H H HO

91

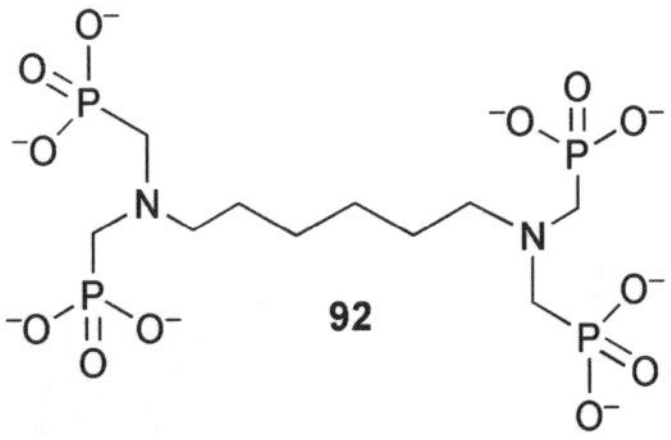

92

OH ^-O P O O^- Me P O O^- O^-

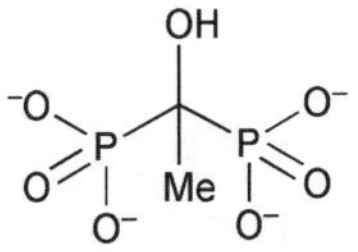

93

Table VIII *Continued*

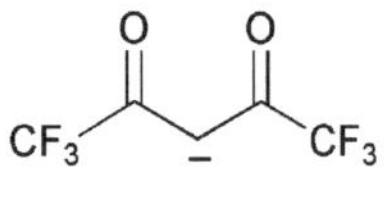

$CF_3C(O)CH^-C(O)CF_3$

94

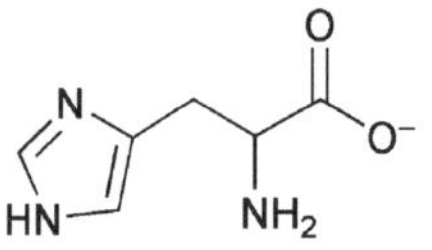

NH_2 O^- HN N

95

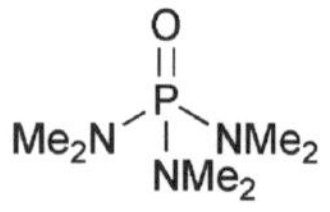

$(Me_2N)_3P=O$

96

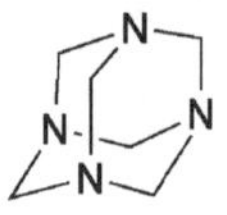

97

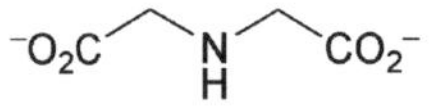

$^-O_2C\ N(H)\ CO_2^-$

98

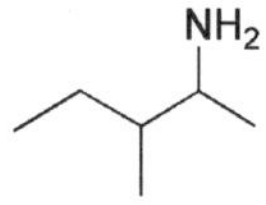

NH_2

99

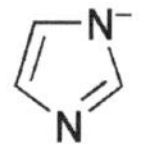

N^- N

100

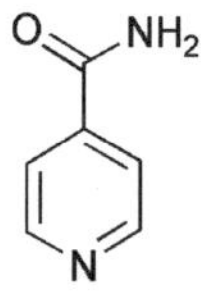

O NH_2 N

101

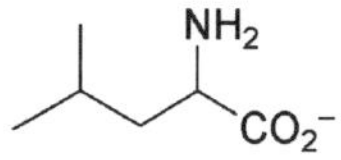

NH_2 CO_2^-

102

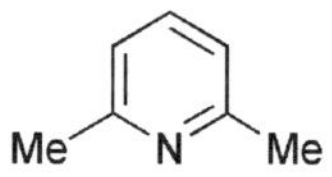

Me N Me

103

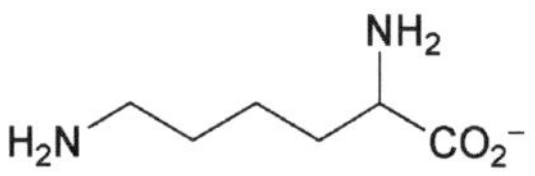

H_2N NH_2 CO_2^-

104

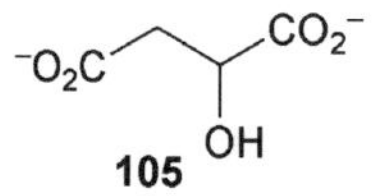

^-O_2C CO_2^- OH

105

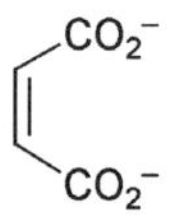

CO_2^- CO_2^-

106

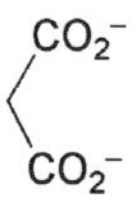

CO_2^- CO_2^-

107

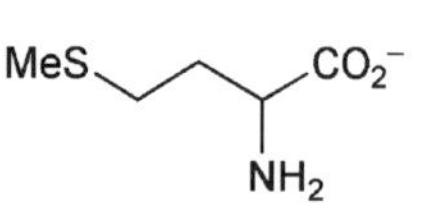

MeS CO_2^- NH_2

110

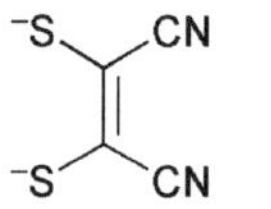

^-S CN ^-S CN

111

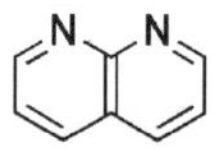

N N

112

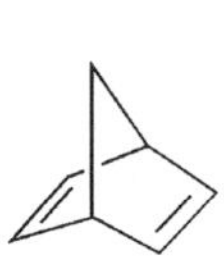

113

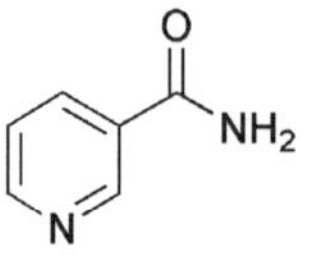

O NH_2 N

114

N Me

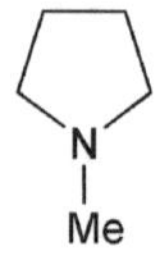

115

Table VIII *Continued*

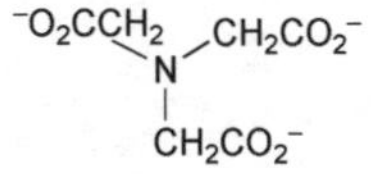

$^{-}O_2CCH_2N(CH_2CO_2^-)_2$

116

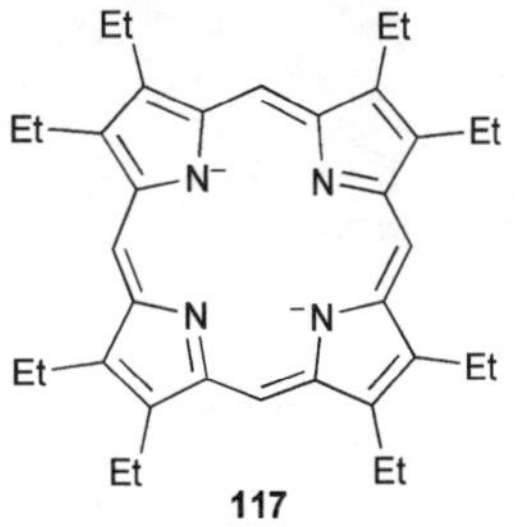

Et Et Et Et N⁻ N N ⁻N Et Et Et Et

117

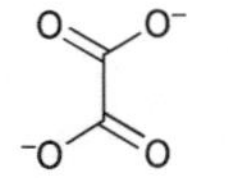

O O⁻ ⁻O O

118

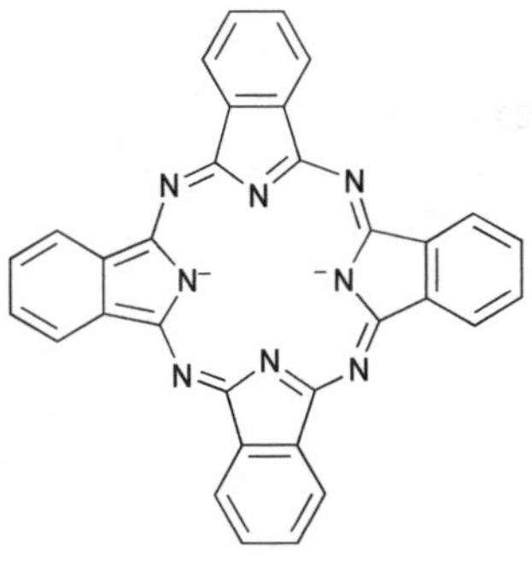

N N N N⁻ ⁻N N N N

119

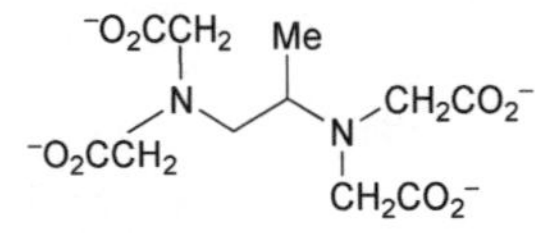

$^{-}O_2CCH_2$ Me N $CH_2CO_2^-$ $^{-}O_2CCH_2$ N $CH_2CO_2^-$

120

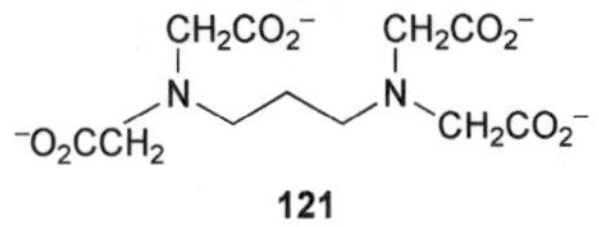

$CH_2CO_2^-$ $CH_2CO_2^-$ N N $^{-}O_2CCH_2$ $CH_2CO_2^-$

121

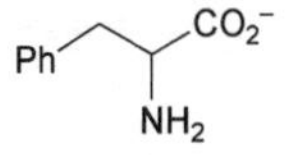

Ph CO_2^- NH_2

123

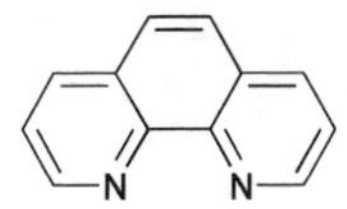

N N

124

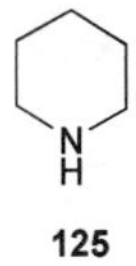

N H

125

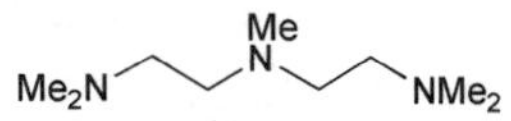

Me Me_2N N NMe_2

126

NH_2 Me NH_2

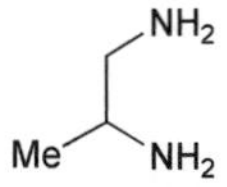

127

Table VIII *Continued*

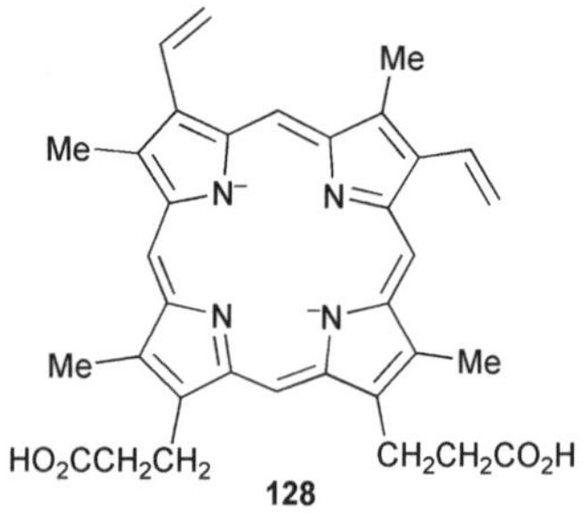

Me Me Me Me N⁻ N N ⁻N $HO_2CCH_2CH_2$ $CH_2CH_2CO_2H$

128

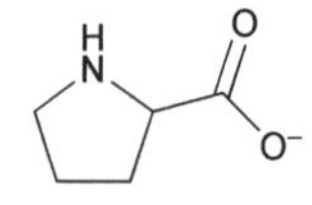

H N O O⁻

129

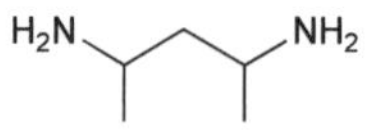

H_2N NH_2

130

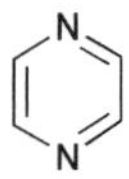

N N

132

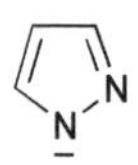

N N

133

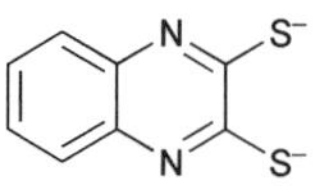

N N S⁻ S⁻

134

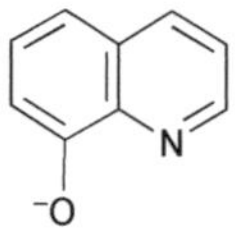

N ⁻O

135

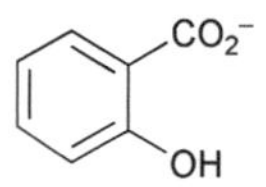

CO_2^- OH

136

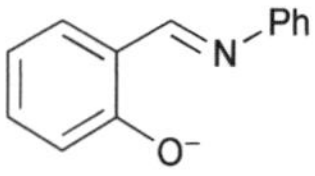

N Ph O⁻

137

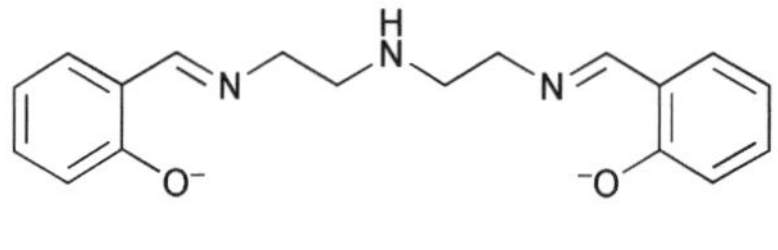

H N N N O⁻ ⁻O

138

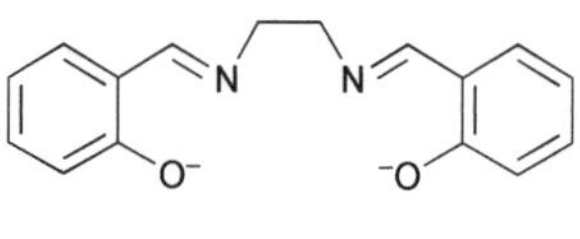

N N O⁻ ⁻O

139

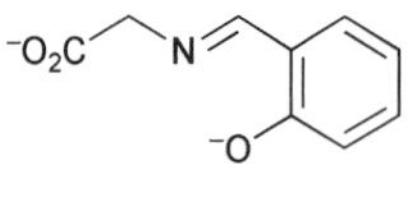

^-O_2C N ⁻O

140

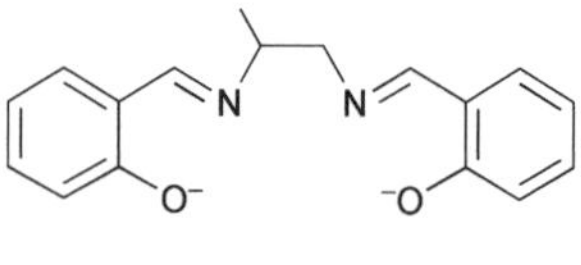

N N O⁻ ⁻O

141

N N O⁻ ⁻O

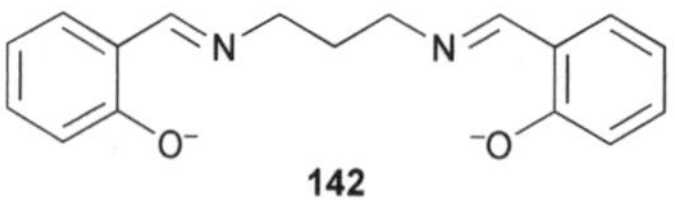

142

Table VIII *Continued*

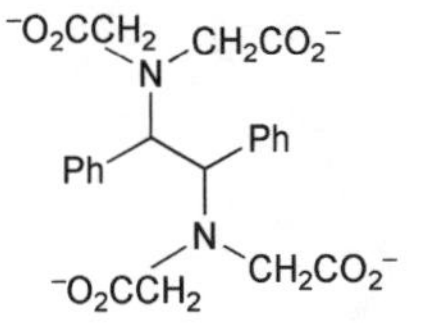

$^{-}O_2CCH_2$ N $CH_2CO_2^{-}$ Ph Ph N $^{-}O_2CCH_2$ $CH_2CO_2^{-}$
143

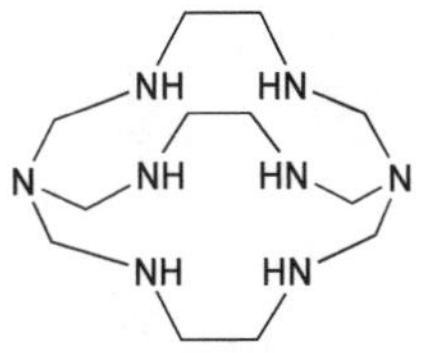

NH HN N NH HN N NH HN
144

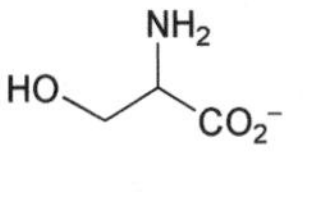

NH_2 HO CO_2^{-}
145

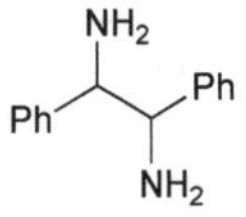

NH_2 Ph Ph NH_2
146

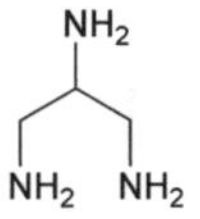

NH_2 NH_2 NH_2
147

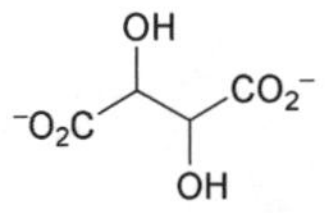

OH $^{-}O_2C$ CO_2^{-} OH
148

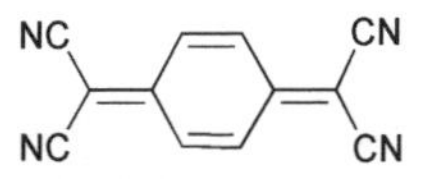

NC CN NC CN
150

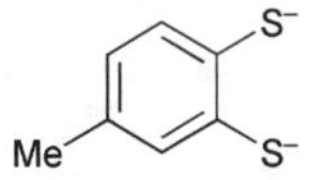

S^{-} Me S^{-}
151

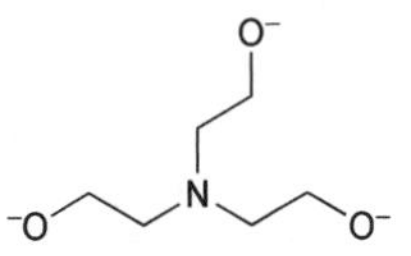

O^{-} ^{-}O N O^{-}
152

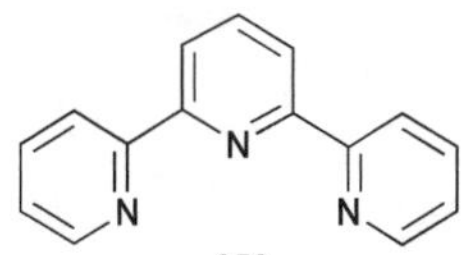

N N N
153

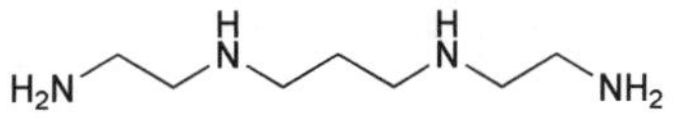

H_2N H N H N NH_2
154

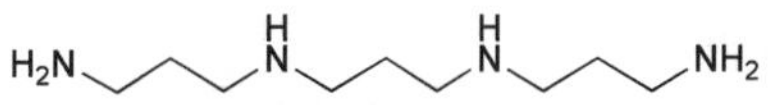

H_2N H N H N NH_2
155

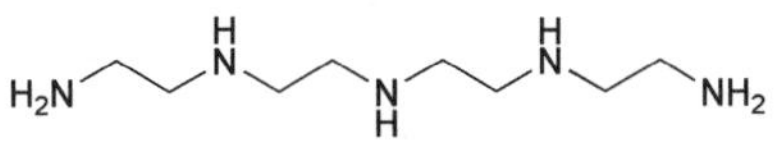

H_2N H N N H H N NH_2
156

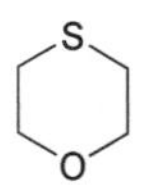

S O
159

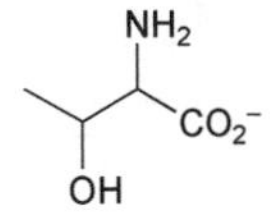

NH_2 CO_2^{-} OH
160

S
161

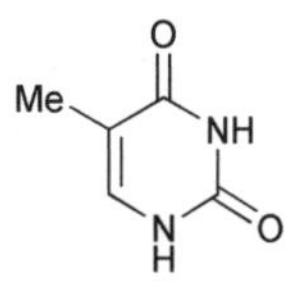

O Me NH N H O
162

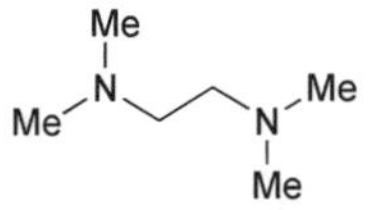

Me Me N N Me Me
163

Table VIII *Continued*

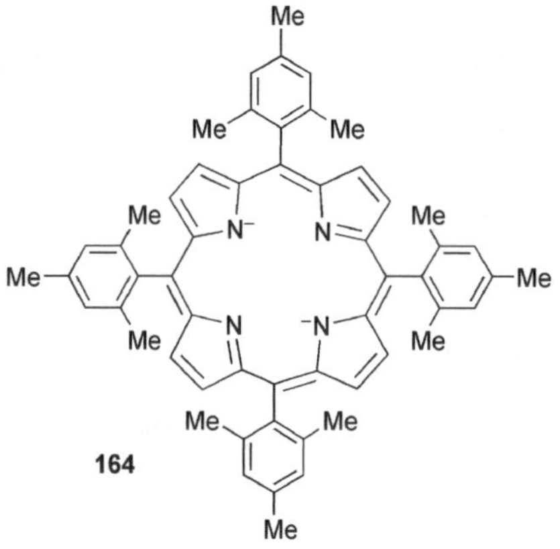

Me
Me Me
Me Me
Me N⁻ N Me
Me N ⁻N Me
Me Me
164 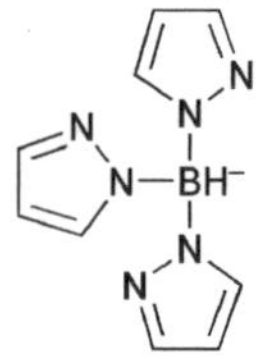
Me

N N
N–BH⁻
N N
N N
167

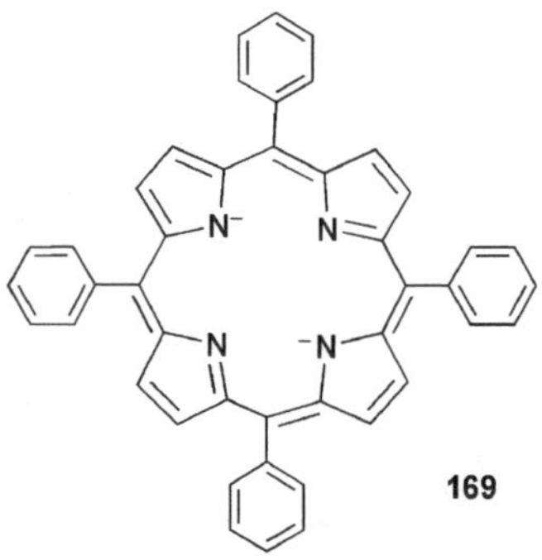

N⁻ N
N ⁻N
169

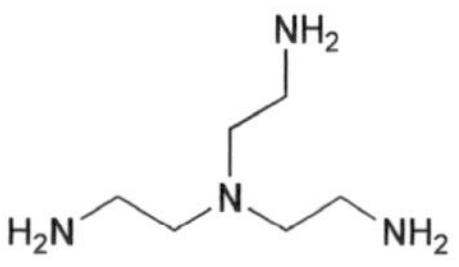

NH_2
H_2N N NH_2
170

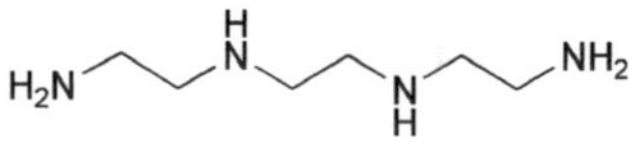

H
H_2N N N NH_2
H
171

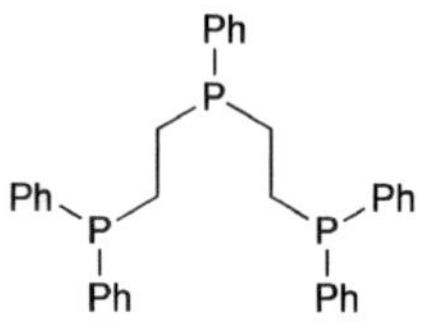

Ph
P
Ph P P Ph
Ph Ph
172

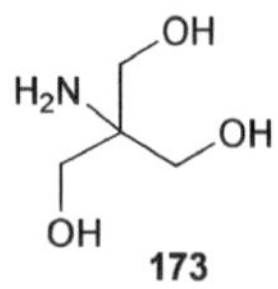

OH
H_2N OH
OH 173

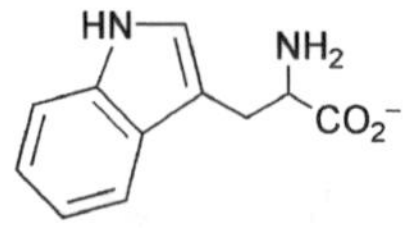

HN NH_2
CO_2^-
174

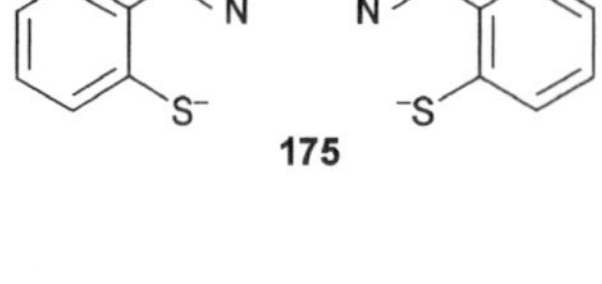

N N
S⁻ ⁻S
175

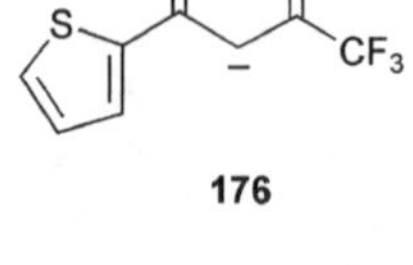

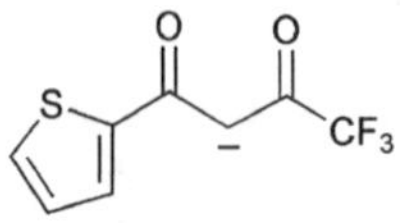

O O
S CF_3
–
176

Table VIII *Continued*

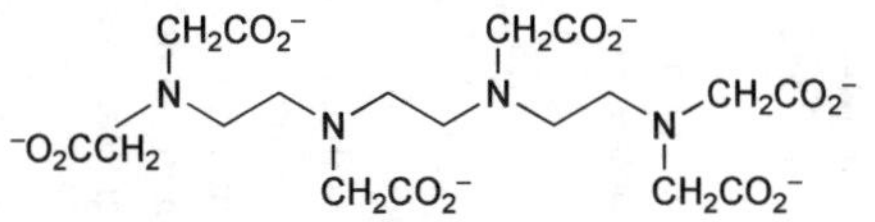

177

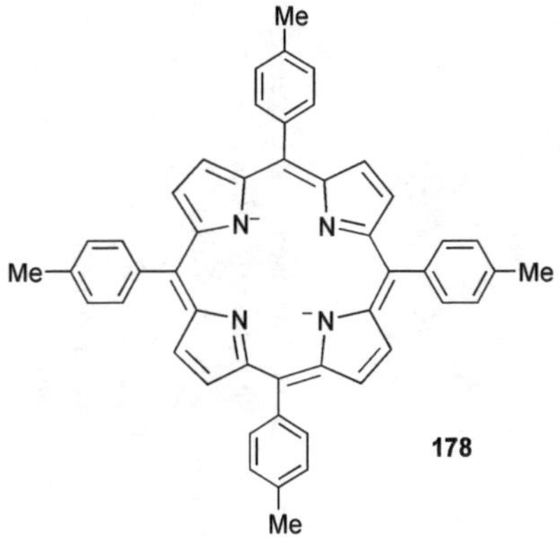

178

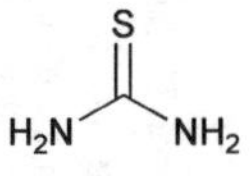

179

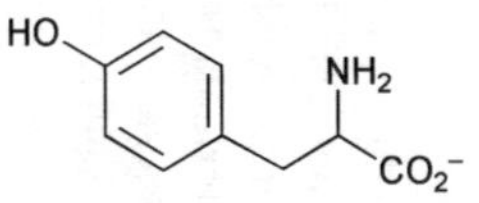

180

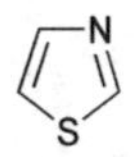

181

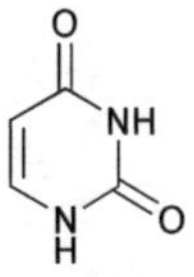

182

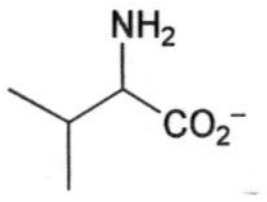

183

Table IX *Names of homoatomic, binary and certain other simple molecules, ions, compounds, radicals and substituent groups*

This Table names a large number of homoatomic and binary compounds and species, and some heteropolynuclear entities, and thus may be used as a reference for names of simple compounds and a source of examples to guide in the naming of further compounds. It may be necessary to browse the Table to find (families of) compounds that match those of interest. For example, all the oxides of potassium are named; corresponding compounds of the other alkali metals, not included here, are named analogously. Several silicon and germanium hydride species are named; names for corresponding tin and lead species are not necessarily included.

Some inorganic acids and their corresponding bases are included in this Table, but more acid names are given in Tables IR-8.1 and IR-8.2. Only a few simple carbon-containing compounds and substituent groups are included. In particular, organic ligands belonging to the general classes alcoholates, thiolates, phenolates, carboxylates, amines, phosphanes and arsanes as well as (partially) dehydronated amines, phosphanes and arsanes are generally *not* included. Their naming is described and exemplified in Section IR-9.2.2.3.

Entries in the first column are ordered alphabetically according to the formulae as they appear here. Formulae for *binary* species are written in this column according to the position of the two elements in Table VI (Section IR-4.4), *e.g.* ammonia is found under 'NH_3', but selane under 'H_2Se' and AlLi under 'LiAl'. In case of doubt, cross references should aid in finding the correct entry in the Table. However, in the first column, formulae of *ternary* and *quaternary* compounds are written strictly alphabetically, *e.g.* '$ClSCN^{\bullet-}$' is found under the entry 'CClNS', and 'HPO_4^{-}' under 'HO_4P', and these formulae are ordered as described in Section IR-4.4.2.2. In the columns to the right of the first column, special formats may be used for formulae in order to stress a particular structure, *e.g.* under the entry '$BrHO_3$' one finds '$HOBrO_2$' rather than '$HBrO_3$' or '$[BrO_2(OH)]$', the two formats presented in Table IR-8.2.

The symbol '⌣' is used for dividing names when this is made necessary by a line break. When the name is reconstructed from the name given in the Table, this symbol should be omitted. Thus, all *hyphens* in the Table are true parts of the names. The symbols '$>$' and '$<$' placed next to an element symbol both denote two single bonds connecting the atom in question to two other atoms.

For a given compound, the various systematic names, if applicable, are given in the order: stoichiometric names (Chapter IR-5), substitutive names (Chapter IR-6), additive names (Chapter IR-7) and hydrogen names (Section IR-8.4). Acceptable names which are not entirely systematic (or not formed according to any of the systems mentioned above) are given at the end after a semicolon. No order of preference is implied by the order in which formulae and names are listed, but in practice it may be useful to select particular formulae and names for particular uses. Thus, for sodium chloride the formula [NaCl] and the additive name 'chloridosodium' may be used specifically for the molecular compound, which can be regarded as a coordination compound, whereas 'sodium chloride' may be used, and traditionally *is* used, for the compound in general and for the solid with the composition NaCl. Corresponding remarks apply to a number of hydrides for which a stoichiometric name may be applied to the compound with the stoichiometry in question without further structural implications (such as 'aluminium trihydride' for AlH_3 or 'dihydrogen disulfide' for H_2S_2), whereas a parent hydride name (needed anyway for naming certain derivatives) or an additive name may be used to denote specifically the molecular compound or entity (such as 'alumane' or 'trihydridoaluminium' for the molecular entity $[AlH_3]$ and 'disulfane' for HSSH).

Note from the examples above that in order to stress the distinctions discussed, the square brackets are sometimes used in the Table to enclose formulae for molecular entities that are otherwise often written with no enclosing marks. When a formula with square brackets is shown, there will also be a coordination-type additive name.

Formula for uncharged atom or group	*Name*			
	Uncharged atoms or molecules (including zwitterions and radicals) or substituent groups[a]	*Cations (including cation radicals) or cationic substituent groups*[a]	*Anions (including anion radicals) or anionic substituent groups*[b]	*Ligands*[c]
Ac	actinium	actinium	actinide[d]	actinido
Ag	silver	silver	argentide	argentido
Al	aluminium	aluminium (general) Al^{+}, aluminium(1+) Al^{3+}, aluminium(3+)	aluminide (general) Al^{-}, aluminide(1−)	aluminido (general) Al^{-}, aluminido(1−)
AlCl	AlCl, aluminium monochloride [AlCl], chloridoaluminium	$AlCl^{+}$, chloridoaluminium(1+)		
$AlCl_3$ (see also Al_2Cl_6)	$AlCl_3$, aluminium trichloride [$AlCl_3$], trichloroalumane, trichloridoaluminium			
$AlCl_4$			$AlCl_4^{-}$, tetrachloroalumanuide, tetrachloridoaluminate(1−)	$AlCl_4^{-}$, tetrachloroalumanuido, tetrachloridoaluminato(1−)
AlH	AlH, aluminium monohydride [AlH], λ^1-alumane (parent hydride name), hydridoaluminium	AlH^{+}, hydridoaluminium(1+)		
AlH_2	$-AlH_2$, alumanyl			
AlH_3	AlH_3, aluminium trihydride [AlH_3], alumane (parent hydride name), trihydridoaluminium	$AlH_3^{\bullet+}$, alumaniumyl, trihydridoaluminium(•1+)	$AlH_3^{\bullet-}$, alumanuidyl, trihydridoaluminate(•1−)[e]	
AlH_4			AlH_4^{-}, alumanuide, tetrahydridoaluminate(1−)	AlH_4^{-}, alumanuido, tetrahydridoaluminato(1−)
AlO	AlO, aluminium mon(o)oxide [AlO], oxidoaluminium	AlO^{+}, oxidoaluminium(1+)	AlO^{-}, oxidoaluminate(1−)	
AlSi	AlSi, aluminium monosilicide [AlSi], silicidoaluminium			

Table IX *Continued*

Formula for uncharged atom or group	*Name*			
	Uncharged atoms or molecules (including zwitterions and radicals) or substituent groups[a]	*Cations (including cation radicals) or cationic substituent groups*[a]	*Anions (including anion radicals) or anionic substituent groups*[b]	*Ligands*[c]
Al_2	Al_2, dialuminium		Al_2^-, dialuminide(1−)	
Al_2Cl_6	$[Cl_2Al(\mu\text{-}Cl)_2AlCl_2]$, di-μ-chlorido-bis(dichlorido-aluminium)			
Al_4			Al_4^{2-}, tetraaluminide(2−)	
Am	americium	americium	americide	americido
Ar	argon	argon (general) Ar^+, argon(1+)	argonide	argonido
ArBe		$ArBe^+$, beryllidoargon(1+)		
ArF	ArF, argon monofluoride [ArF], fluoridoargon	ArF^+, fluoridoargon(1+)		
ArHe		$ArHe^+$, helidoargon(1+)		
ArLi		$ArLi^+$, lithidoargon(1+)		
Ar_2	Ar_2, diargon	Ar_2^+, diargon(1+)		
As	arsenic >As−, arsanetriyl	arsenic	arsenide (general) As^{3-}, arsenide(3−), arsanetriide; arsenide	arsenido (general) As^{3-}, arsanetriido; arsenido
AsH	AsH, arsenic monohydride $AsH^{2\bullet}$, arsanylidene, hydridoarsenic(2•) >AsH, arsanediyl =AsH, arsanylidene	AsH^{2+}, arsanebis(ylium), hydridoarsenic(1+)	AsH^{2-} arsanediide, hydridoarsenate(2−)	AsH^{2-}, arsanediido, hydridoarsenato(2−)

AsHO	>AsH(O), oxo-λ^5-arsanediyl; arsonoyl =AsH(O), oxo-λ^5-arsanylidene; arsonoylidene			
$AsHO_2$	>AsO(OH), hydroxy(oxo)-λ^5-arsanediyl; hydroxyarsoryl =AsO(OH), hydroxy(oxo)-λ^5-arsanylidene; hydroxyarsorylidene		$AsHO_2^{2-}$, hydridodioxidoarsenate(2−); arsonite	$AsHO_2^{2-}$, hydridodioxidoarsenato(2−); arsonito
$AsHO_3$			$AsHO_3^{2-}$, hydridotrioxidoarsenate(2−); arsonate	$AsHO_3^{2-}$, hydridotrioxidoarsenato(2−); arsonato
AsH_2	AsH_2, arsenic dihydride $AsH_2^{\bullet}$, arsanyl, dihydridoarsenic(•) $-AsH_2$, arsanyl	AsH_2^+, arsanylium, dihydridoarsenic(1+)	AsH_2^-, arsanide, dihydridoarsenate(1−)	AsH_2^-, arsanido, dihydridoarsenato(1−)
AsH_2O	$-AsH_2O$, oxo-λ^5-arsanyl; arsinoyl		AsH_2O^-, dihydridooxidoarsenate(1−); arsinite	AsH_2O^-, dihydridooxidoarsenato(1−); arsinito
AsH_2O_2			$AsH_2O_2^-$, dihydridodioxidoarsenate(1−); arsinate	$AsH_2O_2^-$, dihydridodioxidoarsenato(1−); arsinato
AsH_2O_3	$-As(O)(OH)_2$, dihydroxyoxo-λ^5-arsanyl; dihydroxyarsoryl, arsono		$AsO(OH)_2^-$, dihydroxidooxidoarsenate(1−)	$AsO(OH)_2^-$, dihydroxidooxidoarsenato(1−)
AsH_3	AsH_3, arsenic trihydride $[AsH_3]$, arsane (parent hydride name), trihydridoarsenic	$AsH_3^{\bullet+}$, arsaniumyl, trihydridoarsenic(•1+) $-AsH_3^+$, arsaniumyl	$AsH_3^{\bullet-}$, arsanuidyl, trihydridoarsenate(•1−)[e]	
AsH_4	$-AsH_4$, λ^5-arsanyl	AsH_4^+, arsanium, tetrahydridoarsenic(1+)		
AsH_5	AsH_5, arsenic pentahydride $[AsH_5]$, λ^5-arsane (parent hydride name), pentahydridoarsenic			

Table IX *Continued*

Formula for uncharged atom or group	*Name*			
	Uncharged atoms or molecules (including zwitterions and radicals) or substituent groups[a]	*Cations (including cation radicals) or cationic substituent groups*[a]	*Anions (including anion radicals) or anionic substituent groups*[b]	*Ligands*[c]
AsO	>As(O)–, oxo-λ^5-arsanetriyl; arsoryl =As(O)–, oxo-λ^5-arsanylylidene; arsorylidene ≡As(O), oxo-λ^5-arsanylidyne; arsorylidyne			
AsO_3			AsO_3^{3-}, trioxidoarsenate(3−); arsenite, arsorite $-As(=O)(O^-)_2$, dioxidooxo-λ^5-arsanyl; arsonato	AsO_3^{3-}, trioxidoarsenato(3−); arsenito, arsorito
AsO_4			AsO_4^{3-}, tetraoxidoarsenate(3−); arsenate, arsorate	AsO_4^{3-}, tetraoxidoarsenato(3−); arsenato, arsorato
AsS_4			AsS_4^{3-}, tetrasulfidoarsenate(3−)	AsS_4^{3-}, tetrasulfidoarsenato(3−)
As_2H			$HAs{=}As^-$, diarsenide $HAsAs^{3-}$, diarsanetriide	$HAs{=}As^-$, diarsenido $HAsAs^{3-}$, diarsanetriido
As_2H_2	$HAs{=}AsH$, diarsene		H_2AsAs^{2-}, diarsane-1,1-diide $HAsAsH^{2-}$, diarsane-1,2-diide	$HAs{=}AsH$, diarsene H_2AsAs^{2-}, diarsane-1,1-diido $HAsAsH^{2-}$, diarsane-1,2-diido
As_2H_4	H_2AsAsH_2, diarsane			H_2AsAsH_2, diarsane
As_4	As_4, tetraarsenic			As_4, tetraarsenic
At	astatine (general) $At^{\bullet}$, astatine(•), monoastatine	astatine	At^-, astatide(1−); astatide	astatido (general) At^-, astatido(1−); astatido
AtH, see HAt				

At_2	At_2, diastatine			
Au	gold	gold (general) Au^+, gold(1+) Au^{3+}, gold(3+)	auride	aurido
B	boron >B−, boranetriyl ≡B, boranylidyne	boron (general) B^+, boron(1+) B^{3+}, boron(3+)	boride (general) B^-, boride(1−) B^{3-}, boride(3−); boride	borido (general) B^-, borido(1−) B^{3-}, borido(3−); borido
BH	>BH, boranediyl =BH, boranylidene	BH^{2+}, boranebis(ylium), hydridoboron(2+)	BH^{2-}, boranediide, hydridoborate(2−)	BH^{2-}, boranediido, hydridoborato(2−)
BHO_3			$BO_2(OH)^{2-}$, hydroxidodioxidoborate(2−); hydrogenborate	$BO_2(OH)^{2-}$, hydroxidodioxidoborato(2−); hydrogenborato
BH_2	$-BH_2$, boranyl	BH_2^+, boranylium, dihydridoboron(1+)	BH_2^-, boranide, dihydridoborate(2−)	BH_2^-, boranido, dihydridoborato(2−)
BH_2O	−BH(OH), hydroxyboranyl			
BH_2O_2	$-B(OH)_2$, dihydroxyboranyl; borono			
BH_3	BH_3, boron trihydride $[BH_3]$, borane (parent hydride name), trihydridoboron	$BH_3^{\bullet +}$, boraniumyl, trihydridoboron(•1+)	$BH_3^{\bullet -}$ boranuidyl, trihydridoborate(•1−)[e] $-BH_3^-$, boranuidyl	$BH_3^{\bullet -}$, trihydridoborato(•1−)
BH_4		BH_4^+, boranium, tetrahydridoboron(1+)	BH_4^-, boranuide, tetrahydridoborate(1−)	BH_4^-, boranuido, tetrahydridoborato(1−)
BO	BO, boron mon(o)oxide [BO], oxidoboron	BO^+, oxidoboron(1+)	BO^-, oxidoborate(1−)	BO^-, oxidoborato(1−)
BO_2			$(BO_2^-)_n = (OBO)_n^{n-}$, *catena*-poly[(oxidoborate-μ-oxido)(1−)]; metaborate	
BO_3			BO_3^{3-}, trioxidoborate(3−); borate	BO_3^{3-}, trioxidoborato(3−); borato
Ba	barium	barium	baride	barido
BaO	barium oxide			

Table IX *Continued*

Formula for uncharged atom or group	*Name*			
	Uncharged atoms or molecules (including zwitterions and radicals) or substituent groups[a]	*Cations (including cation radicals) or cationic substituent groups*[a]	*Anions (including anion radicals) or anionic substituent groups*[b]	*Ligands*[c]
BaO_2	$Ba^{2+}O_2^{2-}$, barium dioxide(2−); barium peroxide			
Be	beryllium	beryllium (general) Be^{+}, beryllium(1+) Be^{2+}, beryllium(2+)	beryllide	beryllido
BeH	BeH, beryllium monohydride [BeH], hydridoberyllium	BeH^{+}, hydridoberyllium(1+)	BeH^{-}, hydridoberyllate(1−)	BeH^{-}, hydridoberyllato(1−)
Bh	bohrium	bohrium	bohride	bohrido
Bi	bismuth	bismuth	bismuthide (general) Bi^{3-}, bismuthide(3−), bismuthanetriide; bismuthide	bismuthido (general) Bi^{3-}, bismuthido(3−), bismuthanetriido; bismuthido
BiH	>BiH, bismuthanediyl =BiH, bismuthanylidene $BiH^{2\bullet}$, bismuthanylidene, hydridobismuth(2•)	BiH^{2+}, bismuthanebis(ylium), hydridobismuth(2+)	BiH^{2-}, bismuthanediide, hydridobismuthate(2−)	BiH^{2-}, bismuthanediido, hydridobismuthato(2−)
BiH_2	$-BiH_2$, bismuthanyl $BiH_2^{\bullet}$, bismuthanyl, dihydridobismuth(•)	BiH_2^{+}, bismuthanylium, dihydridobismuth(1+)	BiH_2^{-}, bismuthanide, dihydridobismuthate(1−)	BiH_2^{-}, bismuthanido, dihydridobismuthato(1−)
BiH_3	BiH_3, bismuth trihydride $[BiH_3]$, bismuthane (parent hydride name), trihydridobismuth $=BiH_3$, λ^5-bismuthanylidene	$BiH_3^{\bullet +}$, bismuthaniumyl, trihydridobismuth(•1+)	$BiH_3^{\bullet -}$, bismuthanuidyl, trihydridobismuthate(•1−)[e]	

BiH_4		BiH_4^+, bismuthanium, tetrahydridobismuth(1+)		
Bi_5		Bi_5^{4+}, pentabismuth(4+)		
Bk	berkelium	berkelium	berkelide	berkelido
Br	bromine (general) $Br^{\bullet}$, bromine(•), monobromine –Br, bromo	bromine (general) Br^+, bromine(1+)	bromide (general) Br^-, bromide(1−); bromide	bromido (general) Br^-, bromido(1−); bromido
BrCN	BrCN, cyanobromane, bromidonitridocarbon			
BrH, see HBr				
BrHO	HOBr, bromanol, hydroxidobromine[f]; hypobromous acid			
$BrHO_2$	HOBrO, hydroxy-λ^3-bromanone, hydroxidooxidobromine; bromous acid			
$BrHO_3$	$HOBrO_2$, hydroxy-λ^5-bromanedione, hydroxidodioxidobromine; bromic acid			
$BrHO_4$	$HOBrO_3$, hydroxy-λ^7-bromanetrione, hydroxidotrioxidobromine; perbromic acid			
Br_2	Br_2, dibromine	$Br_2^{\bullet+}$, dibromine(•1+)	$Br_2^{\bullet-}$, dibromide(•1−)	Br_2, dibromine
Br_3	Br_3, tribromine		Br_3^-, tribromide(1−); tribromide	Br_3^-, tribromido(1−); tribromido
C	carbon (general) C, monocarbon >C<, methanetetrayl =C=, methanediylidene	carbon (general) C^+, carbon(1+)	carbide (general) C^-, carbide(1−) C^{4-}, carbide(4−), methanetetraide; carbide	carbido (general) C^-, carbido(1−) C^{4-}, carbido(4−), methanetetrayl, methanetetraido
CClNS			$ClSCN^{\bullet-}$, (chloridosulfato)nitrido- carbonate(•1−)	

Table IX *Continued*

Formula for uncharged atom or group	*Name*			
	Uncharged atoms or molecules (including zwitterions and radicals) or substituent groups[a]	*Cations (including cation radicals) or cationic substituent groups*[a]	*Anions (including anion radicals) or anionic substituent groups*[b]	*Ligands*[c]
CH	$CH^{\bullet}$, hydridocarbon(•) $CH^{3\bullet}$, methylidyne, hydridocarbon(3•), carbyne ≡CH, methylidyne −CH=, methanylylidene −CH<, methanetriyl	CH^{+}, λ^2-methanylium, hydridocarbon(1+)	CH^{-}, λ^2-methanide, hydridocarbonate(1−) CH^{3-}, methanetriide, hydridocarbonate(3−)	CH^{-}, λ^2-methanido, hydridocarbonato(1−) CH^{3-}, methanetriyl, methanetriido, hydridocarbonato(3−)
CHN	HCN, hydrogen cyanide HCN = [CHN], methanenitrile, hydridonitridocarbon; formonitrile >C=NH, carbonimidoyl =C=NH, iminomethylidene, carbonimidoylidene			
CHNO	HCNO = [N(CH)O], formonitrile oxide, (hydridocarbonato)oxidonitrogen HNCO = [C(NH)O], (hydridonitrato)oxidocarbon; isocyanic acid HOCN = [C(OH)N], hydroxidonitridocarbon; cyanic acid HONC = [NC(OH)], λ^2-methylidenehydroxylamine, carbidohydroxidonitrogen		$HNCO^{\bullet-}$, (hydridonitrato)oxidocarbonate(•1−) $HOCN^{\bullet-}$, hydroxidonitridocarbonate(•1−)	$HNCO^{\bullet-}$, (hydridonitrato)oxidocarbonato(•1−) $HOCN^{\bullet-}$, hydroxidonitridocarbonato(•1−)

CHNOS			$HONCS^{\bullet -}$, (hydroxidonitrato)sulfidocarbonate(•1−) $HOSCN^{\bullet -}$, (hydroxidosulfato)nitridocarbonate(•1−)	$HONCS^{\bullet -}$, (hydroxidonitrato)sulfidocarbonato(•1−) $HOSCN^{\bullet -}$, (hydroxidosulfato)nitridocarbonato(•1−)
$CHNO_2$			$HOOCN^{\bullet -}$, (dioxidanido)nitridocarbonate(•1−) $HONCO^{\bullet -}$, (hydroxidonitrato)oxidocarbonate(•1−)	$HOOCN^{\bullet -}$, (dioxidanido)nitridocarbonato(•1−) $HONCO^{\bullet -}$, (hydroxidonitrato)oxidocarbonato(•1−)
CHNS	$HCNS = HC{\equiv}N^{+}S^{-}$ = [N(CH)S], (methylidyneammoniumyl)sulfanide, (hydridocarbonato)sulfidonitrogen HNCS = [C(NH)S], (hydridonitrato)sulfidocarbon; isothiocyanic acid HSCN = [CN(SH)], nitridosulfanidocarbon; thiocyanic acid HSNC = [NC(SH)], λ^2-methylidenethiohydroxylamine, carbidosulfanidonitrogen			

Table IX *Continued*

Formula for uncharged atom or group	*Name*			
	Uncharged atoms or molecules (including zwitterions and radicals) or substituent groups[a]	*Cations (including cation radicals) or cationic substituent groups*[a]	*Anions (including anion radicals) or anionic substituent groups*[b]	*Ligands*[c]
CHNSe	HCNSe = HC≡N^+Se^- = [N(CH)Se], (methylidyneammoniumyl)selanide, (hydridocarbonato)selenidonitrogen HNCSe = [C(NH)Se], (hydridonitrato)selenidocarbon; isoselenocyanic acid HSeCN = [CN(SeH)], nitridoselanidocarbon; selenocyanic acid HSeNC = [NC(SeH)], λ^2-methylideneselenohydroxylamine, carbidoselanidonitrogen			
CHO	$HCO^•$, oxomethyl, hydridooxidocarbon(•) –CH(O), methanoyl, formyl			
$CHOS_2$	$HOCS_2^•$, hydroxidodisulfidocarbon(•)			
CHO_2	$HOCO^•$, hydroxidooxidocarbon(•)			

CHO_3	$HOCO_2^{\bullet}$, hydroxidodioxidocarbon(•) $HOOCO^{\bullet}$, (dioxidanido)oxidocarbon(•)		HCO_3^-, hydroxidodioxidocarbonate(1−); hydrogencarbonate	HCO_3^-, hydroxidodioxidocarbonato(1−); hydrogencarbonato
CH_2	CH_2, λ^2-methane $CH_2^{2\bullet}$, methylidene, dihydridocarbon(2•); carbene $>CH_2$, methanediyl, methylene $=CH_2$, methylidene		CH_2^{2-}, methanediide, dihydridocarbonate(2−) $-CH_2^-$, methanidyl	$>CH_2$, methanediyl, methylene $=CH_2$, methylidene CH_2^{2-}, methanediido, dihydridocarbonato(2−)
CH_2N	$H_2CN^{\bullet}$, dihydridonitridocarbon(•)			
CH_2NO	$H_2NCO^{\bullet}$, (dihydridonitrato)oxidocarbon(•) $HNCOH^{\bullet}$, (hydridonitrato)hydroxido- carbon(•)			
CH_3	$CH_3^{\bullet}$, methyl $-CH_3$ *or* −Me, methyl	CH_3^+, methylium, trihydridocarbon(1+)	CH_3^-, methanide, trihydridocarbonate(1−)	CH_3^-, methyl, methanido, trihydridocarbonato(1−)
CH_4	CH_4, methane (parent hydride name), tetrahydridocarbon	$CH_4^{\bullet+}$, methaniumyl, tetrahydridocarbon(•1+)	$CH_4^{\bullet-}$, methanuidyl, tetrahydridocarbonate(•1−)[e]	
CH_5		CH_5^+, methanium, pentahydridocarbon(1+)		
CN	$CN^{\bullet}$, nitridocarbon(•); cyanyl −CN, cyano −NC, isocyano	CN^+, azanylidynemethylium, nitridocarbon(1+)	CN^-, nitridocarbonate(1−); cyanide	nitridocarbonato (general) CN^-, nitridocarbonato(1−); cyanido = [nitridocarbonato(1−)-κ*C*]
CN_2			NCN^{2-}, dinitridocarbonate(2−)	NCN^{2-}, dinitridocarbonato(2−)
CNO	$OCN^{\bullet}$, nitridooxidocarbon(•) −OCN, cyanato −NCO, isocyanato −ONC, λ^2-methylideneazanylylideneoxy −CNO, (oxo-λ^5-azanylidyne)methyl		OCN^-, nitridooxidocarbonate(1−); cyanate ONC^-, carbidooxidonitrate(1−); fulminate $OCN^{\bullet 2-}$, nitridooxidocarbonate(•2−)	OCN^-, nitridooxidocarbonato(1−); cyanato ONC^-, carbidooxidonitrato(1−); fulminato

Table IX *Continued*

Formula for uncharged atom or group	*Name*			
	Uncharged atoms or molecules (including zwitterions and radicals) or substituent groups[a]	*Cations (including cation radicals) or cationic substituent groups*[a]	*Anions (including anion radicals) or anionic substituent groups*[b]	*Ligands*[c]
CNS	$SCN^{\bullet}$, nitridosulfidocarbon(•) –SCN, thiocyanato –NCS, isothiocyanato –SNC, λ^2-methylideneazanylylidenesulfanediyl –CNS, (sulfanylidene-λ^5-azanylidyne)methyl		SCN^-, nitridosulfidocarbonate(1–); thiocyanate SNC^-, carbidosulfidonitrate(1–)	SCN^-, nitridosulfidocarbonato(1–); thiocyanato SNC^-, carbidosulfidonitrato(1–)
CNSe	$SeCN^{\bullet}$, nitridoselenidocarbon(•) –SeCN, selenocyanato –NCSe, isoselenocyanato –SeNC, λ^2-methylideneazanylylideneselanediyl –CNSe, (selanylidene-λ^5-azanylidyne)methyl		$SeCN^-$, nitridoselenidocarbonate(1–); selenocyanate $SeNC^-$, carbidoselenidonitrate(1–)	$SeCN^-$, nitridoselenidocarbonato(1–); selenocyanato $SeNC^-$, carbidoselenidonitrato(1–)
CO	CO, carbon mon(o)oxide >C=O, carbonyl =C=O, carbonylidene	$CO^{\bullet +}$, oxidocarbon(•1+) CO^{2+}, oxidocarbon(2+)	$CO^{\bullet -}$, oxidocarbonate(•1–)	CO, oxidocarbon, oxidocarbonato (general); carbonyl = oxidocarbon-κ*C* (general) $CO^{\bullet +}$, oxidocarbon(•1+) $CO^{\bullet -}$, oxidocarbonato(•1–)
COS	C(O)S, carbonyl sulfide, oxidosulfidocarbon			
CO_2	CO_2, carbon dioxide, dioxidocarbon		$CO_2^{\bullet -}$, oxidooxomethyl, dioxidocarbonate(•1–)	CO_2, dioxidocarbon $CO_2^{\bullet -}$, oxidooxomethyl, dioxidocarbonato(•1–)

CO_3			$CO_3^{\bullet-}$, trioxidocarbonate(•1−), $OCOO^{\bullet-}$, (dioxido)oxidocarbonate(•1−), oxidoperoxidocarbonate(•1−) CO_3^{2-}, trioxidocarbonate(2−); carbonate	CO_3^{2-}, trioxidocarbonato(2−); carbonato
CS	carbon monosulfide >C=S, carbonothioyl; thiocarbonyl =C=S, carbonothioylidene	$CS^{\bullet+}$, sulfidocarbon(•1+)	$CS^{\bullet-}$, sulfidocarbonate(•1−)	CS, sulfidocarbon, sulfidocarbonato, thiocarbonyl (general); $CS^{\bullet+}$, sulfidocarbon(•1+) $CS^{\bullet-}$, sulfidocarbonato(•1−)
CS_2	CS_2, disulfidocarbon, carbon disulfide		$CS_2^{\bullet-}$, sulfidothioxomethyl, disulfidocarbonate(•1−)	CS_2, disulfidocarbon $CS_2^{\bullet-}$· sulfidothioxomethyl, disulfidocarbonato(•1−)
CS_3			CS_3^{2-}, trisulfidocarbonate(2−)	CS_3^{2-}, trisulfidocarbonato(2−)
C_2	C_2, dicarbon	C_2^{+}, dicarbon(1+)	C_2^{-}, dicarbide(1−) C_2^{2-}, dicarbide(2−), ethynediide, acetylenediide; acetylide	dicarbido (general) C_2^{2-}, dicarbido(2−), ethynediido, ethyne-1,2-diyl
C_2H	$HCC^{\bullet}$, ethynyl, hydridodicarbon(•)			
C_2N_2	NCCN, ethanedinitrile, bis(nitridocarbon)(*C—C*); oxalonitrile		$NCCN^{\bullet-}$, bis(nitridocarbonate)(*C—C*)(•1−)	
$C_2N_2O_2$	NCOOCN, dioxidanedicarbonitrile, bis[cyanidooxygen](*O—O*)		$NCOOCN^{\bullet-}$, bis[cyanidooxygenate](*O—O*)(•1−)[e] $OCNNCO^{\bullet-}$, bis(carbonylnitrate)(*N—N*)(•1−)[e]	$NCOOCN^{\bullet-}$, bis[cyanidooxygenato](*O—O*)(•1−) $OCNNCO^{\bullet-}$, bis(carbonylnitrato)(*N—N*)(•1−)
$C_2N_2S_2$	NCSSCN, disulfanedicarbonitrile, bis[cyanidosulfur](*S—S*)		$NCSSCN^{\bullet-}$, bis[cyanidosulfate](*S—S*)(•1−)[e]	$NCSSCN^{\bullet-}$, bis[cyanidosulfato](*S—S*)(•1−)
C_3O_2	C_3O_2, tricarbon dioxide O=C=C=C=O, propa-1,2-diene-1,3-dione			
$C_{12}O_9$	$C_{12}O_9$, dodecacarbon nonaoxide			

Table IX *Continued*

Formula for uncharged atom or group	*Name*			
	Uncharged atoms or molecules (including zwitterions and radicals) or substituent groups[a]	*Cations (including cation radicals) or cationic substituent groups*[a]	*Anions (including anion radicals) or anionic substituent groups*[b]	*Ligands*[c]
Ca	calcium	calcium (general) Ca^{2+}, calcium(2+)	calcide	calcido
Cd	cadmium	cadmium (general) Cd^{2+}, cadmium(2+)	cadmide	cadmido
Ce	cerium	cerium (general) Ce^{3+}, cerium(3+) Ce^{4+}, cerium(4+)	ceride	cerido
Cf	californium	californium	californide	californido
Cl	chlorine (general) $Cl^{\bullet}$, chlorine(•), monochlorine –Cl, chloro	chlorine (general) Cl^{+}, chlorine(1+)	chloride (general) Cl^{-}, chloride(1−); chloride	chlorido (general) Cl^{-}, chlorido(1−); chlorido
ClF	ClF, fluoridochlorine, chlorine monofluoride	ClF^{+}, fluoridochlorine(1+)		
ClF_2			ClF_2^{-}, difluoridochlorate(1−)	ClF_2^{-}, difluoridochlorato(1−)
ClF_4		ClF_4^{+}, tetrafluoridochlorine(1+)	ClF_4^{-}, tetrafluoridochlorate(1−)	ClF_4^{-}, tetrafluoridochlorato(1−)
ClH, see HCl				
ClHN			$NHCl^{-}$, chloroazanide, chloridohydridonitrate(1−)	$NHCl^{-}$, chloroazanido, chloridohydridonitrato(1−)
ClHO	HOCl, chloranol, hydroxidochlorine[f]; hypochlorous acid		$HOCl^{\bullet -}$, hydroxidochlorate(•1−)	
$ClHO_2$	HOClO, hydroxy-λ^3-chloranone, hydroxidooxidochlorine; chlorous acid			

$ClHO_3$	$HOClO_2$, hydroxy-λ^5-chloranedione, hydroxidodioxidochlorine; chloric acid			
$ClHO_4$	$HOClO_3$, hydroxy-λ^7-chloranetrione, hydroxidotrioxidochlorine; perchloric acid			
Cl_2	Cl_2, dichlorine	$Cl_2^{\bullet+}$, dichlorine(•1+)	$Cl_2^{\bullet-}$, dichloride(•1−)	Cl_2, dichlorine $Cl_2^{\bullet-}$, dichlorido(•1−)
Cl_2OP	$-PCl_2(O)$, dichlorooxo-λ^5-phosphanyl, phosphorodichloridoyl			
Cl_4		Cl_4^+, tetrachlorine(1+)		
Cm	curium	curium	curide	curido
Co	cobalt	cobalt (general) Co^{2+}, cobalt(2+) Co^{3+}, cobalt(3+)	cobaltide	cobaltido
Cr	chromium	chromium (general) Cr^{2+}, chromium(2+) Cr^{3+}, chromium(3+)	chromide	chromido
CrO	CrO, chromium mon(o)oxide, chromium(II) oxide			
CrO_2	CrO_2, chromium dioxide, chromium(IV) oxide			
CrO_3	CrO_3, chromium trioxide, chromium(VI) oxide			
CrO_4	$[Cr(O_2)_2]$, diperoxidochromium		CrO_4^{2-}, tetraoxidochromate(2−); chromate CrO_4^{3-}, tetraoxidochromate(3−) CrO_4^{4-}, tetraoxidochromate(4−)	CrO_4^{2-}, tetraoxidochromato(2−); chromato CrO_4^{3-}, tetraoxidochromato(3−) CrO_4^{4-}, tetraoxidochromato(4−)
CrO_5	$[CrO(O_2)_2]$, oxidodiperoxidochromium			

Table IX *Continued*

Formula for uncharged atom or group	*Name*			
	Uncharged atoms or molecules (including zwitterions and radicals) or substituent groups[a]	*Cations (including cation radicals) or cationic substituent groups*[a]	*Anions (including anion radicals) or anionic substituent groups*[b]	*Ligands*[c]
CrO_6			$CrO_2(O_2)_2^{2-}$, dioxidodiperoxidochromate(2−)	
CrO_8			$Cr(O_2)_4^{2-}$, tetraperoxidochromate(2−) $Cr(O_2)_4^{3-}$, tetraperoxidochromate(3−)	
Cr_2O_3	Cr_2O_3, dichromium trioxide, chromium(III) oxide			
Cr_2O_7			$Cr_2O_7^{2-}$, heptaoxidodichromate(2−) $O_3CrOCrO_3^{2-}$, μ-oxido-bis(trioxidochromate)(2−); dichromate	$Cr_2O_7^{2-}$, heptaoxidodichromato(2−) $O_3CrOCrO_3^{2-}$, μ-oxido-bis(trioxidochromato)(2−); dichromato
Cs	caesium	caesium	caeside	caesido
Cu	copper	copper (general) Cu^{+}, copper(1+) Cu^{2+}, copper(2+)	cupride	cuprido
D, see H				
D_2, see H_2				
D_2O, see H_2O				
Db	dubnium	dubnium	dubnide	dubnido
Ds	darmstadtium	darmstadtium	darmstadtide	darmstadtido
Dy	dysprosium	dysprosium	dysproside	dysprosido
Er	erbium	erbium	erbide	erbido

Es	einsteinium	einsteinium	einsteinide	einsteinido
Eu	europium	europium	europide	europido
F	fluorine $F^{\bullet}$, fluorine(•), monofluorine –F, fluoro	fluorine (general) F^{+}, fluorine(1+)	fluoride (general) F^{-}, fluoride(1–); fluoride	F^{-}, fluorido(1–); fluorido
FH, see HF				
FHO	HOF, fluoranol, fluoridohydridooxygen			
FNS	NSF, fluoridonitridosulfur			
FN_3	FNNN, fluorido-1κ*F*-trinitrogen(2 *N*—*N*)			
FO, see OF				
F_2	F_2, difluorine	F_2^{+}, difluorine(•1+)	F_2^{-}, difluoride(•1–)	F_2, difluorine
F_2N_2	FN=NF, difluorido-1κ*F*,2κ*F*-dinitrogen(*N*—*N*), difluorodiazene			
Fe	iron	iron (general) Fe^{2+}, iron(2+) Fe^{3+}, iron(3+)	ferride	ferrido
Fm	fermium	fermium	fermide	fermido
Fr	francium	francium	francide	francido
Ga	gallium	gallium	gallide	gallido
GaH_2	$-GaH_2$, gallanyl			
GaH_3	GaH_3, gallium trihydride $[GaH_3]$, gallane (parent hydride name), trihydridogallium			
Gd	gadolinium	gadolinium	gadolinide	gadolinido
Ge	germanium >Ge<, germanetetrayl =Ge=, germanediylidene	germanium (general) Ge^{2+}, germanium(2+) Ge^{4+}, germanium(4+)	germide (general) Ge^{4-}, germide(4–); germide	germido (general) Ge^{4-}, germido(4–); germido
GeH	>GeH–, germanetriyl =GeH–, germanylylidene ≡GeH, germylidyne			

Table IX *Continued*

Formula for uncharged atom or group	*Name*			
	Uncharged atoms or molecules (including zwitterions and radicals) or substituent groups[a]	*Cations (including cation radicals) or cationic substituent groups*[a]	*Anions (including anion radicals) or anionic substituent groups*[b]	*Ligands*[c]
GeH_2	$>GeH_2$, germanediyl $=GeH_2$, germylidene			
GeH_3	$-GeH_3$, germyl	GeH_3^+, germylium, trihydridogermanium(1+)	GeH_3^-, germanide, trihydridogermanate(1−)	GeH_3^-, germanido, trihydridogermanato(1−)
GeH_4	GeH_4, germane (parent hydride name), tetrahydridogermanium			
Ge_4			Ge_4^{4-}, tetragermide(4−)	
H	hydrogen $H^{\bullet}$, hydrogen(•), monohydrogen (natural or unspecified isotopic composition) $^1H^{\bullet}$, protium(•), monoprotium $^2H^{\bullet} = D^{\bullet}$, deuterium(•), monodeuterium $^3H^{\bullet} = T^{\bullet}$, tritium(•), monotritium	hydrogen (general) H^+, hydrogen(1+), hydron (natural or unspecified isotopic composition) $^1H^+$, protium(1+), proton $^2H^+ = D^+$, deuterium(1+), deuteron $^3H^+ = T^+$, tritium(1+), triton	hydride (general) H^-, hydride (natural or unspecified isotopic composition) $^1H^-$, protide $^2H^- = D^-$, deuteride $^3H^- = T^-$, tritide	hydrido protido deuterido tritido
HAt	HAt, hydrogen astatide [HAt], astatidohydrogen			
HBr	HBr, hydrogen bromide [HBr], bromane (parent hydride name), bromidohydrogen			
HCO, see CHO				

HCl	HCl, hydrogen chloride [HCl], chlorane (parent hydride name), chloridohydrogen	HCl^{+}, chloraniumyl, chloridohydrogen(•1+)		
HF	HF, hydrogen fluoride [HF], fluorane (parent hydride name), fluoridohydrogen	HF^{+}, fluoraniumyl, fluoridohydrogen(•1+)		
HF_2			FHF^{-}, fluorofluoranuide, μ-hydridodifluorate(1−), difluoridohydrogenate(1−)	
HI	HI, hydrogen iodide [HI], iodane (parent hydride name), iodidohydrogen			
HIO	HOI, iodanol, hydroxidoiodine[f]; hypoiodous acid			
HIO_2	HOIO, hydroxy-λ^3-iodanone, hydroxidooxidoiodine; iodous acid			
HIO_3	$HOIO_2$, hydroxy-λ^5-iodanedione, hydroxidodioxidoiodine; iodic acid		$HOIO_2^{\bullet-}$, hydroxidodioxidoiodate(•1−)	
HIO_4	$HOIO_3$, hydroxy-λ^7-iodanetrione, hydroxidotrioxidoiodine; periodic acid			
H_nN_m, see N_mH_n				
$HMnO_4$	$HMnO_4 = [MnO_3(OH)]$, hydroxidotrioxidomanganese		$HMnO_4^{-} = [MnO_3(OH)]^{-}$, hydroxidotrioxidomanganate(1−)	

Table IX *Continued*

Formula for uncharged atom or group	*Name*			
	Uncharged atoms or molecules (including zwitterions and radicals) or substituent groups[a]	*Cations (including cation radicals) or cationic substituent groups*[a]	*Anions (including anion radicals) or anionic substituent groups*[b]	*Ligands*[c]
HNO	HNO = [NH(O)], azanone, hydridooxidonitrogen $HON^{2\bullet}$, hydroxidonitrogen(2•) >NH(O), oxo-λ^5-azanediyl; azonoyl =NH(O), oxo-λ^5-azanylidene; azonoylidene >N–OH, hydroxyazanediyl =N–OH, hydroxyazanylidene; hydroxyimino	$HNO^{\bullet +}$ = $[NH(O)]^{\bullet +}$, hydridooxidonitrogen(•1+)	HON^{2-}, hydroxidonitrate(2–)	HON^{2-}, hydroxidonitrato(2–)
HNO_2	HNO_2 = [NO(OH)], hydroxidooxidonitrogen; nitrous acid >N(O)(OH), hydroxyoxo-λ^5-azanediyl; hydroxyazoryl =N(O)(OH), hydroxyoxo-λ^5-azanylidene; hydroxyazorylidene			
HNO_3	HNO_3 = $[NO_2(OH)]$, hydroxidodioxidonitrogen; nitric acid $HNO(O_2)$ = [NO(OOH)], dioxidanidooxidonitrogen, peroxynitrous acid			

HNO_4	HNO_4 = [NO_2(OOH)], (dioxidanido)dioxidonitrogen; peroxynitric acid			
HNS	>S(=NH), imino-λ^4-sulfanediyl; sulfinimidoyl			
HN_2O_2	−$NHNO_2$, nitroazanyl, nitroamino		[HON=NO]$^-$, 2-hydroxydiazen-1-olate, hydroxido-1κ*O*-oxido-2κ*O*- dinitrate(*N*—*N*)(1−)	
HN_2O_3			$HN_2O_3^-$ = N(H)(O)NO_2^-, hydrido-1κ*H*-trioxido- 1κ*O*,2κ^2*O*-dinitrate(*N*—*N*)(1−)	
HN_3O			$HON_3^{\bullet-}$, hydroxido-1κ*O*- trinitrate(2 *N*—*N*)(•1−)	
HO	$HO^\bullet$, oxidanyl, hydridooxygen(•); hydroxyl −OH, oxidanyl; hydroxy	HO^+, oxidanylium, hydridooxygen(1+); hydroxylium	HO^-, oxidanide, hydridooxygenate(1−); hydroxide	HO^-, oxidanido; hydroxido
HOP	HPO = [P(H)O], phosphanone, hydridooxidophosphorus >PH(O), oxo-λ^5-phosphanediyl; phosphonoyl =PH(O), oxo-λ^5-phosphanylidene; phosphonoylidene =P−OH, hydroxyphosphanylidene			
HOS	−SH(O), oxo-λ^4-sulfanyl −SOH, hydroxysulfanyl −OSH, sulfanyloxy		HSO^-, sulfanolate, hydridooxidosulfate(1−)	HSO^-, sulfanolato, hydridooxidosulfato(1−)
HOSe	−SeH(O), oxo-λ^4-selanyl −SeOH, hydroxyselanyl −OSeH, selanyloxy			

Table IX *Continued*

Formula for uncharged atom or group	*Name*			
	Uncharged atoms or molecules (including zwitterions and radicals) or substituent groups[a]	*Cations (including cation radicals) or cationic substituent groups*[a]	*Anions (including anion radicals) or anionic substituent groups*[b]	*Ligands*[c]
HO_2	$HO_2^{\bullet}$, dioxidanyl, hydridodioxygen(•) −OOH, dioxidanyl; hydroperoxy	HO_2^+, dioxidanylium, hydridodioxygen(1+)	HO_2^-, dioxidanide, hydrogen(peroxide)(1−)	HO_2^-, dioxidanido, hydrogen(peroxido)(1−)
HO_2P	P(O)(OH), hydroxyphosphanone, hydroxidooxidophosphorus >P(O)(OH), hydroxyoxo-λ^5-phosphanediyl; hydroxyphosphoryl =P(O)(OH), hydroxyoxo-λ^5-phosphanylidene; hydroxyphosphorylidene		$HOPO^{\bullet -}$, hydroxidooxidophosphate(•1−) HPO_2^{2-}, hydridodioxidophosphate(2−)	$HOPO^{\bullet -}$, hydroxidooxidophosphato(•1−) HPO_2^{2-}, hydridodioxidophosphato(2−)
HO_2S	$HOOS^{\bullet}$, hydrido-1κ*H*-sulfido-2κ*S*-dioxygen(•) $HOSO^{\bullet}$, hydroxidooxidosulfur(•) $HSOO^{\bullet}$, (hydridosulfato)dioxygen(•) −S(O)(OH), hydroxyoxo-λ^4-sulfanyl; hydroxysulfinyl, sulfino −S(O)$_2$H, dioxo-λ^6-sulfanyl		$HOSO^-$, hydroxysulfanolate, hydroxidooxidosulfate(1−)	$HOSO^-$, hydroxysulfanolato, hydroxidooxidosulfato(1−)

HO_2Se	$-Se(O)(OH)$, hydroxyoxo-λ^4-selanyl; hydroxyseleninyl, selenino $-Se(O)_2H$, dioxo-λ^6-selanyl			
HO_3	$HO_3^{\bullet}$, hydridotrioxygen(•) $HOOO^{\bullet}$, trioxidanyl, hydrido-1κ*H*-trioxygen(2 *O*—*O*)(•) $-OOOH$, trioxidanyl			
HO_3P	$P(O)_2(OH)$, hydroxy-λ^5-phosphanedione, hydroxidodioxidophosphorus		$HOPO_2^{\bullet-}$, hydroxidodioxidophosphate(•1−) PHO_3^{2-}, hydridotrioxidophosphate(2−); phosphonate $HPO_3^{2-} = PO_2(OH)^{2-}$, hydroxidodioxidophosphate(2−); hydrogenphosphite	$HOPO_2^{\bullet-}$, hydroxidodioxidophosphato(•1−) PHO_3^{2-}, hydridotrioxidophosphato(2−); phosphonato $HPO_3^{2-} = PO_2(OH)^{2-}$, hydroxidodioxidophosphato(2−); hydrogenphosphito
HO_3S	$-S(O)_2(OH)$, hydroxydioxo-λ^6-sulfanyl, hydroxysulfonyl; sulfo		HSO_3^-, hydroxidodioxidosulfate(1−), hydrogensulfite	HSO_3^-, hydroxidodioxidosulfato(1−), hydrogensulfito
HO_3Se	$HOSeO_2^{\bullet}$, hydroxidodioxidoselenium(•) $-Se(O)_2(OH)$, hydroxydioxo-λ^6-selanyl, hydroxyselenonyl; selenono		$HSeO_3^-$, hydroxidodioxidoselenate(1−)	$HSeO_3^-$, hydroxidodioxidoselenato(1−)
HO_4P			$HOPO_3^{\bullet-} = PO_3(OH)^{\bullet-}$, hydroxidotrioxidophosphate(•1−) $HPO_4^{2-} = PO_3(OH)^{2-}$, hydroxidotrioxidophosphate(2−); hydrogenphosphate	$HOPO_3^{\bullet-} = PO_3(OH)^{\bullet-}$, hydroxidotrioxidophosphato(•1−) HPO_4^{2-}, hydroxidotrioxidophosphato(2−); hydrogenphosphato
HO_4S	$HOSO_3^{\bullet}$, hydro-xidotrioxidosulfur(•) $-OS(O)_2(OH)$, hydroxysulfonyloxy; sulfooxy		HSO_4^-, hydroxidotrioxidosulfate(1−); hydrogensulfate	HSO_4^-, hydroxidotrioxidosulfato(1−); hydrogensulfato

Table IX *Continued*

Formula for uncharged atom or group	*Name*			
	Uncharged atoms or molecules (including zwitterions and radicals) or substituent groups[a]	*Cations (including cation radicals) or cationic substituent groups*[a]	*Anions (including anion radicals) or anionic substituent groups*[b]	*Ligands*[c]
HO_4Se			$HSeO_4^-$, hydroxidotrioxidoselenate(1−)	$HSeO_4^-$, hydroxidotrioxidoselenato(1−)
HO_5P			$HOPO_4^{\bullet-} = PO_2(OH)(OO)^{\bullet-}$, (dioxido)hydroxidodioxidophosphate(•1−)	$PO_2(OH)(OO)^{\bullet-}$, (dioxido)hydroxidodioxidophosphato(•1−)
HO_5S	$HOSO_4^{\bullet} = [SO_2(OH)(OO)]^{\bullet}$, (dioxido)hydroxidodioxidosulfur(•)			
HS	−SH, sulfanyl $HS^{\bullet}$, sulfanyl, hydridosulfur(•)	HS^+, sulfanylium, hydridosulfur(1+)	HS^-, sulfanide, hydrogen(sulfide)(1−)	HS^-, sulfanido, hydrogen(sulfido)(1−)
HS_2	−SSH, disulfanyl		HSS^-, disulfanide	HSS^-, disulfanido
HS_3	−SSSH, trisulfanyl		$HSSS^-$, trisulfanide	$HSSS^-$, trisulfanido
HS_4	−SSSSH, tetrasulfanyl		$HSSSS^-$, tetrasulfanide	$HSSSS^-$, tetrasulfanido
HS_5	−SSSSSH, pentasulfanyl		$HSSSSS^-$, pentasulfanide	$HSSSSS^-$, pentasulfanido
HSe	$HSe^{\bullet}$, selanyl, hydridoselenium(•) −SeH, selanyl	HSe^+, selanylium, hydridoselenium(1+)	HSe^-, selanide, hydrogen(selenide)(1−)	HSe^-, selanido, hydrogen(selenido)(1−)
HSe_2	−SeSeH, diselanyl		$HSeSe^-$, diselanide	$HSeSe^-$, diselanido
HTe	$HTe^{\bullet}$, tellanyl, hydridotellurium(•) −TeH, tellanyl	HTe^+, tellanylium, hydridotellurium(1+)	HTe^-, tellanide, hydrogen(tellanide)(1−)	
HTe_2	−TeTeH, ditellanyl		$HTeTe^-$, ditellanide	$HTeTe^-$, ditellanido
H_2	H_2, dihydrogen D_2, dideuterium T_2, ditritium	$H_2^{\bullet+}$, dihydrogen(•1+) $^1H_2^{\bullet+}$, diprotium(•1+) $D_2^{\bullet+}$, dideuterium(•1+) $T_2^{\bullet+}$, ditritium(•1+)		

H_2Br	$H_2Br^{\bullet}$, λ[3]-bromanyl, dihydridobromine(•)	H_2Br^+, bromanium, dihydridobromine(1+)		
H_2Cl	$H_2Cl^{\bullet}$, λ[3]-chloranyl, dihydridochlorine(•)	H_2Cl^+, chloranium, dihydridochlorine(1+)		
H_2F	$H_2F^{\bullet}$, λ[3]-fluoranyl, dihydridofluorine(•)	H_2F^+, fluoranium, dihydridofluorine(1+)		
H_2I	$H_2I^{\bullet}$, λ[3]-iodanyl, dihydridoiodine(•)	H_2I^+, iodanium, dihydridoiodine(1+)		
H_2IO_2	$-I(OH)_2$, dihydroxy-λ[3]-iodanyl			
H_2MnO_4	$H_2MnO_4 = [MnO_2(OH)_2]$, dihydroxidodioxidomanganese			
H_2N_m, see N_mH_2				
H_2NO	$H_2NO^{\bullet}$, aminooxidanyl, dihydridooxidonitrogen(•); aminoxyl $HONH^{\bullet}$, hydroxyazanyl, hydridohydroxidonitrogen(•) $-NH(OH)$, hydroxyazanyl, hydroxyamino $-ONH_2$, aminooxy $-NH_2(O)$, oxo-λ[5]-azanyl; azinoyl		$HONH^-$, hydroxyazanide, hydridohydroxidonitrate(1−) H_2NO^-, azanolate, aminooxidanide, dihydridooxidonitrate(1−)	$NHOH^-$, hydroxyazanido, hydridohydroxidonitrato(1−) H_2NO^-, azanolato, aminooxidanido, dihydridooxidonitrato(1−)
H_2NOS	$-S(O)NH_2$, azanyloxo-λ[4]-sulfanyl; aminosulfinyl			
H_2NO_2S	$-S(O)_2NH_2$, azanyldioxo-λ[6]-sulfanyl; aminosulfonyl; sulfamoyl			
H_2NO_3		$[NO(OH)_2]^+$, dihydroxidooxidonitrogen(1+)		
H_2NS	$-SNH_2$, azanylsulfanyl; aminosulfanyl $-NH_2(S)$, sulfanylidene-λ[5]-azanyl; azinothioyl			

Table IX *Continued*

Formula for uncharged atom or group	*Name*			
	Uncharged atoms or molecules (including zwitterions and radicals) or substituent groups[a]	*Cations (including cation radicals) or cationic substituent groups*[a]	*Anions (including anion radicals) or anionic substituent groups*[b]	*Ligands*[c]
H_2N_m, see N_mH_2				
H_2O	H_2O, dihydrogen oxide; water $H_2O = [OH_2]$, oxidane (parent hydride name), dihydridooxygen ^1H_2O, diprotium oxide; $(^1H_2)$water $D_2O = {}^2H_2O$, dideuterium oxide; $(^2H_2)$water $T_2O = {}^3H_2O$, ditritium oxide; $(^3H_2)$water			H_2O, aqua
H_2OP	$-PH_2O$, oxo-λ^5-phosphanyl; phosphinoyl		PH_2O^- dihydridooxidophosphate(1−); phosphinite	PH_2O^-, dihydridooxidophosphato(1−); phosphinito
H_2OSb	$-SbH_2O$, oxo-λ^5-stibanyl, stibinoyl			
H_2O_2	H_2O_2, dihydrogen peroxide; hydrogen peroxide HOOH, dioxidane (parent hydride name), bis(hydridooxygen)(*O—O*)	$HOOH^{\bullet+}$, dioxidaniumyl, bis(hydridooxygen)(*O—O*)(•1+)		HOOH, dioxidane
H_2O_2P	$-P(OH)_2$, dihydroxyphosphanyl $-PH(O)(OH)$, hydroxyoxo-λ^5-phosphanyl		$PH_2O_2^-$, dihydridodioxidophosphate(1−); phosphinate	$PH_2O_2^-$, dihydridodioxidophosphato(1−); phosphinato

H_2O_3B			$H_2BO_3^- = [BO(OH)_2]^-$, dihydroxidooxidoborate(1−); dihydrogenborate	$H_2BO_3^- = [BO(OH)_2]^-$, dihydroxidooxidoborato(1−); dihydrogenborato
H_2O_3P	$-P(O)(OH)_2$, dihydroxyoxo-λ^5-phosphanyl; dihydroxyphosphoryl, phosphono		$[PHO_2(OH)]^-$, hydridohydroxidodioxidophosphate(1−); hydrogenphosphonate $[PO(OH)_2]^-$, dihydroxidooxidophosphate(1−); dihydrogenphosphite	$[PHO_2(OH)]^-$, hydridohydroxidodioxidophosphato(1−); hydrogenphosphonato $[PO(OH)_2]^-$, dihydroxidooxidophosphato(1−); dihydrogenphosphito
H_2O_4P	$(HO)_2PO_2^{\bullet}$, (dihydroxido)dioxidophosphorus(•)		$H_2PO_4^-$, dihydroxidodioxidophosphate(1−); dihydrogenphosphate	$H_2PO_4^-$, dihydroxidodioxidophosphato(1−); dihydrogenphosphato
$H_2O_5P_2$			$P_2H_2O_5^{2-} = [PH(O)_2OPH(O)_2]^{2-}$, μ-oxido- bis(hydridodioxidophosphate)(2−); diphosphonate	$P_2H_2O_5^{2-} = [PH(O)_2OPH(O)_2]^{2-}$, μ-oxido- bis(hydridodioxidophosphato)(2−); diphosphonato
H_2PS	$-PH_2(S)$, sulfanylidene-λ^5-phosphanyl; phosphinothioyl			
H_2Po	H_2Po, dihydrogen polonide $H_2Po = [PoH_2]$, polane (parent hydride name), dihydridopolonium			
H_2S	H_2S, dihydrogen sulfide; hydrogen sulfide $H_2S = [SH_2]$, sulfane (parent hydride name), dihydridosulfur	$H_2S^{\bullet+}$, sulfaniumyl, dihydridosulfur(•1+) $-SH_2^+$, sulfaniumyl	$H_2S^{\bullet-}$, sulfanuidyl, dihydridosulfate(•1−)[e]	H_2S, sulfane
H_2S_2	H_2S_2, dihydrogen disulfide HSSH, disulfane (parent hydride name), bis(hydridosulfur)(*S*—*S*)	$HSSH^{\bullet+}$, disulfaniumyl, bis(hydridosulfur)(*S*—*S*)(•1+)	$HSSH^{\bullet-}$, disulfanuidyl, bis(hydridosulfate)(*S*—*S*)(•1−)[e]	HSSH, disulfane
H_2S_3	H_2S_3, dihydrogen trisulfide HSSSH, trisulfane (parent hydride name)			HSSSH, trisulfane

Table IX *Continued*

Formula for uncharged atom or group	*Name*			
	Uncharged atoms or molecules (including zwitterions and radicals) or substituent groups[a]	*Cations (including cation radicals) or cationic substituent groups*[a]	*Anions (including anion radicals) or anionic substituent groups*[b]	*Ligands*[c]
H_2S_4	H_2S_4, dihydrogen tetrasulfide HSSSSH, tetrasulfane (parent hydride name)			HSSSSH, tetrasulfane
H_2S_5	H_2S_5, dihydrogen pentasulfide HSSSSSH, pentasulfane (parent hydride name)			HSSSSSH, pentasulfane
H_2Se	H_2Se, dihydrogen selenide; hydrogen selenide $H_2Se = [SeH_2]$, selane (parent hydride name), dihydridoselenium	$H_2Se^{\bullet +}$, selaniumyl, dihydridoselenium(•1+) $-SeH_2^+$, selaniumyl	$H_2Se^{\bullet -}$, selanuidyl, dihydridoselenate(•1−)[e]	H_2Se, selane
H_2Se_2	H_2Se_2, dihydrogen diselenide HSeSeH, diselane (parent hydride name), bis(hydridoselenium)(*Se—Se*)	$HSeSeH^{\bullet +}$, diselaniumyl, bis(hydridoselenium)(*Se—Se*)(•1+)	$HSeSeH^{\bullet -}$, diselanuidyl, bis(hydridoselenate)(*Se—Se*)(•1−)[e]	HSeSeH, diselane
H_2Te	H_2Te, dihydrogen tellanide; hydrogen tellanide $H_2Te = [TeH_2]$, tellane (parent hydride name), dihydridotellurium	$H_2Te^{\bullet +}$, tellaniumyl, dihydridotellurium(•1+) $-TeH_2^+$, tellaniumyl	$H_2Te^{\bullet -}$, tellanuidyl, dihydridotellurate(•1−)[e]	H_2Te, tellane
H_3		H_3^+, trihydrogen(1+)		
H_3N_m, see N_mH_3				

H_3NO	$HONH_2$, azanol, dihydridohydroxidonitrogen; hydroxylamine (parent name for organic derivatives)	$HONH_2^{\bullet +}$, hydroxyazaniumyl, dihydridohydroxidonitrogen(•1+)		$HONH_2$, azanol, dihydridohydroxidonitrogen; hydroxylamine
H_3NP	$-PH_2(=NH)$, imino-λ^5-phosphanyl; phosphinimidoyl			
H_3O		H_3O^+, oxidanium, trihydridooxygen(1+), aquahydrogen(1+); oxonium (*not* hydronium)		
H_3OS	$H_3OS^+ = [SH_3(O)]^+$, oxo-λ^5-sulfanylium, trihydridooxidosulfur(1+)			
H_3OSi	$-OSiH_3$, silyloxy			
H_3O_4S		$[SO(OH)_3]^+ = H_3SO_4{}^+$, trihydroxidooxidosulfur(1+), trihydrogen(tetraoxidosulfate)(1+)		
H_3O_5P	$[PO(OH)_2(OOH)]$, (dioxidanido)dihydroxido‿oxidophosphorus; peroxyphosphoric acid, phosphoroperoxoic acid			
H_3S	$H_3S^{\bullet}$, λ^4-sulfanyl, trihydridosulfur(•)	H_3S^+, sulfanium, trihydridosulfur(1+)	H_3S^-, sulfanuide, trihydridosulfate(1−)	
H_3Se	$H_3Se^{\bullet}$, λ^4-selanyl, trihydridoselenium(•)	H_3S^+, selanium, trihydridoselenium(1+)	H_3Se^-, selanuide, trihydridoselenate(1−)	
H_3Te	$H_3Te^{\bullet}$, λ^4-tellanyl, trihydridotellurium(•)	H_3Te^+, tellanium, trihydridotellurium(1+)	H_3Te^-, tellanuide, trihydridotellurate(1−)	
H_4N_m, see N_mH_4				
H_4NO		$NH_2OH_2{}^+$, aminooxidanium, aquadihydridonitrogen(1+) NH_3OH^+, hydroxyazanium, trihydridohydroxidonitrogen(1+); hydroxyammonium		

Table IX *Continued*

Formula for uncharged atom or group	*Name*			
	Uncharged atoms or molecules (including zwitterions and radicals) or substituent groups[a]	*Cations (including cation radicals) or cationic substituent groups*[a]	*Anions (including anion radicals) or anionic substituent groups*[b]	*Ligands*[c]
H_4O		H_4O^{2+}, oxidanediium, tetrahydridooxygen(2+)		
H_5IO_6	$IO(OH)_5$, pentahydroxy-λ^7-iodanone, pentahydroxidooxidoiodine; orthoperiodic acid			
H_5N_2, see N_2H_5				
H_5O_2		$[H(H_2O)_2]^+$, μ-hydrido-bis(dihydridooxygen)(1+), diaquahydrogen(1+)		
H_6N_2, see N_2H_6				
H_nN_m, see N_mH_n				
He	helium	helium (general) $He^{\bullet+}$, helium(•1+)	helide	helido
HeH		HeH^+, hydridohelium(1+)		
He_2		He_2^{+}, dihelium(1+) He_2^{2+}, dihelium(2+)		
Hf	hafnium	hafnium	hafnide	hafnido
Hg	mercury	mercury (general) Hg^{2+}, mercury(2+)	mercuride	mercurido
Hg_2		Hg_2^{2+}, dimercury(2+)		
Ho	holmium	holmium	holmide	holmido

Hs	hassium	hassium	hasside	hassido
I	iodine (general) $I^{\bullet}$, iodine(•), monoiodine –I, iodo	iodine (general) I^{+}, iodine(1+)	iodide (general) I^{-}, iodide(1$^-$); iodide	I^{-}, iodido(1–); iodido
ICl_2	$ICl_2^{\bullet}$, dichloridoiodine(•) $-ICl_2$, dichloro-λ^3-iodanyl	ICl_2^{+}, dichloroiodanium, dichloridoiodine(1+)		
IF	IF, iodine fluoride [IF], fluoridoiodine			
IF_4		IF_4^{+}, tetrafluoro-λ^3-iodanium, tetrafluoridoiodine(1+)	IF_4^{-}, tetrafluoro-λ^3-iodanuide, tetrafluoridoiodate(1–)	IF_4^{-}, tetrafluoro-λ^3-iodanuido, tetrafluoridoiodato(1–)
IF_6			IF_6^{-}, hexafluoro-λ^5-iodanuide, hexafluoridoiodate(1–)	IF_6^{-}, hexafluoro-λ^5-iodanuido, hexafluoridoiodato(1–)
IH, see HI				
I_2	I_2, diiodine	$I_2^{\bullet +}$, diiodine(•1+)	$I_2^{\bullet -}$, diiodide(•1–)	I_2, diiodine
I_3	I_3, triiodine		I_3^{-}, triiodide(1–); triiodide	I_3^{-}, triiodido(1–); triiodido
In	indium	indium	indide	indido
InH_2	$-InH_2$, indiganyl			
InH_3	InH_3, indium trihydride $[InH_3]$, indigane (parent hydride name), trihydridoindium			
Ir	iridium	iridium	iridide	iridido
K	potassium	potassium	potasside	potassido
KO_2	KO_2, potassium dioxide(1–); potassium superoxide			
KO_3	KO_3, potassium trioxide(1–); potassium ozonide			
K_2O	K_2O, dipotassium oxide			
K_2O_2	K_2O_2, dipotassium dioxide(2–); potassium peroxide			
Kr	krypton	krypton	kryptonide	kryptonido
La	lanthanum	lanthanum	lanthanide[d]	lanthanido
Li	lithium	lithium (general) Li^{+}, lithium(1+)	lithide (general) Li^{-}, lithide(1–); lithide	lithido Li^{-}, lithido(1–); lithido

Table IX *Continued*

Formula for uncharged atom or group	*Name*			
	Uncharged atoms or molecules (including zwitterions and radicals) or substituent groups[a]	*Cations (including cation radicals) or cationic substituent groups*[a]	*Anions (including anion radicals) or anionic substituent groups*[b]	*Ligands*[c]
LiAl	[LiAl], aluminidolithium			
LiBe	[LiBe], beryllidolithium			
LiCl	LiCl, lithium chloride [LiCl], chloridolithium	$LiCl^{+}$, chloridolithium(1+)	$LiCl^{-}$, chloridolithate(1−)	$LiCl^{-}$, chloridolithato(1−)
LiH	LiH, lithium hydride [LiH], hydridolithium	LiH^{+}, hydridolithium(1+)	LiH^{-}, hydridolithate(1−)	LiH^{-}, hydridolithato(1−)
LiMg	LiMg, lithium monomagneside	$LiMg^{+}$, magnesidolithium(1+)		
Li_2	Li_2, dilithium	Li_2^{+}, dilithium(•1+)	Li_2^{-}, dilithide(•1−)	Li_2^{-}, dilithido(•1−)
Lr	lawrencium	lawrencium	lawrencide	lawrencido
Lu	lutetium	lutetium	lutetide	lutetido
Md	mendelevium	mendelevium	mendelevide	mendelevido
Mg	magnesium	magnesium (general) Mg^{+}, magnesium(1+) Mg^{2+}, magnesium(2+)	magneside (general) Mg^{-}, magneside(1−)	magnesido Mg^{-}, magnesido(1−)
Mn	manganese	manganese (general) Mn^{2+}, manganese(2+) Mn^{3+}, manganese(3+)	manganide	manganido
MnO	MnO, manganese mon(o)oxide, manganese(II) oxide			
MnO_2	MnO_2, manganese dioxide, manganese(IV) oxide			
MnO_3		MnO_3^{+}, trioxidomanganese(1+)		

MnO_4			MnO_4^-, tetraoxidomanganate(1−), permanganate MnO_4^{2-}, tetraoxidomanganate(2−), manganate(VI) MnO_4^{3-}, tetraoxidomanganate(3−), manganate(V)	MnO_4^-, tetraoxidomanganato(1−), permanganato MnO_4^{2-}, tetraoxidomanganato(2−), manganato(VI) MnO_4^{3-}, tetraoxidomanganato(3−), manganato(V)
Mn_2O_3	Mn_2O_3, dimanganese trioxide, manganese(III) oxide			
Mn_2O_7	Mn_2O_7, dimanganese heptaoxide, manganese(VII) oxide $[O_3MnOMnO_3]$, μ-oxido-bis(trioxidomanganese)			
Mn_3O_4	Mn_3O_4, trimanganese tetraoxide $Mn^{II}Mn^{III}{}_2O_4$, manganese(II,III) tetraoxide			
Mo	molybdenum	molybdenum	molybdenide	molybdenido
Mt	meitnerium	meitnerium	meitneride	meitnerido
Mu	$Mu^{\bullet} = \mu^+e^-$, muonium	$Mu^+ = \mu^+$, muon	$Mu^- = \mu^+(e^-)^2$, muonide	
N	nitrogen $N^{\bullet}$, nitrogen(•), mononitrogen −N<, azanetriyl; nitrilo −N=, azanylylidene ≡N, azanylidyne	nitrogen (general) N^+, nitrogen(1+)	nitride (general) N^{3-}, nitride(3−), azanetriide; nitride =N−, azanidylidene; amidylidene $-N^{2-}$, azanediidyl	N^{3-}, nitrido(3−), azanetriido
NCO, see CNO				
NCS, see CNS				
NCl_2			NCl_2^-, dichloroazanide, dichloridonitrate(1−)	NCl_2^-, dichloroazanido, dichloridonitrato(1−)
NF			NF^{2-}, fluoroazanediide, fluoridonitrate(2−)	NF^{2-}, fluoroazanediido, fluoridonitrato(2−)

Table IX *Continued*

Formula for uncharged atom or group	*Name*			
	Uncharged atoms or molecules (including zwitterions and radicals) or substituent groups[a]	*Cations (including cation radicals) or cationic substituent groups*[a]	*Anions (including anion radicals) or anionic substituent groups*[b]	*Ligands*[c]
NF_3	NF_3, nitrogen trifluoride $[NF_3]$, trifluoroazane, trifluoridonitrogen			NF_3, trifluoroazane, trifluoridonitrogen
NF_4		NF_4^+, tetrafluoroammonium, tetrafluoroazanium, tetrafluoridonitrogen(1+)		
NH	$NH^{2\bullet}$, azanylidene, hydridonitrogen(2•); nitrene >NH, azanediyl =NH, azanylidene; imino	NH^+, azanyliumdiyl, hydridonitrogen(1+) NH^{2+}, azanebis(ylium), hydridonitrogen(2+)	NH^-, azanidyl, hydridonitrate(1−) NH^{2-}, azanediide, hydridonitrate(2−); imide $-NH^-$, azanidyl; amidyl	NH^{2-}, azanediido, hydridonitrato(2−); imido
NH_2	$NH_2^\bullet$, azanyl, dihydridonitrogen(•); aminyl $-NH_2$, azanyl; amino	NH_2^+, azanylium, dihydridonitrogen(1+)	NH_2^-, azanide, dihydridonitrate(1−); amide	NH_2^-, azanido, dihydridonitrato(1−), amido
NH_3	NH_3, azane (parent hydride name), amine (parent name for certain organic derivatives), trihydridonitrogen; ammonia	$NH_3^{\bullet+}$, azaniumyl, trihydridonitrogen(•1+) $-NH_3^+$, azaniumyl; ammonio	$NH_3^{\bullet-}$, azanuidyl, trihydridonitrate(•1−)[e]	NH_3, ammine
NH_4	$NH_4^\bullet$, λ^5-azanyl, tetrahydridonitrogen(•)	NH_4^+, azanium; ammonium		

NO	NO, nitrogen mon(o)oxide (*not* nitric oxide) $NO^{\bullet}$, oxoazanyl, oxidonitrogen(•); nitrosyl –N=O, oxoazanyl; nitroso >N(O)–, oxo-λ^5-azanetriyl; azoryl =N(O)–, oxo-λ^5-azanylylidene; azorylidene ≡N(O), oxo-λ^5-azanylidyne; azorylidyne $-O^{+}=N^{-}$, azanidylideneoxidaniumyl	NO^{+}, oxidonitrogen(1+) (*not* nitrosyl) $NO^{\bullet 2+}$, oxidonitrogen(•2+)	NO^{-}, oxidonitrate(1–) $NO^{(2\bullet)-}$, oxidonitrate(2•1–)	NO, oxidonitrogen (general); nitrosyl = oxidonitrogen-κ*N* (general) NO^{+}, oxidonitrogen(1+) NO^{-}, oxidonitrato(1–)
NO_2	NO_2, nitrogen dioxide $NO_2^{\bullet}$ = $ONO^{\bullet}$, nitrosooxidanyl, dioxidonitrogen(•); nitryl $-NO_2$, nitro –ONO, nitrosooxy	NO_2^{+}, dioxidonitrogen(1+) (*not* nitryl)	NO_2^{-}, dioxidonitrate(1–); nitrite $NO_2^{\bullet 2-}$, dioxidonitrate(•2–)	NO_2^{-}, dioxidonitrato(1–); nitrito $NO_2^{\bullet 2-}$, dioxidonitrato(•2–)
NO_3	NO_3, nitrogen trioxide $NO_3^{\bullet}$ = $O_2NO^{\bullet}$, nitrooxidanyl, trioxidonitrogen(•) $ONOO^{\bullet}$, nitrosodioxidanyl, oxidoperoxidonitrogen(•) $-ONO_2$, nitrooxy		NO_3^{-}, trioxidonitrate(1–); nitrate $NO_3^{\bullet 2-}$, trioxidonitrate(•2–) $[NO(OO)]^{-}$, oxidoperoxidonitrate(1–); peroxynitrite	NO_3^{-}, trioxidonitrato(1–); nitrato $NO_3^{\bullet 2-}$, trioxidonitrato(•2–) $[NO(OO)]^{-}$, oxidoperoxidonitrato(1–); peroxynitrito
NO_4			$NO_2(O_2)^{-}$, dioxidoperoxidonitrate(1–); peroxynitrate	$NO_2(O_2)^{-}$, dioxidoperoxidonitrato(1–); peroxynitrato
NS	NS, nitrogen monosulfide $NS^{\bullet}$, sulfidonitrogen(•) –N=S, sulfanylideneazanyl; thionitroso	NS^{+}, sulfidonitrogen(1+) (*not* thionitrosyl)	NS^{-}, sulfidonitrate(1–)	NS, sulfidonitrogen, sulfidonitrato, thionitrosyl (general) NS^{+}, sulfidonitrogen(1+) NS^{-}, sulfidonitrato(1–)

Table IX *Continued*

Formula for uncharged atom or group	*Name*			
	Uncharged atoms or molecules (including zwitterions and radicals) or substituent groups[a]	*Cations (including cation radicals) or cationic substituent groups*[a]	*Anions (including anion radicals) or anionic substituent groups*[b]	*Ligands*[c]
N_2	N_2, dinitrogen $=N^+=N^-$, (azanidylidene)azaniumylidene; diazo $=NN=$, diazane-1,2-diylidene; hydrazinediylidene $-N=N-$, diazene-1,2-diyl; azo	$N_2^{\bullet+}$, dinitrogen(•1+) N_2^{2+}, dinitrogen(2+) $-N^+{\equiv}N$, diazyn-1-ium-1-yl	N_2^{2-}, dinitride(2−) N_2^{4-}, dinitride(4−), diazanetetraide; hydrazinetetraide	N_2, dinitrogen N_2^{2-}, dinitrido(2−) N_2^{4-}, dinitrido(4−), diazanetetraido; hydrazinetetraido
N_2H		$N{\equiv}NH^+$, diazynium	$N=NH^-$, diazenide NNH^{3-}, diazanetriide, hydrazinetriide	$N=NH^-$, diazenido NNH^{3-}, diazanetriido, hydrazinetriido
N_2H_2	$HN=NH$, diazene $^-N=NH_2^+$, diazen-2-ium-1-ide $H_2NN^{2\bullet}$, diazanylidene, hydrazinylidene $=NNH_2$, diazanylidene; hydrazinylidene $^{\bullet}HNNH^{\bullet}$, diazane-1,2-diyl; hydrazine-1,2-diyl $-HNNH-$, diazane-1,2-diyl; hydrazine-1,2-diyl	$HNNH^{2+}$, diazynediium	$HNNH^{2-}$, diazane-1,2-diide, hydrazine-1,2-diide H_2NN^{2-}, diazane-1,1-diide, hydrazine-1,1-diide	$HN=NH$, diazene $^-N=NH_2^+$, diazen-2-ium-1-ido $HNNH^{2-}$, diazane-1,2-diido, hydrazine-1,2-diido H_2NN^{2-}, diazane-1,1-diido, hydrazine-1,1-diido
N_2H_3	$H_2NNH^{\bullet}$, diazanyl, trihydridodinitrogen(*N*—*N*)(•); hydrazinyl $-NHNH_2$, diazanyl; hydrazinyl $^{2-}NNH_3^+$, diazan-2-ium-1,1-diide	$HN=NH_2^+$, diazenium	H_2NNH^-, diazanide, hydrazinide	$^{2-}NNH_3^+$, diazan-2-ium-1,1-diido H_2NNH^-, diazanido, hydrazinido

N_2H_4	H_2NNH_2, diazane (parent hydride name), hydrazine (parent name for organic derivatives) $^{-}NHNH_3^{+}$, diazan-2-ium-1-ide	$H_2NNH_2^{\bullet+}$, diazaniumyl, bis(dihydridonitrogen)(*N*—*N*)(•1+); hydraziniumyl $H_2N{=}NH_2^{2+}$, diazenediium		H_2NNH_2, diazane, hydrazine $^{-}NHNH_3^{+}$, diazan-2-ium-1-ido
N_2H_5		$H_2NNH_3^{+}$, diazanium, hydrazinium		
N_2H_6		$H_3NNH_3^{2+}$, diazanediium, hydrazinediium		
N_2O	N_2O, dinitrogen oxide (*not* nitrous oxide) NNO, oxidodinitrogen(*N*—*N*) −N(O)=N−, azoxy		$N_2O^{\bullet-}$, oxidodinitrate(•1−)	N_2O, dinitrogen oxide (general) NNO, oxidodinitrogen(*N*—*N*) $N_2O^{\bullet-}$, oxidodinitrato(•1−)
N_2O_2	N_2O_2, dinitrogen dioxide ONNO, bis(oxidonitrogen)(*N*—*N*)		$N_2O_2^{2-}$, diazenediolate, bis(oxidonitrate)(*N*—*N*)(2−)	$N_2O_2^{2-}$, bis(oxidonitrato)(*N*—*N*)(2−)
N_2O_3	N_2O_3, dinitrogen trioxide O_2NNO, trioxido-1κ^2*O*,2κ*O*-dinitrogen(*N*—*N*) $NO^{+}NO_2^{-}$, oxidonitrogen(1+) dioxidonitrate(1−) ONONO, dinitrosooxidane, μ-oxido-bis(oxidonitrogen)		$N_2O_3^{2-} = [O_2NNO]^{2-}$, trioxido-1κ^2*O*,2κ*O*-dinitrate(*N*—*N*)(2−)	
N_2O_4	N_2O_4, dinitrogen tetraoxide O_2NNO_2, bis(dioxidonitrogen)(*N*—*N*) ONOONO, 1,2-dinitrosodioxidane, bis(nitrosyloxygen)(*O*—*O*), 2,5-diazy-1,3,4,6-tetraoxy-[6]catena $NO^{+}NO_3^{-}$, oxidonitrogen(1+) trioxidonitrate(1−)			

Table IX *Continued*

Formula for uncharged atom or group	*Name*			
	Uncharged atoms or molecules (including zwitterions and radicals) or substituent groups[a]	*Cations (including cation radicals) or cationic substituent groups*[a]	*Anions (including anion radicals) or anionic substituent groups*[b]	*Ligands*[c]
N_2O_5	N_2O_5, dinitrogen pentaoxide O_2NONO_2, dinitrooxidane, μ-oxido-bis(dioxidonitrogen)(*N*—*N*) $NO_2^+NO_3^-$, dioxidonitrogen(1+) trioxidonitrate(1−)			
N_3	$N_3^{\bullet}$, trinitrogen(•) $-N{=}N^+{=}N^-$, azido		N_3^-, trinitride(1−); azide	N_3^-, trinitrido(1−); azido
N_3H	N_3H, hydrogen trinitride(1−); hydrogen azide [NNNH], hydrido-1κ*H*-trinitrogen(2 *N*—*N*)			
N_3H_2	−NHN=NH, triaz-2-en-1-yl			
N_3H_4	$-NHNHNH_2$, triazan-1-yl			
N_5		N_5^+, pentanitrogen(1+)		
N_6			$N_6^{\bullet -}$, hexanitride(•1−)	$N_6^{\bullet -}$, hexanitrido(•1−)
Na	sodium	sodium (general) Na^+, sodium(1+)	sodide (general) Na^-, sodide(1−); sodide	sodido Na^-, sodido(1−); sodido
NaCl	NaCl, sodium chloride [NaCl], chloridosodium	$NaCl^+$, chloridosodium(1+)	$NaCl^-$, chloridosodate(1−)	
Na_2	Na_2, disodium	Na_2^+, disodium(1+)	Na_2^-, disodide(1−)	Na_2^-, disodido(1−)
Nb	niobium	niobium	niobide	niobido
Nd	neodymium	neodymium	neodymide	neodymido

Ne	neon	neon (general) Ne^{+}, neon(1+)	neonide	neonido
NeH		NeH^{+}, hydridoneon(1+)		
NeHe		$NeHe^{+}$, helidoneon(1+)		
Ni	nickel	nickel (general) Ni^{2+}, nickel(2+) Ni^{3+}, nickel(3+)	nickelide	nickelido
No	nobelium	nobelium	nobelide	nobelido
Np	neptunium	neptunium	neptunide	neptunido
NpO_2	NpO_2, neptunium dioxide	$NpO_2{}^{+}$, dioxidoneptunium(1+) [*not* neptunyl(1+)] $NpO_2{}^{2+}$, dioxidoneptunium(2+) [*not* neptunyl(2+)]		
O	oxygen (general) O, monooxygen $O^{2\bullet}$, oxidanylidene, monooxygen(2•) >O, oxy, epoxy (in rings) =O, oxo	oxygen (general) $O^{\bullet+}$, oxygen(•1+)	oxide (general) $O^{\bullet-}$, oxidanidyl, oxide(•1−) O^{2-}, oxide(2−); oxide $-O^{-}$, oxido	O^{2-}, oxido
OBr	OBr, oxygen (mono)bromide[f] $OBr^{\bullet}$, bromidooxygen(•)[f]; bromosyl −BrO, oxo-λ^3-bromanyl; bromosyl −OBr, bromooxy	OBr^{+}, bromidooxygen(1+)[f] (*not* bromosyl)	OBr^{-}, bromidooxygenate(1−)[f]; oxidobromate(1−)[f], hypobromite	OBr^{-}, bromidooxygenato(1−)[f]; oxidobromato(1−)[f], hypobromito
OCN, see CNO				
OCl	OCl, oxygen (mono)chloride[f] $OCl^{\bullet}$, chloridooxygen(•)[f]; chlorosyl −ClO, oxo-λ^3-chloranyl; chlorosyl −OCl, chlorooxy		OCl^{-}, chloridooxygenate(1−)[f]; oxidochlorate(1−)[f], hypochlorite	OCl^{-}, chloridooxygenato(1−)[f]; oxidochlorato(1−)[f], hypochlorito
OD_2, see H_2O				

Table IX *Continued*

Formula for uncharged atom or group	*Name*			
	Uncharged atoms or molecules (including zwitterions and radicals) or substituent groups[a]	*Cations (including cation radicals) or cationic substituent groups*[a]	*Anions (including anion radicals) or anionic substituent groups*[b]	*Ligands*[c]
OF	OF, oxygen (mono)fluoride $OF^{\bullet}$, fluoridooxygen(•) –FO, oxo-λ^3-fluoranyl; fluorosyl	OF^+, fluoridooxygen(1+)	OF^-, fluoridooxygenate(1−)	
OF_2	OF_2, oxygen difluoride $[OF_2]$, difluoridooxygen			
OH_n, see H_nO ($n = 1$–4)				
O^1H_2, see H_2O				
OI	OI, oxygen (mono)iodide[f] $OI^{\bullet}$, iodidooxygen(•)[f]; iodosyl –IO, oxo-λ^3-iodanyl; iodosyl –OI, iodooxy	OI^+, iodidooxygen(1+)[f] (*not* iodosyl)	OI^-, iodidooxygenate(1−)[f]; oxidoiodate(1−)[f], hypoiodite $OI^{\bullet 2-}$, iodidooxygenate(•2−)[f]	OI^-, iodidooxygenato(1−)[f]; oxidoiodato(1−)[f], hypoiodito
ONC, see CNO				
OT_2, see H_2O				
O_2	O_2, dioxygen $O_2^{2\bullet}$, dioxidanediyl, dioxygen(2•) –OO–, dioxidanediyl; peroxy	$O_2^{\bullet +}$, dioxidanyliumyl, dioxygen(•1+) O_2^{2+}, dioxidanebis(ylium), dioxygen(2+)	$O_2^{\bullet -}$, dioxidanidyl, dioxide(•1−); superoxide (*not* hyperoxide) O_2^{2-}, dioxidanediide, dioxide(2−); peroxide	dioxido (general) O_2, dioxygen $O_2^{\bullet -}$, dioxido(•1−); superoxido O_2^{2-}, dioxidanediido, dioxido(2−); peroxido
O_2Br	O_2Br, dioxygen bromide[f] $BrO_2^{\bullet}$, dioxidobromine(•) $-BrO_2$, dioxo-λ^5-bromanyl; bromyl –OBrO, oxo-λ^3-bromanyloxy	BrO_2^+, dioxidobromine(1+) (*not* bromyl)	BrO_2^-, dioxidobromate(1−); bromite	BrO_2^-, dioxidobromato(1−); bromito

O_2Cl	O_2Cl, dioxygen chloride[f] $ClO_2^{\bullet}$, dioxidochlorine(•) $ClOO^{\bullet}$, chloridodioxygen(*O—O*)(•) $-ClO_2$, dioxo-λ^5-chloranyl; chloryl $-OClO$, oxo-λ^3-chloranyloxy	ClO_2^+, dioxidochlorine(1+) (*not* chloryl)	ClO_2^-, dioxidochlorate(1−); chlorite	ClO_2^-, dioxidochlorato(1−); chlorito
O_2Cl_2		$O_2Cl_2^+$, (dioxygen dichloride)(1+)[f]		
O_2F_2	O_2F_2, dioxygen difluoride FOOF, difluorodioxidane, bis(fluoridooxygen)(*O—O*)			
O_2I	O_2I, dioxygen iodide[f] $IO_2^{\bullet}$, dioxidoiodine(•) $-IO_2$, dioxo-λ^5-iodanyl; iodyl $-OIO$, oxo-λ^3-iodanyloxy	IO_2^+, dioxidoiodine(1+) (*not* iodyl)	IO_2^-, dioxidoiodate(1−); iodite	IO_2^-, dioxidoiodato(1−); iodito
O_3	O_3, trioxygen; ozone $-OOO-$, trioxidanediyl		$O_3^{\bullet-}$, trioxidanidyl, trioxide(•1−); ozonide	O_3, trioxygen; ozone $O_3^{\bullet-}$, trioxido(•1−); ozonido
O_3Br	O_3Br, trioxygen bromide[f] $BrO_3^{\bullet}$, trioxidobromine(•) $-BrO_3$, trioxo-λ^7-bromanyl; perbromyl $-OBrO_2$, dioxo-λ^5-bromanyloxy	BrO_3^+, trioxidobromine(1+) (*not* perbromyl)	BrO_3^-, trioxidobromate(1−); bromate	BrO_3^-, trioxidobromato(1−); bromato
O_3Cl	O_3Cl, trioxygen chloride[f] $ClO_3^{\bullet}$, trioxidochlorine(•) $-ClO_3$, trioxo-λ^7-chloranyl; perchloryl $-OClO_2$, dioxo-λ^5-chloranyloxy	ClO_3^+, trioxidochlorine(1+) (*not* perchloryl)	ClO_3^-, trioxidochlorate(1−); chlorate	ClO_3^-, trioxidochlorato(1−); chlorato
O_3I	O_3I, trioxygen iodide[f] $IO_3^{\bullet}$, trioxidoiodine(•) $-IO_3$, trioxo-λ^7-iodanyl; periodyl $-OIO_2$, dioxo-λ^5-iodanyloxy	IO_3^+, trioxidoiodine(1+) (*not* periodyl)	IO_3^-, trioxidoiodate(1−); iodate	IO_3^-, trioxidoiodato(1−); iodato
O_4Br	O_4Br, tetraoxygen bromide[f] $BrO_4^{\bullet}$, tetraoxidobromine(•) $-OBrO_3$, trioxo-λ^7-bromanyloxy		BrO_4^-, tetraoxidobromate(1−); perbromate	BrO_4^-, tetraoxidobromato(1−); perbromato

Table IX *Continued*

Formula for uncharged atom or group	*Name*			
	Uncharged atoms or molecules (including zwitterions and radicals) or substituent groups[a]	*Cations (including cation radicals) or cationic substituent groups*[a]	*Anions (including anion radicals) or anionic substituent groups*[b]	*Ligands*[c]
O_4Cl	O_4Cl, tetraoxygen chloride[f] $ClO_4^{\bullet}$, tetraoxidochlorine(•) $-OClO_3$, trioxo-λ^7-chloranyloxy		ClO_4^-, tetraoxidochlorate(1−); perchlorate	ClO_4^-, tetraoxidochlorato(1−); perchlorato
O_4I	O_4I, tetraoxygen iodide[f] $IO_4^{\bullet}$, tetraoxidoiodine(•) $-OIO_3$, trioxo-λ^7-iodanyloxy		IO_4^-, tetraoxidoiodate(1−); periodate	IO_4^-, tetraoxidoiodato(1−); periodato
O_5I			IO_5^{3-}, pentaoxidoiodate(3−)	IO_5^{3-}, pentaoxidoiodato(3−)
O_6I			IO_6^{5-}, hexaoxidoiodate(5−); orthoperiodate	IO_6^{5-}, hexaoxidoiodato(5−); orthoperiodato
O_9I_2			$I_2O_9^{4-}$, nonaoxidodiiodate(4−) $[O_3I(\mu\text{-}O)_3IO_3]^{4-}$, tri-μ-oxido-bis(trioxidoiodate)(4−)	$I_2O_9^{4-}$, nonaoxidodiiodato(4−) $[O_3I(\mu\text{-}O)_3IO_3]^{4-}$, tri-μ-oxido-bis(trioxidoiodato)(4−)
Os	osmium	osmium	osmide	osmido
P	phosphorus (general) $P^{\bullet}$, phosphorus(•), monophosphorus $>P-$, phosphanetriyl	phosphorus (general) P^+, phosphorus(1+)	phosphide (general) P^-, phosphide(1−) P^{3-}, phosphide(3−), phosphanetriide; phosphide	P^{3-}, phosphido, phosphanetriido
PF			PF^{2-}, fluorophosphanediide, fluoridophosphate(2−)	PF^{2-}, fluorophosphanediido, fluoridophosphato(2−)
PF_2			PF_2^-, difluorophosphanide, difluoridophosphate(1−)	PF_2^-, difluorophosphanido, difluoridophosphato(1−)
PF_3	PF_3, phosphorus trifluoride $[PF_3]$, trifluorophosphane, trifluoridophosphorus			

PF_4		PF_4^+, tetrafluorophosphanium, tetrafluoridophosphorus(1+)	PF_4^-, tetrafluorophosphanuide, tetrafluoridophosphate(1−)	PF_4^-, tetrafluorophosphanuido, tetrafluoridophosphato(1−)
PF_5	PF_5, phosphorus pentafluoride $[PF_5]$, pentafluoro-λ^5-phosphane, pentafluoridophosphorus			
PF_6			PF_6^-, hexafluoro-λ^5-phosphanuide, hexafluoridophosphate(1−)	PF_6^-, hexafluoro-λ^5-phosphanuido, hexafluoridophosphato(1−)
PH	$PH^{2\bullet}$, phosphanylidene, hydridophosphorus(2•) >PH, phosphanediyl =PH, phosphanylidene	$PH^{\bullet+}$, phosphanyliumyl, hydridophosphorus(•1+) PH^{2+}, phosphanebis(ylium), hydridophosphorus(2+)	$PH^{\bullet-}$, phosphanidyl, hydridophosphate(•1−) PH^{2-}, phosphanediide, hydridophosphate(2−)	PH^{2-}, phosphanediido, hydridophosphato(2−)
PH_2	$PH_2^{\bullet}$, phosphanyl, dihydridophosphorus(•) $-PH_2$, phosphanyl	PH_2^+, phosphanylium, dihydridophosphorus(1+)	PH_2^-, phosphanide, dihydridophosphate(1−)	PH_2^-, phosphanido, dihydridophosphato(1−)
PH_3	PH_3, phosphorus trihydride $[PH_3]$, phosphane (parent hydride name), trihydridophosphorus	$PH_3^{\bullet+}$, phosphaniumyl, trihydridophosphorus(•1+) $-PH_3^+$, phosphaniumyl	$PH_3^{\bullet-}$, phosphanuidyl, trihydridophosphate(•1−)[e]	PH_3, phosphane
PH_4	$-PH_4$, λ^5-phosphanyl	PH_4^+, phosphanium, tetrahydridophosphorus(1+)	PH_4^-, phosphanuide, tetrahydridophosphate(1−)	PH_4^-, phosphanuido, tetrahydridophosphato(1−)
PH_5	PH_5, phosphorus pentahydride $[PH_5]$, λ^5-phosphane (parent hydride name), pentahydridophosphorus			
PN	P≡N, nitridophosphorus >P≡N, azanylidyne-λ^5-phosphanediyl; phosphoronitridoyl			

Table IX *Continued*

Formula for uncharged atom or group	*Name*			
	Uncharged atoms or molecules (including zwitterions and radicals) or substituent groups[a]	*Cations (including cation radicals) or cationic substituent groups*[a]	*Anions (including anion radicals) or anionic substituent groups*[b]	*Ligands*[c]
PO	$PO^{\bullet}$, oxophosphanyl, oxidophosphorus(•), phosphorus mon(o)oxide; phosphoryl >P(O)–, oxo-λ⁵-phosphanetriyl; phosphoryl =P(O)–, oxo-λ⁵-phosphanylylidene; phosphorylidene ≡P(O), oxo-λ⁵-phosphanylidyne; phosphorylidyne	PO^{+}, oxidophosphorus(1+) (*not* phosphoryl)	PO^{-}, oxidophosphate(1−)	
PO_2	$-P(O)_2$, dioxo-λ⁵-phosphanyl		PO_2^{-}, dioxidophosphate(1−)	PO_2^{-}, dioxidophosphato(1−)
PO_3			PO_3^{-}, trioxidophosphate(1−) $PO_3^{\bullet 2-}$, trioxidophosphate(•2−) PO_3^{3-}, trioxidophosphate(3−); phosphite $(PO_3^{-})_n$ = $\{P(O)_2O\}_n^{n-}$, *catena*-poly[(dioxidophosphate-μ-oxido)(1−)]; metaphosphate $-P(O)(O^{-})_2$, dioxidooxo-λ⁵-phosphanyl; phosphonato	PO_3^{-}, trioxidophosphato(1−) $PO_3^{\bullet 2-}$, trioxidophosphato(•2−) PO_3^{3-}, trioxidophosphato(3−); phosphito

PO_4			$PO_4^{\bullet 2-}$, tetraoxidophosphate(•2−) PO_4^{3-}, tetraoxidophosphate(3−); phosphate	PO_4^{3-}, tetraoxidophosphato(3−); phosphato
PO_5			$PO_5^{\bullet 2-} = PO_3(OO)^{\bullet 2-}$, trioxidoperoxidophosphate(•2−) $PO_5^{3-} = PO_3(OO)^{3-}$, trioxidoperoxidophosphate(3−); peroxyphosphate, phosphoroperoxoate	$PO_5^{3-} = PO_3(OO)^{3-}$, trioxidoperoxidophosphato(3−); peroxyphosphato, phosphoroperoxoato
PS	$PS^{\bullet}$, sulfidophosphorus(•); −PS, thiophosphoryl	PS^{+}, sulfidophosphorus(1+) (*not* thiophosphoryl)		
PS_4			PS_4^{3-}, tetrasulfidophosphate(3−)	PS_4^{3-}, tetrasulfidophosphato(3−)
P_2	P_2, diphosphorus	P_2^{+}, diphosphorus(1+)	P_2^{-}, diphosphide(1−) P_2^{2-}, diphosphide(2−)	P_2, diphosphorus P_2^{-}, diphosphido(1−) P_2^{2-}, diphosphido(2−)
P_2H			$HP{=}P^{-}$, diphosphenide PPH^{3-}, diphosphanetriide	$HP{=}P^{-}$, diphosphenido PPH^{3-}, diphosphanetriido
P_2H_2	$HP{=}PH$, diphosphene (parent hydride name) $H_2PP^{2\bullet}$, diphosphanylidene $=PPH_2$, diphosphanylidene −HPPH−, diphosphane-1,2-diyl		$HPPH^{2-}$, diphosphane-1,2-diide H_2PP^{2-}, diphosphane-1,1-diide	$HP{=}PH$, diphosphene $HPPH^{2-}$, diphosphane-1,2-diido H_2PP^{2-}, diphosphane-1,1-diido
P_2H_3	$H_2PPH^{\bullet}$, diphosphanyl, trihydridodiphosphorus(*P*—*P*)(•) $-HPPH_2$, diphosphanyl		H_2PPH^{-}, diphosphanide	H_2PPH^{-}, diphosphanido
P_2H_4	H_2PPH_2, diphosphane (parent hydride name)			H_2PPH_2, diphosphane
P_2O_6			$O_3PPO_3^{3-}$, bis(trioxidophosphate)(*P*—*P*)(4−); hypodiphosphate	$O_3PPO_3^{3-}$, bis(trioxidophosphato)(*P*—*P*)(4−); hypodiphosphato
P_2O_7			$O_3POPO_3^{4-}$, μ-oxido- bis(trioxidophosphate)(4−); diphosphate	$O_3POPO_3^{4-}$, μ-oxido- bis(trioxidophosphato)(4−); diphosphato

Table IX *Continued*

Formula for uncharged atom or group	*Name*			
	Uncharged atoms or molecules (including zwitterions and radicals) or substituent groups[a]	*Cations (including cation radicals) or cationic substituent groups*[a]	*Anions (including anion radicals) or anionic substituent groups*[b]	*Ligands*[c]
P_2O_8			$O_3POOPO_3{}^{4-}$, μ-peroxido-1κ*O*,2κ*O*′-bis(trioxidophosphate)(4−); peroxydiphosphate	$O_3POOPO_3{}^{4-}$, μ-peroxido-1κ*O*,2κ*O*′-bis(trioxidophosphato)(4−); peroxydiphosphato
P_4	P_4, tetraphosphorus			P_4, tetraphosphorus
Pa	protactinium	protactinium	protactinide	protactinido
Pb	lead	lead (general) Pb^{2+}, lead(2+) Pd^{4+}, lead(4+)	plumbide	plumbido
PbH_4	PbH_4, plumbane (parent hydride name), tetrahydridolead, lead tetrahydride			
Pb_9			$Pb_9{}^{4-}$, nonaplumbide(4−)	
Pd	palladium	palladium (general) Pd^{2+}, palladium(2+) Pd^{4+}, palladium(4+)	palladide	palladido
Pm	promethium	promethium	promethide	promethido
Po	polonium	polonium	polonide	polonido
PoH_2, see H_2Po				
Pr	praseodymium	praseodymium	praseodymide	praseodymido
Pt	platinum	platinum (general) Pt^{2+}, platinum(2+) Pt^{4+}, platinum(4+)	platinide	platinido

Pu	plutonium	plutonium	plutonide	plutonido
PuO_2	PuO_2, plutonium dioxide	PuO_2^{+}, dioxidoplutonium(1+) [*not* plutonyl(1+)] PuO_2^{2+}, dioxidoplutonium(2+) [*not* plutonyl(2+)]		
Ra	radium	radium	radide	radido
Rb	rubidium	rubidium	rubidide	rubidido
Re	rhenium	rhenium	rhenide	rhenido
ReO_4			ReO_4^{-}, tetraoxidorhenate(1−) ReO_4^{2-}, tetraoxidorhenate(2−)	ReO_4^{-}, tetraoxidorhenato(1−) ReO_4^{2-}, tetraoxidorhenato(2−)
Rf	rutherfordium	rutherfordium	rutherfordide	rutherfordido
Rg	roentgenium	roentgenium	roentgenide	roentgenido
Rh	rhodium	rhodium	rhodide	rhodido
Rn	radon	radon	radonide	radonido
Ru	ruthenium	ruthenium	ruthenide	ruthenido
S	sulfur (general) S, monosulfur =S, sulfanylidene; thioxo −S−, sulfanediyl	sulfur (general) S^{+}, sulfur(1+)	sulfide (general) $S^{\bullet-}$, sulfanidyl, sulfide(•1−) S^{2-}, sulfanediide, sulfide(2−); sulfide $-S^{-}$, sulfido	sulfido (general) $S^{\bullet-}$, sulfanidyl, sulfido(•1−) S^{2-}, sulfanediido, sulfido(2−)
SCN, see CNS				
SH, see HS				
SH_2, see H_2S				
SNC, see CNS				
SO	SO, sulfur mon(o)oxide [SO], oxidosulfur >SO, oxo-λ^4-sulfanediyl; sulfinyl	$SO^{\bullet+}$, oxidosulfur(•1+) (*not* sulfinyl or thionyl)	$SO^{\bullet-}$, oxidosulfate(•1−)	[SO], oxidosulfur
SO_2	SO_2, sulfur dioxide [SO_2], dioxidosulfur >SO_2, dioxo-λ^6-sulfanediyl; sulfuryl, sulfonyl		$SO_2^{\bullet-}$, dioxidosulfate(•1−) SO_2^{2-}, dioxidosulfate(2−), sulfanediolate	[SO_2], dioxidosulfur SO_2^{2-}, dioxidosulfato(2−), sulfanediolato

Table IX *Continued*

Formula for uncharged atom or group	*Name*			
	Uncharged atoms or molecules (including zwitterions and radicals) or substituent groups[a]	*Cations (including cation radicals) or cationic substituent groups*[a]	*Anions (including anion radicals) or anionic substituent groups*[b]	*Ligands*[c]
SO_3	SO_3, sulfur trioxide		$SO_3^{\bullet-}$, trioxidosulfate(•1−) SO_3^{2-}, trioxidosulfate(2−); sulfite $-S(O)_2(O^-)$, oxidodioxo-λ[6]-sulfanyl; sulfonato	SO_3^{2-}, trioxidosulfato(2−); sulfito
SO_4	$-OS(O)_2O-$, sulfonylbis(oxy)		$SO_4^{\bullet-}$, tetraoxidosulfate(•1−) SO_4^{2-}, tetraoxidosulfate(2−); sulfate	SO_4^{2-}, tetraoxidosulfato(2−); sulfato
SO_5			$SO_5^{\bullet-} = SO_3(OO)^{\bullet-}$, trioxidoperoxidosulfate(•1−) $SO_5^{2-} = SO_3(OO)^{2-}$, trioxidoperoxidosulfate(2−); peroxysulfate, sulfuroperoxoate	$SO_5^{2-} = SO_3(OO)^{2-}$, trioxidoperoxidosulfato(2−); peroxysulfato, sulfuroperoxoato
S_2	S_2, disulfur $-SS-$, disulfanediyl $>S{=}S$, sulfanylidene-λ[4]-sulfanediyl; sulfinothioyl	$S_2^{\bullet+}$, disulfur(•1+)	$S_2^{\bullet-}$, disulfanidyl, disulfide(•1−) S_2^{2-}, disulfide(2−), disulfanediide $-SS^-$, disulfanidyl	S_2^{2-}, disulfido(2−), disulfanediido
S_2O	$>S({=}O)({=}S)$, oxosulfanylidene-λ[6]-sulfanediyl; sulfonothioyl			

S_2O_2			$S_2O_2^{2-} = OSSO^{2-}$, disulfanediolate, bis(oxidosulfate)(*S*—*S*)(2−) $S_2O_2^{2-} = SOOS^{2-}$, dioxidanedithiolate, peroxybis(sulfanide), bis(sulfidooxygenate)(*O*—*O*)(2−) $S_2O_2^{2-} = SO_2S^{2-}$, dioxido-1κ^2*O*-disulfate(*S*—*S*)(2−), dioxidosulfidosulfate(2−); thiosulfite, sulfurothioite	$S_2O_2^{2-} = OSSO^{2-}$, disulfanediolato, bis(oxidosulfato)(*S*—*S*)(2−) $S_2O_2^{2-} = SOOS^{2-}$, dioxidanedithiolato, peroxybis(sulfanido), bis(sulfidooxygenato)(*O*—*O*)(2−) $S_2O_2^{2-} = SO_2S^{2-}$, dioxido-1κ^2*O*-disulfato(*S*—*S*)(2−), dioxidosulfidosulfato(2−); thiosulfito, sulfurothioito
S_2O_3			$S_2O_3^{\bullet-} = SO_3S^{\bullet-}$, trioxido-1κ^3*O*-disulfate(*S*—*S*)(•1−), trioxidosulfidosulfate(•1−) $S_2O_3^{2-} = SO_3S^{2-}$, trioxido-1κ^3*O*-disulfate(*S*—*S*)(2−), trioxidosulfidosulfate(2−); thiosulfate, sulfurothioate	$S_2O_3^{2-} = SO_3S^{2-}$, trioxido-1κ^3*O*-disulfato(*S*—*S*)(2−), trioxidosulfidosulfato(2−); thiosulfato, sulfurothioato
S_2O_4			$S_2O_4^{2-} = O_2SSO_2^{2-}$, bis(dioxidosulfate)(*S*—*S*)(2−); dithionite	$S_2O_4^{2-} = O_2SSO_2^{2-}$, bis(dioxidosulfato)(*S*—*S*)(2−); dithionito
S_2O_5			$S_2O_5^{2-} = O_3SSO_2^{2-}$, pentaoxido-1κ^3*O*,2κ^2*O*-disulfate(*S*—*S*)(2−) $S_2O_5^{2-} = O_2SOSO_2^{2-}$, μ-oxido-bis(dioxidosulfate)(2−)	$S_2O_5^{2-} = O_3SSO_2^{2-}$, pentaoxido-1κ^3*O*,2κ^2*O*-disulfato(*S*—*S*)(2−) $S_2O_5^{2-} = O_2SOSO_2^{2-}$, μ-oxido-bis(dioxidosulfato)(2−)
S_2O_6			$S_2O_6^{2-} = O_3SSO_3^{2-}$, bis(trioxidosulfate)(*S*—*S*)(2−); dithionate	$S_2O_6^{2-} = O_3SSO_3^{2-}$, bis(trioxidosulfato)(*S*—*S*)(2−); dithionato
S_2O_7			$S_2O_7^{2-} = O_3SOSO_3^{2-}$, μ-oxido-bis(trioxidosulfate)(2−); disulfate	$S_2O_7^{2-} = O_3SOSO_3^{2-}$, μ-oxido-bis(trioxidosulfato)(2−); disulfato

Table IX *Continued*

Formula for uncharged atom or group	*Name*			
	Uncharged atoms or molecules (including zwitterions and radicals) or substituent groups[a]	*Cations (including cation radicals) or cationic substituent groups*[a]	*Anions (including anion radicals) or anionic substituent groups*[b]	*Ligands*[c]
S_2O_8			$S_2O_8^{2-} = O_3SOOSO_3^{2-}$, μ-peroxido-1κ*O*,2κ*O*′- bis(trioxidosulfate)(2−); peroxydisulfate	$S_2O_8^{2-} = O_3SOOSO_3^{2-}$, μ-peroxido-1κ*O*,2κ*O*′- bis(trioxidosulfato)(2−); peroxydisulfato
S_3	S_3, trisulfur −SSS−, trisulfanediyl >S(=S)$_2$, bis(sulfanylidene)-λ^6-sulfanediyl; sulfonodithioyl, dithiosulfonyl	S_3^{2+}, trisulfur(2+)	$S_3^{\bullet-}$, trisulfide(•1−) $SSS^{\bullet-}$, trisulfanidyl S_3^{2-}, trisulfide(2−) SSS^{2-}, trisulfanediide	$S_3^{\bullet-}$, trisulfido(•1−) $SSS^{\bullet-}$, trisulfanidyl S_3^{2-}, trisulfido(2−) SSS^{2-}, trisulfanediido
S_4	S_4, tetrasulfur −SSSS−, tetrasulfanediyl	S_4^{2+}, tetrasulfur(2+)	S_4^{2-}, tetrasulfide(2−) $SSSS^{2-}$, tetrasulfanediide	S_4^{2-}, tetrasulfido(2−) $SSSS^{2-}$, tetrasulfanediido
S_4O_6			$S_4O_6^{2-} = O_3SSSSO_3^{2-}$, disulfanedisulfonate, bis[(trioxidosulfato)sulfate]⊂ (*S*—*S*)(2−); tetrathionate $S_4O_6^{\bullet 3-} = O_3SSSSO_3^{\bullet 3-}$, bis[(trioxidosulfato)sulfate]⊂ (*S*—*S*)(•3−)	$S_4O_6^{2-} = O_3SSSSO_3^{2-}$, disulfanedisulfonato, bis[(trioxidosulfato)sulfato]⊂ (*S*—*S*)(2−); tetrathionato $S_4O_6^{\bullet 3-} = O_3SSSSO_3^{\bullet 3-}$, bis[(trioxidosulfato)sulfato]⊂ (*S*—*S*)(•3−)
S_5	S_5, pentasulfur		S_5^{2-}, pentasulfide(2−) $SSSSS^{2-}$, pentasulfanediide	S_5^{2-}, pentasulfido(2−) $SSSSS^{2-}$, pentasulfanediido
S_8	S_8, octasulfur	S_8^{2+}, octasulfur(2+)	S_8^{2-}, octasulfide(2−) $S[S]_6S^{2-}$, octasulfanediide	S_8, octasulfur S_8^{2-}, octasulfido(2−) $S[S]_6S^{2-}$, octasulfanediido

Sb	antimony >Sb–, stibanetriyl	antimony	antimonide (general) Sb^{3-}, antimonide(3–), stibanetriide; antimonide	antimonido (general) Sb^{3-}, antimonido, stibanetriido
SbH	$SbH^{2\bullet}$, stibanylidene, hydridoantimony(2•) >SbH, stibanediyl =SbH, stibanylidene	SbH^{2+}, stibanebis(ylium), hydridoantimony(2+)	SbH^{2-}, stibanediide, hydridoantimonate(2–)	SbH^{2-}, stibanediido, hydridoantimonato(2–)
SbH_2	$SbH_2^{\bullet}$, stibanyl, dihydridoantimony(•) $-SbH_2$, stibanyl	SbH_2^+, stibanylium, dihydridoantimony(1+)	SbH_2^-, stibanide, dihydridoantimonate(1–)	SbH_2^-, stibanido, dihydridoantimonato(1–)
SbH_3	SbH_3, antimony trihydride $[SbH_3]$, stibane (parent hydride name), trihydridoantimony	$SbH_3^{\bullet+}$, stibaniumyl, trihydridoantimony(•1+) $-SbH_3^+$, stibaniumyl	$SbH_3^{\bullet-}$, stibanuidyl, trihydridoantimonate(•1–)[e]	SbH_3, stibane
SbH_4	$-SbH_4$, λ^5-stibanyl	SbH_4^+, stibanium, tetrahydridoantimony(1+)		
SbH_5	SbH_5, antimony pentahydride $[SbH_5]$, λ^5-stibane (parent hydride name), pentahydridoantimony			
Sc	scandium	scandium	scandide	scandido
Se	Se (general) Se, monoselenium >Se, selanediyl =Se, selanylidene; selenoxo	selenium	selenide (general) $Se^{\bullet-}$, selanidyl, selenide(•1–) Se^{2-}, selanediide, selenide(2–); selenide	selenido (general) $Se^{\bullet-}$, selanidyl, selenido(•1–) Se^{2-}, selanediido, selenido(2–)
SeCN, see CNSe				
SeH, see HSe				
SeH_2, see H_2Se				
SeO	SeO, selenium mon(o)oxide [SeO], oxidoselenium >SeO, seleninyl			[SeO], oxidoselenium

Table IX *Continued*

Formula for uncharged atom or group	*Name*			
	Uncharged atoms or molecules (including zwitterions and radicals) or substituent groups[a]	*Cations (including cation radicals) or cationic substituent groups*[a]	*Anions (including anion radicals) or anionic substituent groups*[b]	*Ligands*[c]
SeO_2	SeO_2, selenium dioxide $[SeO_2]$, dioxidoselenium $>SeO_2$, selenonyl		SeO_2^{2-}, dioxidoselenate(2−)	$[SeO_2]$, dioxidoselenium SeO_2^{2-}, dioxidoselenato(2−)
SeO_3	SeO_3, selenium trioxide		$SeO_3^{\bullet-}$, trioxidoselenate(•1−) SeO_3^{2-}, trioxidoselenate(2−); selenite	SeO_3^{2-}, trioxidoselenato(2−); selenito
SeO_4			SeO_4^{2-}, tetraoxidoselenate(2−); selenate	SeO_4^{2-}, tetraoxidoselenato(2−); selenato
Sg	seaborgium	seaborgium	seaborgide	seaborgido
Si	silicon $>Si<$, silanetetrayl $=Si=$, silanediylidene	silicon (general) Si^{+}, silicon(•1+) Si^{4+}, silicon(4+)	silicide (general) Si^{-}, silicide(•1−) Si^{4-}, silicide(4−); silicide	silicido (general) Si^{-}, silicido(•1−) Si^{4-}, silicido(4−); silicido
SiC	SiC, silicon carbide [SiC], carbidosilicon	SiC^{+}, carbidosilicon(1+)		
SiH		SiH^{+}, silanyliumdiyl, hydridosilicon(1+)	SiH^{-}, silanidediyl, hydridosilicate(1−)	
SiH_2	$SiH_2^{2\bullet}$, silylidene, dihydridosilicon(2•) $>SiH_2$, silanediyl $=SiH_2$, silylidene			
SiH_3	$SiH_3^{\bullet}$, silyl, trihydridosilicon(•) $-SiH_3$, silyl	SiH_3^{+}, silylium, trihydridosilicon(1+)	SiH_3^{-}, silanide, trihydridosilicate(1−)	SiH_3^{-}, silanido

SiH_4	SiH_4, silicon tetrahydride $[SiH_4]$, silane (parent hydride name), tetrahydridosilicon			
SiO	SiO, oxidosilicon, silicon mon(o)oxide	SiO^+, oxidosilicon(1+)		
SiO_2	SiO_2, silicon dioxide			
SiO_3			$SiO_3^{\bullet -}$, trioxidosilicate(•1−) $(SiO_3^{2-})_n = (Si(O)_2O)_n^{2n-}$, *catena*-poly[(dioxidosilicate-μ-oxido)(1−)]; metasilicate	$SiO_3^{\bullet -}$, trioxidosilicato(•1−)
SiO_4			SiO_4^{4-}, tetraoxidosilicate(4−); silicate	SiO_4^{4-}, tetraoxidosilicato(4−); silicato
Si_2	Si_2, disilicon	Si_2^+, disilicon(1+)	Si_2^-, disilicide(1−)	
Si_2H_4	$>SiHSiH_3$, disilane-1,1-diyl $-SiH_2SiH_2-$, disilane-1,2-diyl $=SiHSiH_3$, disilanylidene			
Si_2H_5	$Si_2H_5^{\bullet}$, disilanyl, pentahydridodisilicon(*Si—Si*)(•) $-Si_2H_5$, disilanyl	$Si_2H_5^+$, disilanylium	$Si_2H_5^-$, disilanide	$Si_2H_5^-$, disilanido
Si_2H_6	Si_2H_6, disilane (parent hydride name)			Si_2H_6, disilane
Si_2O_7			$Si_2O_7^{6-}$, μ-oxido-bis(trioxidosilicate)(6−); disilicate	$Si_2O_7^{6-}$, μ-oxido-bis(trioxidosilicato)(6−); disilicato
Si_4			Si_4^{4-}, tetrasilicide(4−)	
Sm	samarium	samarium	samaride	samarido
Sn	tin	tin (general) Sn^{2+}, tin(2+) Sn^{4+}, tin(4+)	stannide	stannido
$SnCl_3$			$SnCl_3^-$, trichloridostannate(1−)	$SnCl_3^-$, trichloridostannato(1−)
SnH_4	SnH_4, tin tetrahydride $[SnH_4]$, stannane (parent hydride name), tetrahydridotin			

Table IX *Continued*

Formula for uncharged atom or group	*Name*			
	Uncharged atoms or molecules (including zwitterions and radicals) or substituent groups[a]	*Cations (including cation radicals) or cationic substituent groups*[a]	*Anions (including anion radicals) or anionic substituent groups*[b]	*Ligands*[c]
Sn_5			Sn_5^{2-}, pentastannide(2−)	Sn_5^{2-}, pentastannido(2−)
Sr	strontium	strontium	strontide	strontido
T, see H				
T_2, see H_2				
T_2O, see H_2O				
Ta	tantalum	tantalum	tantalide	tantalido
Tb	terbium	terbium	terbide	terbido
Tc	technetium	technetium	technetide	technetido
TcO_4			TcO_4^{-}, tetraoxidotechnetate(1−) TcO_4^{2-}, tetraoxidotechnetate(2−)	TcO_4^{-}, tetraoxidotechnetato(1−) TcO_4^{2-}, tetraoxidotechnetato2−)
Te	tellurium >Te, tellanediyl =Te, tellanylidene; telluroxo	tellurium	telluride (general) $Te^{\bullet-}$, tellanidyl, telluride(•1−) Te^{2-}, tellanediide, telluride(2−); telluride	tellurido (general) $Te^{\bullet-}$, tellanidyl, tellurido(•1−) Te^{2-}, tellanediido, tellurido(2−)
TeH, see HTe				
TeH_2, see H_2Te				
TeO_3			$TeO_3^{\bullet-}$, trioxidotellurate(•1−) TeO_3^{2-}, trioxidotellurate(2−)	$TeO_3^{\bullet-}$, trioxidotellurato(•1−) TeO_3^{2-}, trioxidotellurato(2−)
TeO_4			TeO_4^{2-}, tetraoxidotellurate(2−); tellurate	TeO_4^{2-}, tetraoxidotellurato(2−); tellurato
TeO_6			TeO_6^{6-}, hexaoxidotellurate(6−); orthotellurate	TeO_6^{6-}, hexaoxidotellurato(6−); orthotellurato
Th	thorium	thorium	thoride	thorido

Ti	titanium	titanium	titanide	titanido
TiO	TiO, titanium(II) oxide	TiO^{2+}, oxidotitanium(2+)		
Tl	thallium	thallium	thallide	thallido
TlH_2	$-TlH_2$, thallanyl			
TlH_3	TlH_3, thallium trihydride $[TlH_3]$, thallane (parent hydride name), trihydridothallium			
Tm	thulium	thulium	thulide	thulido
U	uranium	uranium	uranide	uranido
UO_2	UO_2, uranium dioxide	UO_2^+, dioxidouranium(1+) [*not* uranyl(1+)] UO_2^{2+}, dioxidouranium(2+) [*not* uranyl(2+)]		
V	vanadium	vanadium	vanadide	vanadido
VO	VO, vanadium(II) oxide, vanadium mon(o)oxide	VO^{2+}, oxidovanadium(2+) (*not* vanadyl)		
VO_2	VO_2, vanadium(IV) oxide, vanadium dioxide	VO_2^+, dioxidovanadium(1+)		
W	tungsten	tungsten	tungstide	tungstido
Xe	xenon	xenon	xenonide	xenonido
Y	yttrium	yttrium	yttride	yttrido
Yb	ytterbium	ytterbium	ytterbide	ytterbido
Zn	zinc	zinc	zincide	zincido
Zr	zirconium	zirconium	zirconide	zirconido
ZrO	ZrO, zirconium(II) oxide	ZrO^{2+}, oxidozirconium(2+)		

[a] Where an element symbol occurs in the first column, the unmodified element name is listed in the second and third columns. The unmodified name is generally used when the element appears as an electropositive constituent in the construction of a stoichiometric name (Sections IR-5.2 and IR-5.4). Names of homoatomic cations consisting of the element are also constructed using the element name, adding multiplicative prefixes and charge numbers as applicable (Sections IR-5.3.2.1 to IR-5.3.2.3). In selected cases, examples are given in the Table of specific cation names, such as gold(1+), gold(3+); mercury(2+), dimercury(2+). In such cases, the unmodified element name appears with the qualifier '(general)'.

[b] Where an element symbol occurs in the first column, the fourth column gives the element name appropriately modified with the ending 'ide' (argentide, americide, ferride, *etc.*). The 'ide' form of the element name is generally used when the element appears as an electronegative constituent in the construction of a stoichiometric name (Sections IR-5.2 and IR-5.4). Names of homoatomic anions consisting of the element in question are also constructed using this modified form, adding multiplicative prefixes and charge numbers as applicable (Sections IR-5.3.3.1 to IR-5.3.3.3). Examples are given in the Table of names of some specific anions, *e.g.* arsenide(3−), chloride(1−), oxide(2−), dioxide(2−). In certain cases, a particular anion has the 'ide' form itself as an accepted short name, *e.g.* arsenide, chloride, oxide. If specific anions are named, the 'ide' form of the element name with no further modification is given as the first entry in the fourth column, with the qualifier '(general)'.

[c] Ligand names must be placed within enclosing marks whenever necessary to avoid ambiguity, *cf.* Section IR-9.2.2.3. Some ligand names must always be enclosed. For example, if 'dioxido' is cited as is it must be enclosed so as to distinguish it from two 'oxido' ligands; if combined with a multiplicative prefix it must be enclosed because it starts with a multiplicative prefix itself. A ligand name such as 'nitridocarbonato' must always be enclosed to avoid interpreting it as two separate ligand names, 'nitrido' and 'carbonato'. In this table, however, these enclosing marks are omitted for the sake of clarity. Note that the ligand names given here with a charge number can generally also be used without if it is not desired to make any implication regarding the charge of the ligand. For example, the ligand name '[dioxido(•1−)]' may be used if one wishes explicitly to consider the ligand to be the species dioxide(•1−), whereas the ligand name '(dioxido)' can be used if no such implications are desirable.

[d] The ending 'ide' in 'actinide' and 'lanthanide' indicates a negative ion. Therefore, 'actinoid' should be used as the collective name for the elements Ac, Th, Pa, U, Np, Pu, Am, Cm, Bk, Cf, Es, Fm, Md, No, Lr, and 'lanthanoid' as a collective name for the elements La, Ce, Pr, Nd, Pm, Sm, Eu, Gd, Tb, Dy, Ho, Er, Tm, Yb, Lu (*cf.* Section IR-3.5).

[e] The radical names in the present recommendations sometimes differ from those given in "Names for Inorganic Radicals", W.H. Koppenol, *Pure Appl. Chem.*, **72**, 437–446 (2000). Firstly, the exceptional status of anion radicals consisting of hydrogen and only one other element has been lifted. For example, the coordination-type additive name of $BH_3^{\bullet-}$ is 'trihydridoborate(•1−)' (*not* '-boride'). Secondly, concatenation of ligand names, such as in 'hydridodioxido' (meaning the ligand 'dioxidanido'), which is otherwise never used in additive nomenclature, is not recommended here. Thirdly, additive names of dinuclear compounds are based here on selecting the most centrally placed atoms in the molecule as central atoms (see the general principles described in Section IR-7.1.2), *e.g.* $NCSSCN^{\bullet-}$ is named here 'bis[cyanidosulfur](*S*—*S*)(•1−)' rather than 'bis(nitridosulfidocarbonate)(*S*—*S*)(•1−)'.

[f] Due to the strict adherence in these recommendations to the element sequence in Table VI, the order of oxygen and the elements chlorine, bromine and iodine, respectively, has been reversed relative to traditional names. This applies to binary stoichiometric names such as dioxygen chloride (*cf.* Section IR-5.2) and to additive names for the hypohalites, where the rules for selecting central atoms (Section IR-7.1.2) dictate the selection of oxygen rather than the halide. However, because of the additive names for the last three members of the series OX^-, XO_2^-, XO_3^-, XO_4^- (X = Cl, Br, I), namely dioxidohalogenate(1−), trioxidohalogenate(1−) and tetraoxidohalogenate(1−) (the halogen is chosen as the central atom because it has the central position in the structure), the additive names oxidochlorate(1−), oxidobromate(1−) and oxidoiodate(1−) are acceptable alternatives to the systematic 'oxygenate' names. Similar remarks apply to HOCl, $HOCl^{\bullet-}$, *etc.*

Table X *Anion names, 'a' terms used in substitutive nomenclature and 'y' terms used in chains and rings nomenclature*

Element name	*Anion name*[a]	*'a' term*	*'y' term*
actinium	actinate	actina	actiny
aluminium	aluminate	alumina	aluminy
americium	americate	america	americy
antimony	antimonate	stiba[b]	stiby[b]
argon	argonate	argona	argony
arsenic	arsenate	arsa	arsy
astatine	astatate	astata	astaty
barium	barate	bara	bary
berkelium	berkelate	berkela	berkely
beryllium	beryllate	berylla	berylly
bismuth	bismuthate	bisma	bismy
bohrium	bohrate	bohra	bohry
boron	borate	bora	bory
bromine	bromate	broma	bromy
cadmium	cadmate	cadma	cadmy
caesium	caesate	caesa	caesy
calcium	calcate	calca	calcy
californium	californate	californa	californy
carbon	carbonate	carba	carby
cerium	cerate	cera	cery
chlorine	chlorate	chlora	chlory
chromium	chromate	chroma	chromy
cobalt	cobaltate	cobalta	cobalty
copper	cuprate[c]	cupra[c]	cupry[c]
curium	curate	cura	cury
darmstadtium	darmstadtate	darmstadta	darmstadty
deuterium	deuterate	deutera	deutery
dubnium	dubnate	dubna	dubny
dysprosium	dysprosate	dysprosa	dysprosy
einsteinium	einsteinate	einsteina	einsteiny
erbium	erbate	erba	erby
europium	europate	europa	europy
fermium	fermate	ferma	fermy
fluorine	fluorate	fluora	fluory
francium	francate	franca	francy
gadolinium	gadolinate	gadolina	gadoliny
gallium	gallate	galla	gally
germanium	germanate	germa	germy
gold	aurate[d]	aura[d]	aury[d]
hafnium	hafnate	hafna	hafny
hassium	hassate	hassa	hassy
helium	helate	hela	hely
holmium	holmate	holma	holmy
hydrogen	hydrogenate	–	hydrony
indium	indate	inda	indy
iodine	iodate	ioda	iody
iridium	iridate	irida	iridy
iron	ferrate[e]	ferra[e]	ferry[e]
krypton	kryptonate	kryptona	kryptony
lanthanum	lanthanate	lanthana	lanthany
lawrencium	lawrencate	lawrenca	lawrency

Table X *Continued*

Element name	*Anion name*[a]	*'a' term*	*'y' term*
lead	plumbate[f]	plumba[f]	plumby[f]
lithium	lithate	litha	lithy
lutetium	lutetate	luteta	lutety
magnesium	magnesate	magnesa	magnesy
manganese	manganate	mangana	mangany
meitnerium	meitnerate	meitnera	meitnery
mendelevium	mendelevate	mendeleva	mendelevy
mercury	mercurate	mercura	mercury
molybdenum	molybdate	molybda	molybdy
neodymium	neodymate	neodyma	neodymy
neon	neonate	neona	neony
neptunium	neptunate	neptuna	neptuny
nickel	nickelate	nickela	nickely
niobium	niobate	nioba	nioby
nitrogen	nitrate	aza[g]	azy[g]
nobelium	nobelate	nobela	nobely
osmium	osmate	osma	osmy
oxygen	oxygenate	oxa	oxy
palladium	palladate	pallada	pallady
phosphorus	phosphate	phospha	phosphy
platinum	platinate	platina	platiny
plutonium	plutonate	plutona	plutony
polonium	polonate	polona	polony
potassium	potassate	potassa	potassy
praseodymium	praseodymate	praseodyma	praseodymy
promethium	promethate	prometha	promethy
protactinium	protactinate	protactina	protactiny
protium	protate	prota	proty
radium	radate	rada	rady
radon	radonate	radona	radony
rhenium	rhenate	rhena	rheny
rhodium	rhodate	rhoda	rhody
roentgenium	roentgenate	roentgena	roentgeny
rubidium	rubidate	rubida	rubidy
ruthenium	ruthenate	ruthena	rutheny
rutherfordium	rutherfordate	rutherforda	rutherfordy
samarium	samarate	samara	samary
scandium	scandate	scanda	scandy
seaborgium	seaborgate	seaborga	seaborgy
selenium	selenate	selena	seleny
silicon	silicate	sila	sily
silver	argentate[h]	argenta[h]	argenty[h]
sodium	sodate	soda	sody
strontium	strontate	stronta	stronty
sulfur	sulfate	thia[i]	sulfy
tantalum	tantalate	tantala	tantaly
technetium	technetate	techneta	technety
tellurium	tellurate	tellura	tellury
terbium	terbate	terba	terby
thallium	thallate	thalla	thally
thorium	thorate	thora	thory
thulium	thulate	thula	thuly

Table X *Continued*

Element name	*Anion name*[a]	*'a' term*	*'y' term*
tin	stannate[j]	stanna[j]	stanny[j]
titanium	titanate	titana	titany
tritium	tritate	trita	trity
tungsten	tungstate	tungsta	tungsty[k]
uranium	uranate	urana	urany
vanadium	vanadate	vanada	vanady
xenon	xenonate	xenona	xenony
ytterbium	ytterbate	ytterba	ytterby
yttrium	yttrate	yttra	yttry
zinc	zincate	zinca	zincy
zirconium	zirconate	zircona	zircony

[a] Modified element name used in additive names for heteroatoamic anions containing the element as the central atom.
[b] From the name stibium.
[c] From the name cuprum.
[d] From the name aurum.
[e] From the name ferrum.
[f] From the name plumbum.
[g] From the name azote.
[h] From the name argentum.
[i] From the name theion.
[j] From the name stannum.
[k] 'Wolframy' was used in "Nomenclature of Inorganic Chains and Ring Compounds", E.O. Fluck and R.S. Laitinen, *Pure Appl. Chem.*, **69**, 1659–1692 (1997)" and in Chapter II-5 of *Nomenclature of Inorganic Chemistry II, IUPAC Recommendations 2000*, eds. J.A. McCleverty and N.G. Connelly, Royal Society of Chemistry, 2001.

Subject Index

Element names, parent hydride names and systematic names derived using any of the nomenclature systems described in this book are, with very few exceptions, not included explicitly in this index. If a name or term is referred to in several places in the book, only the more informative references may be indexed.

F

G

N

O